Fourth Edition

Policing America

Methods, Issues, Challenges

KENNETH J. PEAK

University of Nevada, Reno

Upper Saddle River, New Jersey 07458

Library of Congress Cataloging-in-Publication Data

Peak, Kenneth J.
  Policing America : methods, issues, challenges / Kenneth J. Peak.—4th ed.
    p. cm.
  Includes bibliographical references and index.
  ISBN 0-13-094099-2
  1. Police—United States. 2. Law enforcement—United States. I. Title.
  HV8141 .P33 2003
  363.2'0973—dc21

                                                    2002022089

Publisher: Jeff Johnston
Executive Editor: Kim Davies
Assistant Editor: Sarah Holle
Managing Editor: Mary Carnis
Production Editor: Linda B. Pawelchak
Production Liaison: Barbara Marttine Cappuccio
Director of Manufacturing and Production: Bruce Johnson
Manufacturing Buyer: Cathleen Petersen
Art Director: Cheryl Asherman
Cover Design Coordinator: Miguel Ortiz
Cover Design: Scott Garrison
Cover Illustration: Corbis Digital Stock
Marketing Manager: Jessica Pfaff
Composition: Clarinda Company
Printing and Binding: Phoenix Book Tech Park

Pearson Education LTD., *London*
Pearson Education Australia PTY. Limited, *Sydney*
Pearson Education Singapore, Pte. Ltd.
Pearson Education North Asia Ltd., *Hong Kong*
Pearson Education Canada, Ltd., *Toronto*
Pearson Educación de Mexico, S.A. de C.V.
Pearson Education—Japan, *Tokyo*
Pearson Education Malaysia, Pte. Ltd.
Pearson Education, *Upper Saddle River, NJ*

10 9 8 7 6 5
ISBN 0-13-094099-2

*To **Jack Spencer Sr.,** with whom I studied the finer points of policing at several colleges and universities; we served together in both civilian and military police organizations and literally chased bad guys together—even while serving as police executives in different agencies. Jack lost his life in the line of duty in 1998 as an officer of the U.S. Bureau of Indian Affairs (BIA). No more dedicated, innovative protector of the people ever wore the uniform.*

*And to **Creighton Spencer,** Jack's son, one of my former students and also an officer with the BIA; Creighton lost his life while on duty in March 2001.*

May God grant their surviving family members—wife and mother Kay, and son and brother Jack Jr.—eternal blessings and solace.

Contents
in Brief

Contents

2

POLICING LEVELS, ROLES, AND FUNCTIONS   36

3

POLICE SUBCULTURE: THE MAKING OF A COP   62

4

ORGANIZATION AND ADMINISTRATION   88

5

ON PATROL: THE BEAT, DEPLOYMENT, AND DISCRETION   116

8 EXTRAORDINARY PROBLEMS AND METHODS 206

Foreword

(Courtesy Kris Solow)

Policing in America has seen enormous change over the past 30 years. During that time, more police officers entered the profession with a college background than ever before. Most police executives possess advanced degrees and apply sound management principles to their work. Prior to 1970, research on the impact of police methods was virtually nonexistent. Today, police methods and philosophy have been heavily influenced by the research of the past 30 years. Technology has brought advances in policing that could hardly be imagined in 1970. We have seen the development of automated fingerprint identification systems with which a single latent print can be searched against large criminal history databases. DNA has advanced the science of identifying criminal offenders to new levels and has set those free who were innocent of the crime they were accused of committing. Patrol vehicles have computers that give officers access to information to help them clear crimes and solve neighborhood problems. Tremendous strides have been made in radio and telephone communications that have contributed to improved service to the community and greater safety for officers. Relationships with the community have changed as well.

The police have enhanced their relationships with the public as they have reached out through community policing and problem solving to develop partnerships to prevent crime and improve safety. Thirty years ago there were high levels of tension between the police and minority communities. The police struggled with maintaining the peace as large numbers of Americans—particularly students—protested the war in Vietnam. The police were (and continue to be) on the cutting edge of dramatic changes in the American way of life and find it difficult to understand why citizens lash out at them when they were just doing their job. The police have learned the value of problem-solving partnerships. They have formed strong bonds with neighborhood leaders intent on creating and maintaining safe communities. Relationship problems continue to exist to be sure—but most close observers would acknowledge that things are much better today than they have ever been.

As the police have improved their knowledge and the practice of their craft and have developed stronger relationships and seen sharp declines in reported crime over the past eight to ten years, they have come to understand how complex policing has become. The terrorist attacks of September 11, 2001, in New York City and Washington, D.C., have had a massive impact on America and policing. Because of those events and the new bioterrorist attacks using anthrax, Americans now cope with new types of fear on a daily basis. They are looking to the police to provide them with some sense of security from these mysterious enemies and events. The already multifaceted tasks of the police have been increased tenfold as they try to sort out how to respond to these new policing challenges.

Ken Peak's fourth edition of *Policing America* helps practitioners, students, and those with an interest in policing bring some clarity to the complex job of policing. He provides a solid foundation by tracking the historical development of policing in America from its English roots to the first organized municipal police departments in the 1830s. He helps us understand how policing evolved by the 1930s into a corrupt politically directed institution when the first reforms were launched with a view toward creating professional police. He then turns his attention to describing the various federal law enforcement organizations and how they relate to state and local police. This is followed by an insightful examination of the police subculture. The foundation is complete with an explanation of the manner in which police agencies are organized and managed.

Throughout the rest of the text, one finds the complexities of policing addressed in a way that helps build knowledge and understanding of the challenges of policing along with some approaches to meeting those challenges. The most significant and primary service delivery system—patrol—is addressed in a straightforward manner that allows the reader to gain helpful insight into the primary police function. Community policing and problem solving is appropriately sandwiched between patrol and criminal investigations. Dr. Peak explains these new approaches to policing and the important links between the patrol and investigative functions of policing.

These essential chapters are followed by a number of chapters addressing particular aspects of policing. One chapter deals with the many special problems the police are expected to address—gang, organized crime, terrorism, and the unique demands of rural areas. The importance of the rule of law, accountability, and civil liability are addressed as well. An understanding of these areas is vital to effective policing. The police cannot successfully make a full contribution to creating a safe community without the trust of the people they serve. More than a positive relationship with the public is necessary to engender trust in the police; the law must also be applied in a manner that is believed to be fair. At times, police must take stands that are not popular with some members of the community. The rule of law provides a basis for making these difficult decisions.

To help the reader understand American law enforcement, a chapter is devoted to policing in other countries. These comparisons offer insight into the significant differences in police responsibility in a democracy. This insight into other approaches to policing helps one see the strengths and limitations of the justice system in America.

Given the tremendous impact that technology has on the police, and on America as a whole, an entire chapter is devoted to this subject. The application of wireless technology is discussed along with the advances in the forensic sciences. The application of "less-than-lethal" police weapons is addressed as well. For much of the history of policing, the primary tools available for controlling a person who resisted arrest or wanted to harm a citizen or officer was the use of hands, a baton, or a firearm. There are other options available today that have made a difference—mace was largely ineffective, but many departments have adopted pepper spray, and it has proved a useful tool in controlling suspects without injuring them. Electronic shocking tools have evolved over the years to the stage where they have been effectively deployed to help control violent subjects. Beanbag rounds and rubber bullets have also been used to control individuals and crowds. These weapons have helped decrease the tension and difficulty the police face when they have to use "deadly" force in fulfilling their responsibilities. Other important uses of technology are described as well—including uses for traffic control and training simulations.

This excellent text is rounded out in the final chapter with a discussion of the future. As America continues to change and face new challenges, so must the police. If the recent history of the police is any indication, the police will continue to progress toward building true problem-solving partnerships with the community. These partnerships are critical to controlling crime, violence, disorder, and drug abuse.

The fourth edition of *Policing America* is a continuation and update of a volume that captures the important issues in policing. Ken Peak has done a marvelous job of explaining an extremely complex profession that operates in an ever-changing world. Those who read it will be well informed on the issues and challenges of policing in America.

Darrel W. Stephens, Chief of Police,
Charlotte, North Carolina

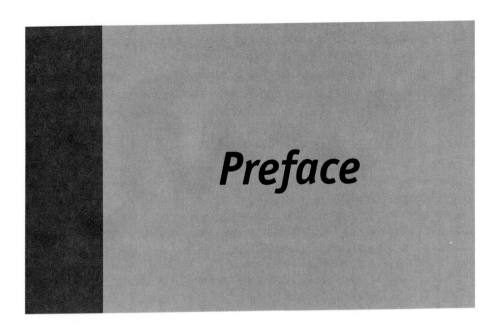

Preface

Author Ken Peak believes that this, the fourth edition of *Policing America*, is by far "bigger and better" than its three predecessors, providing a comprehensive view of the largely misunderstood, often obscure world of policing. A new chapter has been added concerning community oriented policing and problem solving. New materials have been added to other chapters as well, including discussions of terrorism, less-than-lethal weapons, hate crimes, stalking, and updated court decisions. Meanwhile, this edition continues to provide in-depth coverage of such topics as patrol, the police subculture, accountability, civil liability, extraordinary problems and practices, the rule of law, investigations, policing in selected foreign venues, and policing in the future.

The author brings more than 30 years of both scholarly and policing backgrounds to this effort; as a result, the chapters contain a "real world" flavor not found in most policing textbooks. Disseminated throughout the book are several "Practitioner's Perspectives"—short essays written by selected individuals who have expertise in particular areas of policing.

From its introduction, written by Darrel W. Stephens, police chief of Charlotte, North Carolina, through the final chapter, the reader is provided with a penetrating view of what is certainly one of the most difficult and challenging occupations in America.

Pedagogical Attributes

To make this textbook more reader-friendly, each chapter in this fourth edition begins with a listing of its key terms and concepts and an overview of the chapter. The textbook also includes "Items for Review" sections at the end of each chapter;

it is recommended that the reader examine these items prior to reading the chapter to get a feel for the chapter's contents and to obtain some insight as to its more substantive aspects.

Other instructional aids include the "Practitioner's Perspectives," tables and figures, boxes with recent news items, and photographs to aid readers in understanding the work of policing. A listing of relevant Web sites is provided in Appendix A for readers who wish to independently obtain more information about various aspects of policing. Finally, a detailed index at the end of the book facilitates the reader's ability to locate specific topics more quickly.

CHAPTER ORGANIZATION AND OVERVIEW

Chapter 1 discusses the history of policing, and Chapter 2 examines the contemporary status of federal, state, and local law enforcement agencies and their roles and functions. Chapter 3 examines the police subculture and how ordinary citizens are socialized to the role. The next chapter considers how police organizations are organized and administered and how administrators, middle managers, and supervisors perform their functions. Chapter 5 explores the very important function of patrolling and includes a discussion of the concepts of community policing and community problem solving.

Chapter 6, a new chapter, focuses on a rapidly spreading form of policing that is being embraced by thousands of police agencies across the United States and around the world: community oriented policing and problem solving—COPPS. Chapter 7 focuses on criminal investigation, including the highly progressive fields of forensic science and criminalistics, and Chapter 8 looks at several extraordinary police problems and methods: policing terrorism, hate crimes, and militias; the mafia; gangs; small jurisdictions; the homeless; and the nation's borders. The "rule of law" is discussed in Chapter 9, which delineates the constitutional guidelines that direct and constrain police actions. Chapter 10 looks at police accountability to the public, including the issues of police ethics, use of force, and corruption.

Police civil liability is examined in Chapter 11. Chapter 12 describes a number of trends and issues, including rights of police officers, women and minorities in policing, the private police, unionization, contract and consolidated policing, civilianization and accreditation of police agencies, higher education for police, and police stress. Then, to better understand policing in this country, Chapter 13 analyzes policing in four international venues: China, Mexico, Northern Ireland, and Saudi Arabia. Interpol, the international crime-fighting organization, is also discussed.

Chapter 14 examines police technology, including the myriad uses of computers, electronics, and imaging and communications systems. Developments with firearms and other tools are also discussed. Finally, Chapter 15 looks at the police of the future and how predictions are made.

The entire text provides the reader with a comprehensive and penetrating view of what is certainly one of the most difficult, challenging, and obscure occupations in America.

ACKNOWLEDGMENTS

This edition, like its three predecessors, is the result of the professional assistance of several practitioners and publishing people at Prentice Hall. First, I continue to benefit from my friendships and professional associations with Kim Davies, Executive Editor, and production editor Linda Pawelchak. The author also wishes to acknowledge the invaluable assistance of Steven Brandl, University of Wisconsin, William E. Kelly, Auburn University; Charles Martin, Marshall University; and George Rush, California State University–Long Beach, whose review of this fourth edition resulted in many beneficial changes. Also, David Balleau and Michael Goo, Washoe County (Nevada) Sheriff's Office, provided photographic assistance.

About the Author

Ken Peak is a full professor and former chairman of the Department of Criminal Justice, University of Nevada, Reno, where he was named Teacher of the Year by the university's Honor Society. He entered municipal policing in Kansas in 1970 and subsequently held positions as a nine-county criminal justice planner in Kansas; director of a four-state Technical Assistance Institute for the Law Enforcement Assistance Administration; director of university of police at Pittsburg State University (Kansas); acting director of public safety, University of Nevada, Reno; and assistant professor of criminal justice at Wichita State University. His textbooks include *Community Policing and Problem Solving: Strategies and Practices* (3d ed., with Ronald W. Glensor); *Justice Administration: Police, Courts, and Corrections Management* (3d ed.); *Police Supervision* (with Ronald W. Glensor and Larry K. Gaines); and *Policing Communities: Understanding Crime and Solving Problems* (an anthology, with R. Glensor and M. Correia). He has published two historical books: *Kansas Temperance: Much Ado About Booze, 1870–1920* (with P. Peak), and *Kansas Bootleggers* (with Patrick G. O'Brien). He also has published more than 50 journal articles and book chapters. He served as chairman of the Police Section of the Academy of Criminal Justice Sciences from 1997–1999 and recently served as president of the Western and Pacific Association of Criminal Justice Educators. His teaching interests include policing, administration, victimology, and comparative justice systems. He received two gubernatorial appointments to statewide criminal justice committees while residing in Kansas and holds a doctorate from the University of Kansas.

CHAPTER 1

Historical Development

Key Terms and Concepts

August Vollmer
Constable
Coroner
Justice of the Peace
Modus operandi
President's Crime Commission

Professional era of policing
Republicanism
Sheriff
Team policing
Wickersham Commission
William H. Parker

The farther back you can look, the farther forward you are likely to see.—*Winston Churchill*

Human history becomes more and more a race between education and catastrophe.—*H. G. Wells*

Fellow citizens, we cannot escape history.—*Abraham Lincoln*

INTRODUCTION

To understand contemporary policing in America, it is necessary to understand its antecedents. The police, it has been said, are "to a great extent, the prisoners of the past. Day-to-day practices are influenced by deeply ingrained traditions."[1] Another reason for analyzing historical developments and trends is that several discrete legacies have been transmitted to modern police agencies. In view of the significant historical impact on modern policing, it is necessary to turn back the clock to about A.D. 900.

Therefore, we begin with a brief history of the evolution of four primary criminal justice officers—sheriff, constable, coroner, and justice of the peace—from early England to the 1900s in America. We then examine policing from its early beginnings in England to the American colonial period, when volunteers watched over their "human flock." The concepts of patrol, crime prevention, authority, professionalism, and discretion may be traced to the colonial period. We move on to the adoption of full-time policing in American cities (with its predominant issues, political influences, and other problems) and on the western frontier. Then we consider the movement to professionalize the police by removing them from politics (and, at the same time, the citizenry) and casting them as crime fighters. Next, we discuss the movement away from the professional model, centering on the influence of the President's Crime Commission. The chapter concludes with an overview of the current era to which policing has evolved: community oriented policing and problem solving (COPPS).

ENGLISH AND COLONIAL OFFICERS OF THE LAW

All four of the primary criminal justice officials of early England—the sheriff, constable, coroner, and justice of the peace—either still exist or until recently existed in the United States. Accordingly, it is important to grasp a basic understanding of these offices, including their early functions in England and, later, in America. Following is a brief discussion of each.

SHERIFF

The word **sheriff** is derived from *shire reeve*—*shire* meaning county and *reeve* meaning agent of the king. The shire reeve appeared in England before the Norman Conquest of 1066. His job was to maintain law and order in the tithings. The office survives in England, but since the nineteenth century, the sheriff has had no police powers. When the office began, however, the sheriff virtually exercised the powers of a viceroy in his county. He assisted the king in fiscal, military, and judicial affairs and was referred to as the "king's steward," but the sheriff was never a popular officer in England. As men could buy their appointment from the Crown, the office was often held by nonresidents of the county who seemed intent only upon fattening their purses and abusing the public. Further, English sheriffs were often charged with being lazy in the pursuit of criminals. Indeed, by the late 1200s, sheriffs were forbidden to act as justices. The position of coroner was created, and coroners acted as monitors over the sheriff. Thereafter, the status and responsibility of the position began to diminish. Coroners were locally elected officials, and their existence prompted the public to seek similar selection and control over sheriffs. In response, just before his death, Edward I granted to the counties the right to select their sheriffs. Then, with the subsequent appearance of the justice of the peace, the sheriff's office declined in power even further. At the present time in England, a sheriff's only duties are to act as officer of the court, summon juries, and enforce civil judgments.[2]

The first sheriffs in America appeared in the early colonial period. By contrast to the status of the office in England, control over sheriffs has rested with the county electorate since 1886. Today, the American sheriff remains the basic source of rural crime control. When the office appeared in the colonies, it was little changed from the English model. However, the power of appointment was originally vested in the governor, and the sheriff's duties included apprehending criminals, caring for prisoners, executing civil process, conducting elections, and collecting taxes. In keeping with English tradition, and fearing oppression and extortion by the sheriff, colonists generally limited the sheriff's term of office; sometimes the term was as short as a single year. The duties of collecting taxes and conducting elections alone accorded the sheriffs tremendous power and influence.[3]

In the late 1800s, the sheriff became a popular figure in the legendary Wild West (discussed later in this chapter). The frontier sheriffs often used the concept of *posse comitatus*, an important part of the criminal justice machine that allowed the sheriff to deputize common citizens to assist in the capture of outlaws, among other tasks. The use of the posse declined after 1900, because the enlistment of untrained people did not meet the requirements of a more complex society. Overall, by the turn of the century, the powers and duties of the sheriff in America had changed very little in status or function. In fact, the office has not changed much today; the sheriff continues to enjoy the authorization to utilize police powers.

CONSTABLE

Like the sheriff, the **constable** may be traced back to Anglo-Saxon times. The office began during the reign of Edward I, when every parish or township had a constable. As the county militia turned more and more to matters of defense, the constable alone pursued felons. Hence, the ancient custom of citizens raising a loud "hue and cry" and joining in pursuit of criminals lapsed into disuse. During the Middle Ages, there was as yet no high degree of specialization. The constable had a variety of duties, including collecting taxes, supervising highways, and serving as magistrate. The office soon became subject to election and was conferred upon local persons of prominence. However, the creation of the office of justice of the peace around 1200 quickly changed this trend forever; soon he was limited to making arrests only with warrants issued by a justice of the peace. As a result, the office, deprived of social and civic prestige, was no longer attractive. It carried no salary and the duties often were dangerous. Additionally, there was heavy attrition in the office, and so its term was limited to one year in an attempt to attract officeholders. Patrick Colquhoun, a London magistrate and noted police author, disposed of the parish constables altogether, and in 1856 Parliament completely discarded the office.[4]

The office of constable experienced a similar process of disintegration in the colonies. As elected officials, American constables had a variety of duties similar to those of their counterparts in England. However, the American constables, usually two in each town, were given control over the night watch. By the 1930s, state constitutions in 21 states provided for the office of constable. However, constables still received no pay in the early part of this century, and, again like their British colleagues, they have enjoyed little prestige or popularity since the early 1900s. The position fell into disfavor largely because most of its occupants were untrained. By the early twentieth century in America, the constables were believed to be wholly inadequate as officials of the law.[5]

CORONER

The office of **coroner** is more difficult to describe. It has been used to fulfill many different roles throughout its history and has steadily changed over the centuries. There is no agreement concerning the date when the coroner first appeared in Eng-

A leatherhead and his sentry box. Called leatherheads because of their distinctive leather helmets, constables patrolled the city from scattered sentry boxes in the early decades of the 1800s. They wore helmets for protection against falling debris from fires—a constant danger in cities with many wooden buildings. (*Courtesy NYPD Photo Unit*)

land, but there is general consensus that the office was functioning by the end of the twelfth century. The reason is that both the Crown and the property holders were anxious to increase the prestige of this office at the expense of the sheriff. From the beginning, the coroner was elected; his duties included oversight of the interests of the Crown, not only in criminal matters but in fiscal matters as well. In felony cases, the coroner could conduct a preliminary hearing, and the sheriff often came to the coroner's court to preside over the coroner's jury. The coroner's inquest provided another means of power and prestige. The inquest determined the cause of death and the party responsible for it. Initially, coroners were given no compensation, yet they were elected for life. Soon, however, it became apparent that officials holding this office were unhappy with the burdensome tasks and the absence of compensation. Therefore, they were given the right to charge fees for their work.[6]

As was true of sheriffs and constables, at first the office of the coroner in America was only slightly changed from what it had been in England. The office was slow in gaining recognition in America, as many of the coroners' duties were already being performed by the sheriffs and justices of the peace. By 1933, the coroner was recognized as a separate office in two-thirds of the states. Tenure was generally limited to two years. By the 1930s, however, the office had been stripped of many of its original functions, especially those of a fiscal nature. Today, in many states, the coroner legally serves as sheriff when the elected sheriff is disabled or disqualified. However, since the early part of this century, the coroner has basically performed a single function: determining the causes of all deaths by violence or suspicious circumstances. The coroner or his or her assistants are expected to determine the causes and effects of wounds, lesions, contusions, fractures, poisonous substances, and other fatal elements. The coroner's inquest resembles a grand jury, at which the coroner serves as a kind of presiding magistrate. If the inquest determines that the deceased came to his or her death through criminal means, the coroner may issue a warrant for the arrest of the accused party.[7]

The primary debate regarding the office of coroner has centered on the qualifications needed to hold the office. Many states have traditionally allowed laypeople, as opposed to physicians, to be coroners. Thus, people of all backgrounds—ranging from butchers to musicians—have occupied this powerful office. America still clings to the old English philosophy that the chief qualifications for a person to be a coroner are "the possession of tact, sound discretion, practical sense, sympathy, quick perception and a knowledge of human nature."[8]

JUSTICE OF THE PEACE

The **justice of the peace (JP)** may be traced back as far as 1195 in England; by 1264 the *custos pacis*, or conservator of the peace, nominated by the king for each county, presided over criminal trials. Soon, in recognition of its new status, the term *custos* was dropped and *justice* substituted for it. Early JPs were wealthy landholders. They allowed constables to make arrests by issuing them warrants. Over time, this practice removed power from constables and sheriffs. The duties of JPs eventually included the granting of bail to felons, which led to corruption and criticism as the justices bailed people who clearly should not have been released into the community. By the sixteenth century, the office came under criticism, again because of the caliber of people holding it. Officeholders were often referred to as "boobies" and "scum of the earth."[9] The only qualification necessary was being a wealthy landowner who was able to buy his way into office.

By the early 1900s, England had abolished the property-holding requirement, and many of the medieval functions of the JP office were removed. Thereafter, the office possessed strictly extensive criminal jurisdiction but no jurisdiction whatsoever in civil cases. This contrasts with the American system, which gives JPs limited jurisdiction in both criminal and civil cases. Interestingly, even today relatively few English JPs have been trained in the law.

The JP office in the colonies was a distinct change from the position as it existed in England. JPs were elected to office and given jurisdiction in both civil and criminal cases. By 1930, the office had constitutional status in all of the states. JPs have long been allowed to collect fees for their services. For example, in the 1930s the JP typically received 10 cents for administering an oath, 50 cents for preparing information, 50 cents for issuing a warrant for arrest, 1 dollar per day for attendance in court, 25 cents for issuing a subpoena, and 5 cents for filing each paper required by law. As in its English form, typically it is not necessary to hold a law degree or to have pursued legal studies in order to be a JP. Thus, tradespeople and laborers have long held the position, and American JPs come from all walks of life. In the past, they have held court in their saddle shops, kitchens, and flour mills.[10]

Perhaps the most colorful justice was Roy Bean, popularized in the movies as the sole peace officer in a 35,000–square-mile area west of the Pecos River, near Langtry, Texas. Bean was known to hold court in his shack, where three signs hung on the front porch. One read "Justice Roy Bean, Notary Public," another read "Law West of the Pecos," and the third read "Beer Saloon." Cold beer and the law undoubtedly shared many quarters on the western frontier.

A familiar complaint about modern-day JPs has been that they often operate in collusion with police officers, who set up speed traps in order to collect and share the fines. Indeed, many JPs have been known to complain when the police bring them few cases, or even none at all, or when the police take their "business" to other JPs, thereby reducing their incomes. This reputation, coupled with the complaints about their qualifications, has caused the office of *justice of the peace* to be ridiculed for some time. JPs are today what they perhaps were intended to be— lay and inexpert upholders of the law. On the whole, the office has declined from dignity to obscurity and ridicule. As one observer noted, this loss of prestige can never be recovered.[11]

THE OLD ENGLISH SYSTEM OF POLICING

Like much of the American criminal justice system, modern American policing can be traced directly to its English heritage. Ideas concerning community policing, crime prevention, the posse, constables, and sheriffs were developed from English policing. Beginning at about 900, the role of law enforcement was placed in the hands of common citizens. Each citizen was responsible for aiding neighbors who might become victimized by outlaws.[12] No formal mechanism existed with which to police the villages, so this voluntary, informal model that developed was referred to as "kin police."[13] Slowly this model developed into a more formalized, community-based system.

After the Norman Conquest of 1066, a community-based system called "frankpledge" was established. This system required that every male above the age of 12 form a group with nine of his neighbors. This group, called a tithing, was sworn to help protect fellow citizens and to apprehend and deliver any of its members who committed a crime. Tithingmen were not paid salaries for their work, but they were required to perform certain duties under penalty of law.[14] Ten tithings were grouped into a hundred, directed by a constable who was appointed by a nobleman and who, in effect, became the first police officer. The constable was the first police official with law enforcement responsibility greater than simply protecting his neighbors. As the tithings were grouped into hundreds, the hundreds were grouped into shires, which are similar to today's counties.

By the late 1500s, however, wealthier merchants and farmers became reluctant to take their turn in the rotating job of constable. The office was still unpaid, and the duties were numerous. Wealthier men paid the less fortunate to serve in their place until there came a point at which no one but the otherwise unemployable would serve as constable. Thus, from about 1689 on, the demise of the once-powerful office was swift. All who could afford to pay their way out of service as constable to King George I did so. Daniel Defoe, the noted author, spoke for many when he wrote in 1714 that the office of constable represented "an unsupportable hardship; it takes up so much of a man's time that his own affairs are frequently totally neglected, too often to his ruin."[15]

Another cause of the decline of the old system, as mentioned earlier, was the corruption of the justices of the peace. Following the Glorious Revolution in 1688, many families whose members had once filled the office became disaffected with the Crown and so began to refuse the position. Soon, the people who did become justices of the peace were inspired primarily by the potential for profit the office presented. The unsavory magistrates became known as the "justices of mean degree," and the "trading justice" of the first half of the eighteenth century emerged in a criminal justice system where anything was possible for a fee.

The potential for corruption in this system is obvious. The justice of the peace was rewarded in proportion to the number of people he convicted, so extortion was rampant. Ingenious criminals were able to exploit this state of affairs to great advantage. One such criminal was Jonathan Wild, who, for seven years before his execution in 1725, obtained single-handed control over most of London's criminals. Wild's system was simple: After ordering his men to commit a burglary, he would meet the victim courteously and offer to return the stolen goods for a commission. Wild was so successful in fencing stolen property that he found it necessary to transport his booty to warehouses abroad. That he could have operated such a business for so long is a testimony to the corrupt nature of the magistrates of the "trading justice" period.[16]

This early English system, in large measure voluntary and informal, continued with some success well into the 1700s. By the end of the eighteenth century, however, the collapse of its two primary offices and the growth of large cities, crime, and civil disobedience required that changes be made in the system. Soon the British Parliament would be forced to consider and adopt a more dependable system.

POLICING IN COLONIAL AMERICA

The first colonists transplanted the English policing system, with all of its virtues and faults, to seventeenth- and eighteenth-century America. Most of the time the colonies were crime free as their inhabitants busied themselves carving out a farm and a living. Occasionally colonists ran afoul of the law by violating or neglecting some moral obligation—often finding themselves in court for working on the Sabbath, cursing in public, failing to pen animals properly, or begetting children out of wedlock. Only two "crime waves" of note occurred during the seventeenth-century period, both in Massachusetts. In one case, between 1656 and 1665, Quakers who dared challenge the religion of the Puritan colony were whipped, banished, and in three instances, hanged. The second "crime wave" involved witchcraft. Several alleged witches were hanged in 1692 in Salem; dozens more languished in prison before the hysteria abated.[17]

Once colonists settled into villages, including Boston (1630), Charleston (1680), and Philadelphia (1682), local ordinances provided for the appointment of constables, whose duties were much like those of their English predecessors. County governments, again drawing upon English precedent, appointed sheriffs as well. The county sheriff, appointed by a governor, became the most important law enforcement official, particularly when the colonies were small and rural. The sheriff apprehended criminals, served subpoenas, appeared in court, and collected taxes. The sheriff was also paid a fixed amount for each task performed; the more taxes he collected, the higher his pay.[18]

Criminal acts were so infrequent as to be largely ignored; in fact, law enforcement was given low priority. Service as a constable or watchman was obligatory, and for a few years citizens did not seem to mind this duty. But as towns grew and the task of enforcing the laws became more difficult and time-consuming, the colonists, like their English counterparts, began to evade the duty when possible. The "watch-and-ward" responsibility of citizens became more of a comical "snooze-and-snore" system. In Boston, the citizens were so evasive about performing police services that in 1650 the government was moved to threaten citizens who refused to serve with heavy fines. New Amsterdam's Dutch officials introduced a paid watch in 1658, and Boston tried the concept in 1663, but the expense involved quickly forced both cities to discontinue the practice.[19] Apparently, paid policing might have been a good substitute for half-hearted volunteerism, but only if the citizens did not have to pay for it.

Unfortunately for these colonists, their refusal to provide a dependable voluntary policing system into the eighteenth century came at a time when economic, population, and crime growth required reliable police. But the most reliable citizens continued to refuse the duty, and watchmen were hardly able to stay awake at night. The citizen participation model of policing was breaking down, and something had to be done, especially in the larger colonies. Philadelphia devised a plan, enacted into law, that restructured the way in which the watch was performed. City officials hoped that it would solve the problem of enforcing the laws. The law empowered officials, called wardens, to hire as many watchmen as needed; the powers of the watch were increased; and the legislature levied a tax to pay for the watch. Instead of requiring all males to participate, only male citizens interested in making money needed to join the watch. This Philadelphia plan was moderately successful; other cities were soon inspired to follow its example and offer tax-supported wages for watches.[20]

From the middle to the late 1700s, massive social and political unrest caused police problems to increase still further. From 1754 to 1763, the French and Indian War disrupted colonial society. When the war ended, a major depression impoverished many citizens. That depression ended with the American Revolution. In 1783, after the revolution ended, another depression struck; this one lasted until about 1790. Property and street crimes continued to flourish, and the constabulary and the watches were unable to cope with it. Soon it became evident that, like the English, the American people needed a more dependable, formal system of policing.

LEGACIES OF THE COLONIAL PERIOD

As uncomplicated and sedate as colonial law enforcement seems, especially when compared to contemporary police problems, the colonial period is very important to the history of policing because many of the basic ideas that influence modern policing were developed during that era. Specifically, the colonial period transmitted three legacies to contemporary policing.[21]

First, as was shown in the foregoing discussion, the colonists committed themselves to local (as opposed to centralized) policing. Second, the colonists reinforced that commitment by creating a theory of government called **republicanism**. Republicanism asserted that power can be divided, and it relied upon local interests to promote the general welfare. Police chiefs and sheriffs might believe that they alone know how to address crime and disorder, but under republicanism, neighborhood groups and local interest blocs have input with respect to crime control policy. Republicanism thus established the controversial political framework within which the police would develop during the next 200 years.[22]

Finally, the colonial period witnessed the onset of the theory of crime prevention. This legacy would alter the shape of policing after 1800 and eventually bring about the emergence of modern police agencies. The population of England doubled between 1700 and 1800. Parliament, however, took no measures to help solve the problems that arose from social change. Each municipality or county, therefore, was forced to attempt to solve its problems in piecemeal fashion. After 1750, practically every English city increased the number of watchmen and constables, hoping to address the problem of crime and disorder but not giving any thought to whether or not this ancient system of policing still actually worked. However, the cities did convert to providing paid, rather than voluntary, watches.[23]

London probably suffered the most from this general inattention to social problems; awash in crime, whole districts had become criminal haunts where no watchmen visited and no honest citizens frequented. Thieves became very bold, robbing their victims in broad daylight on busy streets. In the face of this situation, English officials still continued to prefer the existing policing arrangements over any new ideas. However, three men—Henry Fielding, his half-brother John Fielding, and Patrick Colquhoun—began to experiment with possible solutions. And although their efforts would have no major effect in the short term in England, they laid the foundation upon which later reformers would build new innovations.

Henry Fielding's acute interest in, and knowledge of, policing led to his 1748 appointment as chief magistrate of Bow Street in London. He soon became one of England's most acclaimed theorists in the area of crime and punishment. Fielding's primary argument was that the severity of the English penal code, which provided for the death penalty for a large number of offenses, including the theft of a handkerchief, did not work in controlling criminals. He believed the country should reform the criminal code in order to deal more with the origins of crime. In 1750, Fielding made the pursuit of criminals more systematic by creating a small group of "thief-takers." Victims of crime paid handsome rewards for the capture of their assailants, so these volunteers stood to profit nicely by pursuing criminals.[24]

When Henry Fielding died in 1754, John Fielding succeeded him as Bow Street magistrate. By 1785, his thief-takers had evolved into the Bow Street Runners—some of the most famous policemen in English history. While the Fieldings started to think about ways of creating a police force that could deal with the changing English society, horrible punishments and incompetent policing continued throughout England.

Patrick Colquhoun was independently wealthy and sincerely interested in improving social conditions in England. In 1792, Colquhoun was appointed London magistrate, and for the next quarter of a century he focused on police reform. He, like the Fieldings, wrote lengthy treatises on the police, and soon he established himself as an authority on police reform. Colquhoun believed that government could, and should, regulate people's behavior. This notion contradicted tradition and even constitutional ideals, undermining the old principle that the residents of local communities, through voluntary watchmen and constables, should police the conduct of their neighbors. Colquhoun also endorsed three ideas originally set forth by the Fieldings: (1) the police should have an intelligence service for gathering information about offenders; (2) a register of known criminals and unlawful groups should be maintained; and (3) a police gazette should be published to assist in the apprehension of criminals and to promote the moral education of the public by publicizing punishments such as whipping, the pillory, and public execution. To justify these reforms, Colquhoun's research estimated that in 1800 there were 10,000 thieves, prostitutes, and other criminals in London, who stole more than a half-million pounds of goods annually from the riverside docks alone.[25]

Colquhoun also believed that policing should maintain the public order, prevent and detect crime, and correct bad manners and morals. He did not agree with the centuries-old notion that watchmen—who, after all, were amateurs—could adequately police the communities. Thus, Colquhoun favored a system of paid professional police officers who would be recruited and maintained by a centralized governmental authority. Colquhoun believed that potential criminals could be identified before they did their unlawful deeds.[26] Thus began the notion of proactive policing, that is, preventing the crime before it occurs. Colquhoun died before his proposals were adopted, and as the eighteenth century ended in England and America, the structure of policing was largely unchanged. However, both nations had experienced the inadequacies of the older form of policing. Although new ideas had emerged, loyalties to the old system of policing would remain for some time.

POLICE REFORM IN ENGLAND AND AMERICA, 1829–1860

Two powerful trends in England and America brought about changes in policing in both countries. First was urbanization, and second was industrialization. These developments generally increased the standard of living for both Americans and western Europeans. Suddenly, factories needed sober, dependable people who could be trusted with machines. To create a reliable workforce, factory owners began advocating temperance. While many workers resented this attempt at social control and reform, clearly a new age, a new way of thinking, had begun. Crime also increased during this period. Thus, social change, crime, and unrest made the old system of policing obsolete. A new policing system was needed, one that could effectively deal with criminals, maintain order, and prevent crime.[27]

In England, after the end of the Napoleonic Wars in 1815, workers protested against new machines, food riots, and an ongoing increase in crime. The British army, traditionally used to disperse rioters, was becoming less effective as people began resisting its commands. In 1822, England's ruling party, the Tories, moved to consider new alternatives. The prime minister appointed Sir Robert Peel to establish a police force to quell the problems. Peel, a wealthy member of Parliament who was familiar with the reforms suggested by the Fieldings and Colquhoun, found that many English people objected to the idea of a professional police force, thinking it a possible restraint on their liberty. They also feared a stronger police organization because the criminal law was already quite harsh, as it had been for many years. By the early nineteenth century, there were 223 crimes in England for which a person could be hanged. Because of these two obstacles, Peel's efforts to gain support for full-time, paid police officers failed for seven years.[28]

Peel finally succeeded in 1829. He had established a base of support in Parliament and tried to reform only the metropolitan police of London rather than trying to create policing for the entire country. Peel submitted a bill to Parliament. This bill, which was very vague about details, was called "An Act for Improving the Police In and Near the Metropolis." Parliament passed the Metropolitan Police Act of 1829. The *General Instructions* of the new force stressed its preventive nature, saying that "the principal object to be attained is 'the prevention of crime.' The security of person and property will thus be better effected, than by the detection and punishment of the offender after he has succeeded in committing the crime."[29] The act called for the home secretary to appoint two police commissioners to command the new organization. These two men were to recruit "a sufficient number of fit and able men" as constables.[30] Peel chose a former military colonel, Charles Rowan, as one commissioner, and a barrister (attorney), Richard Mayne, as the other. Both would turn out to be excellent choices. They divided London into 17 divisions, using crime data as the primary basis for creating the boundaries. Each division had a commander called a superintendent; each superintendent had a force of four inspectors, 16 sergeants, and 165 constables. Thus, London's Metropolitan Police immediately consisted of nearly 3,000 officers. The commissioners also decided to put their constables in a uniform (blue coat, blue pants, and a black top hat) and to arm them with a short baton (known as a truncheon) and a rattle (for raising an alarm). Each constable was to wear his own identifying number on his collar, where it could be easily seen.[31]

Interestingly, the London police (nicknamed "bobbies" after Sir Robert Peel) quickly met with tremendous public hostility. Wealthy people resented their very existence and became particularly incensed at their attempts to control the movements of their horse-drawn coaches. Several aristocrats ordered their coachmen to whip the officers or simply drive over them. Juries and judges refused to punish those persons who assaulted the police. Defendants acquitted by a hostile judge would often sue the officer for false arrest. Policing London's streets in the early 1830s proved to be a very dangerous and lonely business. The two commissioners, Rowan and Mayne, fearing that public hostility might kill off the police force, moved to counter it. The bobbies were constantly told to be respectful yet firm when dealing with the public. Citizens were invited to lodge complaints if their officers were truly unprofessional. This policy of creating public support gradually worked; as the police became more moderate in their conduct, public hostility also declined.[32]

Peel, too, proved to be very farsighted and keenly aware of the needs of both a professional police force and the public that would be asked to maintain it. Indeed, Peel saw that the poor quality of policing contributed to social disorder. Accordingly, he drafted several guidelines for the force, many of which focused on community relations. He wrote that the power of the police to fulfill their duties depended on public approval of their actions; that as public cooperation

A "Peeler," c. 1829. "Peeler," "Robert," or "Bobby" were all early names for a police officer, the latter remaining as a nickname today. All are memorials to Sir Robert Peel.

increased, the need for physical force by the police would decrease; that officers needed to display absolutely impartial service to law; and that force should be employed by the police only when attempts at persuasion and warning had failed, and then they should use only the minimal degree of force possible. Peel's remark that "the police are the public, and the public are the police" emphasized his belief that the police are first and foremost members of the larger society.[33]

Peel's attempts to appease the public were necessary; during the first three years of his reform effort, he encountered strong opposition. He was denounced as a potential dictator, the *London Times* urged revolt, and *Blackwood's Magazine* referred to the bobbies as "general spies" and "finished tools of corruption." A secret national group, called the "Blue Devils" and the "Raw Lobsters," was organized to combat the police. During this initial five-year period, Peel endured the largest police turnover rate in history. Estimates range widely, but these are thought to be fairly accurate: 1,341 constables resigned from London's Metropolitan Police from 1829–1834, or roughly one-half of the constables on the force. The pay of three shillings a day was meager, and probably few of the officers ever considered the position as a career. In fact, many of the men, who had been laborers, took jobs as bobbies to tide them through the "stress of weather," waiting until the inclement weather passed and they could resume their trades.[34]

Peel drafted what have become known as "Peel's Principles" of policing. Most, if not all, of them are still relevant to today's police community:

1. The police must be stable, efficient, and organized along military lines.
2. The police must be under governmental control.
3. The absence of crime will best prove the efficiency of the police.
4. The distribution of crime news is absolutely essential.
5. The deployment of police strength both by time and area is essential.
6. No quality is more indispensable to a policeman than a perfect command of temper; a quiet, determined manner has more effect than violent action.
7. Good appearance commands respect.
8. The securing and training of proper persons is at the root of efficiency.
9. Public security demands that every police officer be given a number.
10. Police headquarters should be centrally located and easily accessible to the people.
11. Policemen should be hired on a probationary basis.
12. Police records are necessary to the correct distribution of police strength.[35]

London's experiment in full-time policing did not bring about the instant expansion of the model across England. It would take many years for other English communities to replace their stubborn reliance on the watchman system.

FULL-TIME POLICING COMES TO THE UNITED STATES

IMITATING PEEL

Americans, meanwhile, were observing Peel's successful experiment with the bobbies on the patrol beat. However, industrialization and social upheaval had not reached the proportions here that they had in England, so there was not the urgent need for full-time policing that had been felt there. Yet by the 1840s, when industrialization began in earnest in America, U.S. officials began to watch the police reform movement in England more closely.

When the movement to improve policing did begin in America in the 1840s, it occurred in New York City. (Philadelphia, with a private bequest of $33,000, actually began a paid daytime police force in 1833; however, it was disbanded in three years.) The police reform movement had actually begun in New York in 1836, when the mayor advocated a new police organization that could deal with civil disorders. The city council denied the mayor's request, saying that the doctrine of republicanism prevented it and that, instead, citizens should simply aid one another in the combatting of crime.

Efforts at police reform thus fell dormant until 1841, when a highly publicized murder case resurrected the issue, showing again the incompetence of the officers under the old system of policing. One Mary Cecilia Rogers left her New

York home one day and disappeared; three days later, her body was discovered in the Hudson River. The public and newspapers clamored for police to solve the crime. The police appeared unwilling to investigate the crime until an adequate reward was offered.[36] Edgar Allen Poe's 1850 short story "The Mystery of Marie Roget" was based on this case. The Rogers case and the police response did more to encourage police reorganization than all of the previous cries for change.

In 1844, the New York state legislature passed a law establishing a full-time preventive police force for New York City. However, this new body came into being in a very different form than what was current in Europe. The American version, as begun in New York City, was deliberately placed under the control of the city government and city politicians. The American plan required that each ward in the city be a separate patrol district, unlike the European model, which divided the districts along the lines of criminal activity. The process for selecting officers was also different. The mayor chose the recruits from a list of names submitted by the aldermen and tax assessors of each ward; the mayor then submitted his choices to the city council for approval. This system adhered to the principles of republicanism and resulted in most of the power over the police going to the ward aldermen. Those politicians were seldom concerned about selecting the best people for the job; instead, the system allowed and even encouraged political patronage and rewards for friends.[37]

The law also provided for the hiring of 800 officers—not nearly enough to cover the city—and for the hiring of a chief of police, who had no power to hire officers, assign them to duties, or fire them. Furthermore, the law did not require the officers to wear uniforms; instead, they were to carry a badge or other emblem for identification. Citizens would be hard pressed to recognize an officer when they needed one. As a result of the law, New York's officers would be patrolling a beat around the clock, and pay scales were high enough to attract good applicants. At the same time, the position of constable was dissolved. Overall, these were important reforms over the old system and provided the basis for continued improvements that were supported by the public.[38]

It did not take long for other cities to adopt the general model of the New York City police force. New Orleans and Cincinnati adopted plans for a new police force in 1852, Boston and Philadelphia followed in 1854, Chicago in 1855, and Baltimore and Newark in 1857.[39] By 1880, virtually every major American city had a police force based on Peel's model.

EARLY ISSUES AND NEW TRADITIONS

Three important issues confronted these early American police officers as they took to the streets between 1845 and 1869: whether the police should be in uniform, whether they should be armed, and whether they should use force.

The issue of a police uniform was important for several reasons. First, the lack of a uniform negated one of the basic principles of crime prevention—that police officers be visible. Crime victims wanted to find a police officer in a hurry.

Further, uniforms would make it difficult for officers to avoid their duties, since it would strip them of their anonymity. Interestingly, police officers themselves tended to prefer not to wear a uniform. They contended that the uniform would hinder their work because criminals would recognize them and flee and that the uniform was demeaning and would destroy their sense of manliness and democracy. One officer went so far as to argue that the sun reflecting off his badge would warn criminals of his approach; another officer hired an attorney and threatened to sue if he were compelled to don a uniform. To remedy the problem, New York City officials took advantage of the fact that their officers served four-year terms of office; when those terms expired in 1853, the city's police commissioners announced they would not rehire any officer who refused to wear a uniform. Thus, New York became the first American city with a uniformed police force. In 1860, it was followed by Philadelphia, where there was also strong police objection to the policy. In Boston (1858) and Chicago (1861), police accepted the adoption of uniforms more easily.[40]

A more serious issue confronting politicians and the new police officers was the carrying of arms. At stake was the personal safety of the officers and the citizens they served. Nearly everyone viewed an armed police force with considerable suspicion. However, after some surprisingly calm objections by members of the public, who noted that the London police had no need to bear arms, it was agreed that an armed police force was unavoidable. Of course, America had a long tradition that citizens had the right—sometimes even the duty—to own firearms. And armed only with nightsticks, the new police could hardly withstand attacks by armed assailants. The public allowed officers to carry arms simply because there was no alternative, which was a significant change in American policing and a major point of departure from its English model. Practically from the first day, then, the American police have been much more physical and violent than their English counterparts.[41]

Eventually the use of force, the third issue, would become necessary and commonplace for the American officers. Indeed, the uncertainty about whether or not an offender was armed perpetuated the need for an officer to rely on physical prowess for survival on the streets. The issue of use of force will be discussed further in Chapter 9.

ATTEMPTS AT REFORM IN DIFFICULT TIMES

By 1850, American police officers still faced a difficult task. In addition to maintaining order and coping with vice and crime, they would, soon after putting on the uniform, be separated from their old associates and viewed with suspicion by most citizens. With few exceptions, the work was steady and layoffs were uncommon. The nature of the work and the possibility of a retirement pension tied officers closely to their work and their colleagues. By 1850, there was a surplus of unskilled labor, particularly in the major eastern cities. The desire for economic security was reason enough for many able-bodied men to try to enter police service. New York City paid its police officers about twice as much as the wages paid unskilled laborers. Police departments had about twice as many applicants as positions. The system of political patronage prevailed in most cities, even after civil service laws attempted to introduce merit systems for hiring police.[42]

In New York, the police reform board was headed by Theodore Roosevelt, who sought applications for the department from residents in upstate areas. When these officers, later called "bushwhackers," were appointed, they were criticized by disgruntled Tammanyites (corrupt New York City politicians) who favored the political patronage system. The Tammanyites complained that the bushwhackers "could not find their way to a single station house."[43] Roosevelt's approach violated the American tradition of hiring local boys for local jobs. In England, meanwhile, police officials purposely sought applicants from outside the London area, believing it more advantageous to hire young men not too wise to the local ways or involved with local people.[44]

Citizens saw these new uniformed anomalies as people who wanted to spoil their fun or close their saloons on Sunday. In addition to police officers' geographical and social isolation, they became isolated in other ways, most of which still prevail today. For example, from the onset of professional American policing, there was little or no lateral movement from one department to another; the officer typically spent his entire career in one city, unable to transfer seniority

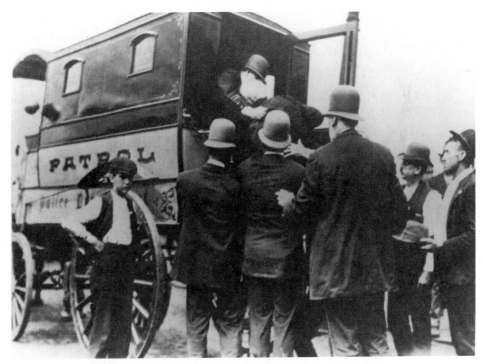

An example of a horse-drawn paddy wagon in use in many cities at the turn of the century. (*Courtesy IACP*)

or knowledge for use in a promotion in another city. Consequently, police departments soon became very inbred; new blood entered only at the lowest level. Tradition became the most important determinant of police behavior: A major teaching tool was the endless string of war stories the recruit heard, and the emphasis in most departments was upon doing things as they had always been done. Innovation was frowned upon, and the veterans impressed upon the rookies the reasons why things had to remain the same.[45]

The police officers of the late nineteenth century kept busy with riots, strikes, parades, and fires. These occurrences often made for hostile interaction between citizens and the police. Labor disputes often meant long hours of extra duty for the officers, for which no extra pay was received. This, coupled with the fact that the police did not engage in collective bargaining, resulted in the police having little empathy or identification with strikers or strikebreakers. Therefore, the use of the baton to put down riots, known as the "baton charge," was not uncommon. On New York's Lower East Side—where labor conflict was frequent—Jewish spokespeople called the police "Black Hundreds" in memory of conditions in Czarist Russia.[46]

During the late nineteenth century, large cities gradually became more orderly places. The number of riots dropped. In the post–Civil War period, however, ethnic group conflict sometimes resulted in individual and group acts of violence and disorder. Hatred of Catholics and Irish Protestants led to the killing and wounding of over 100 people in large eastern cities. Vaudeville and boxing served as safety valves for venting the tensions of urban life. Still, American cities were more orderly in 1900 than they had been in 1850. The possibility of violence involving labor disputes remained, and race riots increased in number and intensity after 1900, but daily urban life became more predictable and controlled. American cities absorbed millions of newcomers after 1900 without the social strains that attended the Irish immigration of the 1830s to 1850s.[47]

INCREASED POLITICS AND CORRUPTION

A more developed urban life also promoted order. Work groups and social clusters provided a sense of integration and belonging. Immigrants established benefit societies, churches, synagogues, and social clubs. At the same time, the police had acquired experience in dealing with potential sources of violence. Police in northern and western cities reflected the times. Irish Americans constituted a heavy proportion of the police departments by the 1890s; they made up more than one-fourth of the New York police force as early as the 1850s. Huge proportions of Irish officers were also found in Boston, Chicago, Cleveland, and San Francisco.[48]

Ethnic and religious disputes were found in many police departments. In Cleveland, for example, Catholics and Masons distrusted one another, while in New York, the Irish officers controlled many hirings and promotions. And there were still strong political influences at work. George Walling, a New York Police Department superintendent for 11 years, lamented after his retirement that he was largely a figurehead. Politics were played to such an extent that even nonranking patrol officers used political backers to obtain promotions, desired assignments, and transfers.

Police corruption also surfaced at this time. Corrupt officers wanted beats close to the gamblers, saloonkeepers, madams, and pimps—people who could not operate if the officers were "untouchable" or "100 percent coppers."[49] Political pull for corrupt officers could work for or against them; the officer who incurred the wrath of his superiors could be transferred to the outposts, where he would have no chance for financial advancement.

While police departments often had strong rivalries and political and religious factions, the officers banded together against outside attack. In New York, officers routinely committed perjury to protect each other against civilian complaints. An early form of "Internal Affairs" thus developed in the 1890s: the "shoo-fly," a plainclothes officer who checked on the performance of the patrol officers. When Theodore Roosevelt served as police commissioner in New York, he frequently made clandestine trips to the beats to check his officers; those malingerers found in the saloons were summoned to headquarters in the morning.[50]

Turn-of-the-century police equipment. (*Courtesy NYPD Photo Unit*)

MEANWHILE, ON THE AMERICAN FRONTIER . . .

While large cities in the east were struggling to overcome social problems and establish preventive police forces, the western half of America was anything but passive. Many historians believe that the true character of Americans developed on the frontier. Rugged individualism, independence, and simplicity of manners and behavior lent dignity to American life.

For most Americans, this period of police history holds a fascination, a time when heroic marshals had gunfights in Dodge City and other wild cowboy towns. But this period is also riddled with exaggerated legends and half-truths. During the latter half of the nineteenth century, the absence of government created a confusing variety of forms of policing in the West. Large parts of the West were under federal control, some had been organized into states, and still others were under Indian control, at least on paper. Law enforcement was performed largely by federal marshals and their deputies. Once a state was created within a territory, its state legislature had the power to attempt to deal with crime by appointing county sheriffs. Otherwise, there was no uniform method or standard for attempting to control the problems of the West.

When the people left the wagon trains and their relatively law-abiding ways, they attempted to live together in communities. Many different ethnic groups— Anglo-Americans, Mexicans, Chinese, Indians, freed blacks, Australians, Scandinavians, and others—competed for often scarce resources and fought one another violently, often with mob attacks. Economic conflicts were frequent between cattlemen and sheepherders, and they often led to major range wars. There was constant labor strife in the mines. The bitterness of the slavery issue remained, and many men with firearms skills learned during the Civil War turned to outlawry after leaving the service. (Jesse James was one such person.) In spite of these difficulties, westerners did manage to establish peace by relying on a combination of four groups who assumed responsibility for law enforcement: private citizens, U.S. marshals, businessmen, and town police officers.[51]

Private citizens usually helped to enforce the law by joining a posse or through individual efforts. Such was the case with the infamous Dalton Gang in Coffeyville, Kansas. The five gang members attempted to rob a bank in Coffeyville in 1892. However, private citizens, seeing what was occurring, armed themselves and shot at the Daltons as they attempted to escape, killing four of the gang members. Another example of citizen policing was the formation of vigilante committees. Between 1849 and 1902, there were 210 vigilante movements in the United States, most of them in California.[52] While it is true that they occasionally hung outlaws, they also performed valuable work by ridding their communities of dangerous criminals.

Federal marshals were created by congressional legislation in 1789. As marshals began to appear on the frontier, the vigilantes tended to disappear. U.S. marshals enforced federal laws, so they had no jurisdiction over matters not involving a federal offense. Marshals could act only in cases involving theft of mail, crimes against railroad property, murder on federal lands (much of the West

---

What Science Has Done for the Police

Chief Francis O'Neill, Chicago, 1903

The watchman of a century ago with his lantern and staff who called out the passing hours in stentorian tones during the night is now but a tradition. He has been succeeded by a uniformed constabulary and police who carry arms and operate under semimilitary discipline. The introduction of electricity as a means of communication between stations was the first notable advance in the improvement of police methods. I remember the time when the manipulation of the dial telegraph by the station keeper while sending messages excited the greatest wonder and admiration. The adoption of the Morse system of telegraphy was a long step forward and proved of great advantage. In 1876, all desk sergeants were required to take up the immediate study of the Morse system of telegraphy. Scarcely one-fourth of them became proficient before modern science, advancing in leaps and bounds, brought forth that still more modern miracle—the telephone. Less than one-quarter century ago the policeman on post had no aid from science in communicating with his station or in securing assistance in case of need. When required by duty to care for the sick and injured or to remove a dead body, an appeal to the owner of some suitable vehicle was his only resource. These were desperate times for policemen in a hostile country with unpaved streets. The patrol wagon and signal service have effected a revolution in police methods. The forward stride from the lanterned night watch, with staff, to the uniformed and disciplined police officer of the present, equipped with telegraph, telephone, signal service, and the Bertillon system of identification, is indeed an interesting one to contemplate.*

*Alphonse Bertillon (1853–1914) developed a system for criminal identification based on precise measurements of the human body. The system was used in Paris in 1882 and was officially adopted for all of France in 1888.

Proceedings of the International Chiefs of Police, 10th Annual convention, May 12–14, 1903, p. 67.

---

was federal property for many decades), and a few other crimes. Their primary responsibility was in civil matters arising from federal court decisions. Federal marshals obtained their office through political appointment; therefore, they did not need any prior experience and were politically indebted. Initially, they received no salary but were instead compensated from fees and rewards. Because chasing outlaws did not pay as much as serving civil process papers, the marshals tended to prefer the more lucrative, less dangerous task of serving the courts' paperwork. Congress saw the folly in this system and, in 1896, enacted legislation providing regular salaries for marshals.[53]

When a territory became a state, the primary law enforcement functions usually fell to local sheriffs and marshals. Train robbers such as Jesse James and the Dalton Gang tended to be the most famous outlaws to violate federal laws. Many train robbers became legendary for having the courage to steal from the despised railroad owners. What is often overlooked in the tales of these legends is their often total disregard for the safety and lives of their victims. In order to combat these criminals, federal marshals found their hideouts, and railroad companies and other businesses often offered rewards for information leading to their capture. Occasionally, as in the case of Jesse James and the Daltons, the marshals' work was done for them—outlaws were often killed by friends (usually for a reward) or by private citizens.[54]

Gunfights in the West actually occurred very rarely; few individuals on either side of the law actually welcomed stand-up gunfights. It was infinitely more sensible to find cover from which to have a shootout. Further, handguns were not the preferred weapon; a double-barreled shotgun could do far more damage than a handgun at close range. Local law enforcement occurred as people settled into communities. Town meetings were held where a government was established and local officials were elected. Sheriffs quickly became important officials, but they spent more time collecting taxes, inspecting cattle brands, maintaining jails, and serving civil papers than they did actually dealing with outlaws. In fact, "Wild Bill" Hickok only killed two men while marshal of Abilene, Kansas; William "Bat" Masterson killed no one while living near Dodge City; and Wyatt Earp, who was never actually a marshal, may have killed one.[55]

Only 45 violent deaths from all causes can be found in western cowtowns from 1870 to 1885, when they were thriving. This low figure reflects the real nature of the cowtowns. Businessmen had a vested interest in preventing crime from occurring and not hiring a trigger-happy sheriff or marshal. They tended to avoid hiring individuals such as John Slaughter, sheriff of Cochise County, Arizona, for eight years, who never returned from a chase with a live prisoner. Too much violence ruined a town's reputation and harmed the local economy.[56]

THE ENTRENCHMENT OF POLITICAL INFLUENCE

Partly because of their closeness to politicians, police during the early 1900s began providing a wide array of services to citizens. Many police departments were involved in the prevention of crime and the maintenance of order as well as a variety of social services. In some cities, they operated soup lines, helped find lost children, and found jobs and temporary lodging in station houses for newly

Roll call in Chicago (1904) shows the uniform and equipment of the police officer of the day.

arrived immigrants.[57] Police organizations were typically quite decentralized, with cities being divided into precincts and run like small-scale departments, hiring, firing, managing, and assigning personnel as necessary. Officers were often recruited from the same ethnic stock as the dominant groups in the neighborhoods and lived in the beats they patrolled, with considerable discretion in handling their individual beats. Detectives operated from a caseload of "persons" rather than offenses, relying on their charges to inform on other criminals.[58]

The strength of the local political influence over the police was that they were integrated into neighborhoods. This strategy proved useful, as it helped contain riots and the police assisted immigrants in establishing themselves in communities and finding jobs. There were weaknesses as well: The intimacy with the community, closeness to politicians, and decentralized organizational structure (and its inability to provide supervision of officers) also led to police corruption. The close identification of police with neighborhoods also resulted in discrimination against strangers, especially minority ethnic and racial groups. Police often ruled their beats with the "end of their nightsticks" and practiced "curbside justice."[59] The lack of organizational control over officers also caused some inefficiencies and disorganization; thus, the image of Keystone Kops—bungling police—was widespread.

THE MOVEMENT TOWARD PROFESSIONALIZATION

ATTEMPTS TO THWART POLITICAL PATRONAGE

During the early nineteenth century, reformers sought to reject political involvement by the police, and civil service systems were created to eliminate patronage and ward influences in hiring and firing police officers. In some cities, officers could not live in the same beat they patrolled in order to isolate them as completely as possible from political influences. Police departments became one of the most autonomous agencies in urban government.[60] However, policing also became a matter viewed as best left to the discretion of police executives. Police organizations became *law enforcement* agencies with the sole goal of controlling crime. Any noncrime activities they were required to do were considered "social work." The professional era of policing would soon be in full bloom.

The scientific theory of administration was adopted, as advocated by Frederick Taylor during the early twentieth century. Taylor first studied the work process, breaking down jobs to their basic steps and emphasizing time and motion studies, all with the goal of maximizing production. From this emphasis on production and unity of control flowed the notion that police officers were best managed by a hierarchical pyramid of control. Police leaders routinized and standardized police work; officers were to enforce laws and make arrests whenever they could. Discretion was limited to the extent it was possible to do so. When special problems arose, special units (e.g., vice, juvenile, drugs, tactical) were created rather than assigning problems to patrol officers.

THE ERA OF AUGUST VOLLMER

August Vollmer's career in policing has been established as one of the most important periods in the development of police professionalism. In April 1905 at age 29, Vollmer became the town marshal in Berkeley, California. At that time, policing had become a major issue all across America. Big-city police departments had become notorious for their corruption, and politics rather than professional principles dominated most police departments.[61]

Vollmer commanded a force of only three deputies; his first act as town marshal was to request an increase in his force from 3 to 12 deputies in order to form a day and night patrol. Obtaining that, he soon won national publicity for being the first chief to order his men to patrol on bicycles. Time checks he had run demonstrated that bicycles would allow officers to respond three times more quickly to calls than men on foot possibly could. His confidence growing, Vollmer next persuaded the Berkeley City Council to purchase a system of red lights. The lights, hung at each street intersection, served as an emergency notification system for police officers—the first such signal system in the country.[62]

In 1906, Vollmer, curious about the methods criminals used to commit their crimes, began to question the criminals he arrested. He found that nearly all criminals used their own peculiar method of operation, or **modus operandi**. In 1907, following an apparent suicide case that Vollmer suspected of being murder, Vollmer sought the advice of a professor of biology at the University of California. He became convinced of the value of scientific knowledge in criminal investigation.[63]

Vollmer's most daring innovation came in 1908—the idea of a police school. The first formal training program for police officers in the country drew upon the expertise of university professors as well as police officers. The school included courses on police methods and procedures, fingerprinting, first aid, criminal law, anthropometry, photography, public health, and sanitation. In 1917, the curriculum was expanded from one to three years.[64]

In 1916, Vollmer persuaded a professor of pharmacology and bacteriology to become a full-time criminalist in charge of the department's criminal investigation laboratory. By 1917, Vollmer had his entire patrol force operating out of automobiles—the first completely mobile patrol force in the country. And in 1918, to improve the quality of police recruits in his department, he began to hire college students as part-time officers and to administer a set of intelligence, psychiatric, and neurological tests to all applicants. Out of this group of "college cops" came several outstanding and influential police leaders (including O. W. Wilson, who served as police chief in Wichita and Chicago and as the first dean of the school of

August Vollmer as town marshal, police chief, and criminalist.
(*Courtesy Samuel G. Chapman*)

criminology at the University of California). Then, in 1921, while also experimenting with the lie detector, two of his officers installed a crystal set and earphones in
a Model T touring car, thus creating the first radio car.

These and other innovations at Berkeley had begun to attract attention from
municipal police departments across the nation, including Los Angeles, which
persuaded Vollmer to serve a short term as chief of police beginning in August
1923. Gambling, the illicit sale of liquor (Prohibition was then in effect), and police
corruption were major problems in Los Angeles. Vollmer hired ex-criminals to
gather intelligence information on the criminal network. He also promoted honest
officers, required 3,000 patrol officers to take an intelligence test, and, using those
tests, reassigned personnel.[65] Already unpopular with crooks and corrupt politicians, these personnel actions made Vollmer very unpopular within the department as well. When he returned to Berkeley in 1924, he had made many enemies
and his attempts at reform had met with too much opposition to have any lasting
effect. It would not be until the 1950s, under Chief William Parker, that the Los
Angeles Police Department (LAPD) would become a leader in what is now known
as the **professional era of policing**.[66]

Vollmer, although a leading proponent of police professionalism, also advocated the idea that the police should function as social workers. He believed the
police should do more than merely arrest offenders; they should also seek to prevent crime by "saving" offenders.[67] He suggested that police work closely with
existing social welfare agencies, inform voters about overcrowded schools, and
support the expansion of recreational facilities, community social centers, and
anti-delinquency agencies. Basically, he was suggesting that the police play an
active part in the life of the community. These views were very prescient; today,
his ideas are being implemented in the contemporary movement toward community policing and problem oriented policing (discussed in Chapter 5). Yet the
major thrust of police professionalization had been to insulate the police from politics. This contradiction illustrated one of the fundamental ambiguities of the
whole notion of professionalism.[68]

In the late 1920s, Vollmer was appointed the first professor of police administration in the country at the University of Chicago. Upon returning to Berkeley
in 1931, he received a similar appointment at the University of California, a position he held concurrently with the office of chief of police until his retirement from
the force in 1932. He continued to serve as university professor until 1938.[69]

THE "CRIME-FIGHTER" IMAGE

The 1930s marked an important turning point in the history of police reform.
O. W. Wilson emerged as the leading authority on police administration, and a
redefinition of the police role and the ascendancy of the crime-fighter image
occurred.

Wilson, who learned from J. Edgar Hoover's transformation of the Federal
Bureau of Investigation (FBI) into one of high prestige, became the principal architect of the police reform strategy.[70] Hoover, appointed FBI director in 1924, had

The Crib of Modern Law Enforcement: A Chronology of August Vollmer and the Berkeley Police Department

1905: Vollmer elected Berkeley Town Marshal. Town trustees appoint six police officers @ $70 monthly salary.

1906: Trustees create detective rank. Vollmer initiates a red light signal system to reach officers on beats from headquarters; telephones installed in boxes. A police records system is created.

1908: Two motorcycles added to the department, and Vollmer begins a police school.

1909: Vollmer appointed Berkeley chief of police under new charter form of government. Trustees approve appointment of Bertillon expert and the purchase of fingerprinting equipment. A "Modus Operandi" file is created, modeled from British system.

1911: All patrol officers are using bicycles.

1914: Three privately owned autos authorized for patrol use. (By 1917, Vollmer had the first completely motorized force; officers furnished their own automobiles.)

1915: A central office is established for police reports.

1916: Vollmer urges Congress to establish a national fingerprint bureau (later created by the FBI in Washington, D.C.), begins annual lectures on police procedures, and persuades biochemist Albert Schneider to install and direct a crime laboratory at headquarters.

1917: Vollmer recruits college students for part-time police jobs. He also began consulting with police, reorganizing departments around the country.

1918: Entrance examinations are initiated to measure mental, physical, emotional fitness of recruits; a part-time police psychiatrist is employed.

1919: Vollmer begins testing delinquents and using psychology to anticipate criminal behavior. He implements juvenile program to reduce child delinquency.

1921: Vollmer guides the development of the first lie detector and begins developing radio communications between patrol cars, handwriting analysis, and use of business machine equipment (a Hollerith tabulator).

Following his retirement from active law enforcement in 1932, Vollmer traveled around the world to study police methods. He continued serving as professor of police administration at U.C. Berkeley until 1938, authoring or co-authoring four books on police and crime from 1935 to 1949. He died in Berkeley in 1955. [The author is indebted to Professor Samuel G. Chapman for contributing much of the chapter information and photographs from his personal files and correspondence.]

raised eligibility and training standards of recruits and had developed an incorruptible crime-fighting organization. Municipal police found Hoover's path a compelling one.

Professionalism came to mean a combination of managerial efficiency and technological sophistication and an emphasis on crime fighting. The social work aspects of the policing movement fell into almost total eclipse. In sum, under the professional model of policing, officers were to remain in their "rolling fortresses," going from one call to the next with all due haste. As Mark Moore and George Kelling observed, "In professionalizing crime fighting . . . citizens on whom so much used to depend [were] removed from the fight."[71]

THE WICKERSHAM COMMISSION

Another important development in policing, one that was strongly influenced by August Vollmer, was the creation of the **Wickersham Commission**. President Coolidge appointed the first National Crime Commission in 1925, in an admission that crime control had become a national problem. This commission was criticized for working neither through the states nor with professionals in criminal justice, psychiatry, social work, and the like. Nevertheless, coming on the heels of World War I, the crime commission took advantage of FBI director J. Edgar Hoover's popular "war on crime" slogan to enlist public support. Political leaders and police officials also loudly proclaimed the "war on crime" concept; it continued the push for police professionalism.

Coolidge's successor, President Herbert Hoover, became concerned about the lax enforcement of Prohibition, which took effect in 1920. It was common knowledge that an alarming number of American police chiefs and sheriffs were accepting graft money in exchange for overlooking moonshiners; other types of police corruption were occurring as well. Hoover replaced the National Crime Commission with the National Commission on Law Observance and Enforcement—popularly known as the Wickersham Commission, after its chairman, former U.S. Attorney General George W. Wickersham. This presidential commission completed the first national study of crime and criminal justice, issuing 14 reports. Two of those reports, the "Report on Lawlessness in Law Enforcement" and the "Report on Police," represented a call by the national government for increased police professionalism.

The "Report on Police" was written in part by August Vollmer, but his imprint on the remainder of this and other reports is evident. The second report, on "Lawlessness in Law Enforcement," concerned itself with police misconduct and has received the greatest public attention, both then and today. The report indicated that the use of third-degree interrogation methods of suspects by the police (including the infliction of physical or mental pain to extract confessions) was widespread in America. This report, through its recommendations, mapped out a path of professionalism in policing for the next two generations. The Wickersham Commission recommended, for example, that the corrupting influence of politics should be removed from policing. Police chief executives should be selected on merit, and patrol officers should be tested and should meet minimal physical standards. Police salaries, working conditions, and benefits should be decent, the commission stated, and there should be adequate training for both pre-service and in-service officers. The commission also called for the use of policewomen (in juvenile and female cases), crime-prevention units, and bureaus of criminal investigation.

Many of these recommendations represented what progressive police reformers had been wanting for the previous 40 years; unfortunately, President Hoover and his administration could do little more than report the Wickersham Commission's recommendations before leaving office. Franklin Roosevelt's administration provided the funding and leadership necessary for implementing Wickersham's suggestions in the states.

POLICE AS THE "THIN BLUE LINE": WILLIAM H. PARKER

The movement to transform the police into professional crime fighters found perhaps its most staunch champion in **William H. Parker**, who began as a patrol officer with the Los Angeles Police Department in 1927. Parker used his law degree to advance his career, and by 1934 he was the LAPD's trial prosecutor and an assistant to the chief.[72]

Parker became police chief in 1950 and, following an uproar over charges of police brutality in 1951, he conducted an extensive investigation that resulted in the dismissal or punishment of over 40 officers. Following this incident, he launched a campaign to transform the LAPD. His greatest success, typical of the new professionalism, came in administrative reorganization. The command structure was simplified, as Parker aggressively sought ways to free every possible officer for duty on the streets, including forcing the county sheriff's office to guard prisoners and adopting one-person patrol cars. Parker also made rigorous selection and training of personnel a major characteristic of the LAPD. Higher standards of physical fitness, intelligence, and scholastic achievement weeded out many applicants, while others failed the psychiatric examinations.

Once accepted, recruits attended a 13-week academy that included a rigorous physical program, rigid discipline, and intensive study. Parker thus molded an image of a tough, competent, polite, and effective crime fighter by controlling recruitment. During the 1950s, this image made the LAPD the model for reform across the nation; thus, the 1950s marked a turning point in the history of professionalism.[73]

But Parker's impact on the police shows the very real limitations of the professional style of policing. Parker conceived of the police as a thin blue line, protecting society from barbarism and Communist subversion. He viewed urban society as a jungle, needing the restraining hand of the police; only the law and law enforcement saved society from the horrors of anarchy. The police had to enforce the law without fear or favor. Parker opposed any restrictions on police methods. The law, he believed, should give the police wide latitude to use wiretaps and to conduct search and seizure. For him, the Bill of Rights was not absolute but relative. Any conflict between effective police operation and individual rights should be resolved in favor of the police, and the rights of society took precedence over the rights of the individual. He thought that evidence obtained illegally should still be admitted in court and that the police could not do their jobs if they were going to be continually second-guessed by the courts and other civilians.

Basically, Parker believed that some "wicked men with evil hearts" preyed upon society and that the police must protect society from attack by them. But Parker's brand of professional police performance lacked total public support. Voters often supported political machines that controlled and manipulated the police in anything but a professional manner; the public demanded a police department subject to political influence and manipulation and then condemned the force for its crookedness. The professional police officer was in the uncomfortable position of offering a service that society required for its very survival, but that many people did not want at all.[74]

Still, Parker's influence over police administration has not perished altogether. And to the extent that his influence remains today, it may hinder the spread of the concepts of community oriented policing and problem oriented policing (discussed more thoroughly in Chapter 6).

A RETREAT FROM THE PROFESSIONAL MODEL

COMING FULL CIRCLE TO PEEL: THE PRESIDENT'S CRIME COMMISSION

The 1960s were a time of explosion and turbulence. Inner-city residents rioted in several major cities, protestors denounced military involvement in Vietnam, and assassins ended the lives of President John F. Kennedy and Dr. Martin Luther King Jr. The country was witnessing tremendous upheaval, and such incidents as the so-called police riot at the 1968 Democratic National Convention in Chicago raised many questions about the police and their function and role.

To this point, there had been few inquiries concerning police functions and methods,[75] for two reasons. First, was a tendency on the part of the police to resist outside scrutiny. Functioning in a bureaucratic environment, they, like other bureaucrats, were sensitive to outside research. Many police administrators perceived a threat to personal careers and to the image of the organization, as well as a concern about the legitimacy of the research itself. There was a natural reluctance to invite trouble. Second, few people in policing perceived a need to challenge traditional methods of operation. The "if it ain't broke, don't fix it" attitude prevailed, particularly among old-school administrators. Some ideas were etched in stone, such as the belief that more police personnel and vehicles equalled more patrolling and, therefore, less crime, a quicker response rate, and a happier private citizen. A corollary belief is that the more officers riding in the patrol car, the better. The methods and effectiveness of detectives and their investigative techniques were not even open to debate. As Herman Goldstein has stated, however,

Crises stimulate progress. The police came under enormous pressure in the late 1960s and early 1970s, confronted with concern about crime, civil rights demonstrations, racial conflicts, riots and political protests.[76]

Five national studies looked into police practices during the 1960s and 1970s, each with a different focus: the President's Commission on Law Enforcement and the Administration of Justice (1967), the National Advisory Commission on Civil Disorders (1968), the National Commission on the Causes and Prevention of Violence (1968), the President's Commission on Campus Unrest (1970), and the National Advisory Commission on Criminal Justice Standards and Goals (1973).

Of particular note was a commission whose findings are still widely cited today and that provided the impetus to return the police to the community: the President's Commission on Law Enforcement and the Administration of Justice. Termed the **President's Crime Commission**, this body was charged by President Lyndon Johnson to find solutions to America's internal crime problems, including the root causes of crime, the workings of the justice system, and the hostile, antagonistic relations between the police and civilians. Among the commission's recommendations for the police were hiring more minority members as officers to improve police-community relations, upgrading the quality of police officers through better educated officers, promoting to supervisory positions college-educated individuals, screening applicants more rigorously, and providing intensive pre-service training for new recruits. It was believed that a higher caliber of recruits would raise police service delivery, promote tranquility within the community, and relegate police corruption to a thing of the past.[77]

The President's Crime Commission brought policing full circle, restating several of the same principles that were laid out by Sir Robert Peel in 1829: that the police should be close to the public, that poor quality of policing contributed to social disorder, and that the police should focus on community relations. Thus, by 1970, there had been what was termed a systematic demolition of the assumptions underlying the professional era of policing.[78] Few authorities on policing today could endorse the basic approaches to police management that were propounded by O. W. Wilson or William Parker. We now know much that was still unknown by the staff of the President's Crime Commission in 1967. For example, as will be seen in Chapter 5, we later learned that adding more police or intensifying patrol coverage does not reduce crime and that neither faster response time nor additional detectives will improve clearance rates.

COMMUNITY ORIENTED POLICING AND PROBLEM SOLVING ERA

In the early 1970s, it was suggested that the performance of patrol officers would improve by redesigning their job based on motivators.[79] This suggestion later evolved into a concept known as **team policing**, which sought to restructure police departments, improve police-community relations, enhance police officer morale, and facilitate change within the police organization. Its primary element was a decentralized neighborhood focus to the delivery of police services. Officers were to be generalists, trained to investigate crimes and basically attend to all of the problems in their area; a team of officers would be assigned to a particular neighborhood and would be responsible for all police services in that area.

In the end, however, team policing failed, for several reasons. Most of the experiments were poorly planned and hastily implemented, resulting in street officers who did not understand what they were supposed to do. Many midmanagement personnel felt threatened by team policing and did not support the experiment.

There were other developments for the police during the late 1970s and early 1980s. Foot patrol became more popular, and many jurisdictions (such as Newark, New Jersey; Boston; and Flint, Michigan) even demanded it. In Newark, an evaluation found that officers on foot patrol were easily seen by residents and produced a significant increase in the level of satisfaction with police service, led to a significant reduction of perceived crime problems, and resulted in a significant increase in the perceived level of safety of the neighborhood.[80]

These findings and others discussed later shattered several long-held myths about measures of police effectiveness. In addition, research conducted during the 1970s suggested that *information* could help police improve their ability to deal with crime. These studies, along with studies of foot patrol and fear reduction, created new opportunities for police to understand the increasing concerns of citizens' groups about problems (e.g., gangs, prostitutes) and to work with citizens to do something about them. Police discovered that when they asked citizens about their priorities, citizens appreciated their asking and often provided useful information.

Simultaneously, the problem-oriented approach to policing was being tested in Madison, Wisconsin; Baltimore County, Maryland; and Newport News, Virginia. Studies there found that police officers have the capacity to do problem solving successfully and can work well with citizens and other agencies. Also, citizens seemed to appreciate working with police. Moreover, this approach gave officers more autonomy to analyze the underlying causes of problems and to find creative solutions. Crime control remains an important function, but equal emphasis is given to *prevention*.

In sum, following are some of the factors that set the stage for the emergence of what has been termed the community oriented policing and problem solving (COPPS) era:

- The narrowing of the police mission to crime fighting
- Increased cultural diversity in our society
- The detachment of patrol officers in patrol vehicles
- Increased violence in our society
- A scientific view of management, stressing efficiency more than effectiveness, quantitative policing more than qualitative policing
- A downturn in the economy and, subsequently, a "do more with less" philosophy toward the police

- Increased dependence on high-technology equipment rather than contact with the public
- The emphasis on organizational change, including decentralization and greater officer discretion
- Isolation of police administration from community and officer input
- Concern about police violation of the civil rights of minorities
- A yearning for personalization of government services
- Burgeoning attempts by the police to adequately reach the community through crime prevention, team policing, and police-community relations

Most of these elements contain a common theme: the isolation of the police from the public.

COPPS is now recognized as being on the cutting edge of what is new in policing.[81] Indeed, the Violent Crime Control and Law Enforcement Act of 1994 authorized $8.8 billion over six years to create the Office of Community Oriented Policing Services (COPS) in the U.S. Department of Justice, add 100,000 more police officers to communities across the country, and create 28 regional community policing institutes (RCPIs) to provide training and technical assistance for implementation and technology throughout the nation.

SUMMARY

This chapter has presented the evolution of policing and some of the individuals, events, and national commissions that were instrumental in taking policing through several eras. It has also shown how the history of policing may be said to have come full circle to its roots, wherein it was intended to operate with the consent and assistance of the public. Policing is now attempting to throw off the shackles of tradition and become more community oriented.

This historical overview also reveals that many of today's policing issues and problems actually began surfacing many centuries ago: graft and corruption, negative community relations, police use of force, coping with public unrest and rioting, general accountability, struggles to establish the proper roles and functions of the police, the police subculture, and the tendency to withdraw from the public, cling to tradition, and be inbred.

ITEMS FOR REVIEW

1. What were the major police-related offices and their functions during the early English and colonial periods?
2. What legacies of colonial policing remained intact and were brought to the United States?
3. Describe the three early issues of American policing and their present status.

4. What unique characteristics of law enforcement existed in the Wild West? What myths concerning early western law enforcement continue today?
5. What were some of the advantages of the political and professional eras of policing?
6. What led to the development of the contemporary community oriented policing and problem solving era, and what are some of its main features?
7. Explain how it can be said that policing has come full circle, returning to its origins.

NOTES

1. Samuel Walker, *The Police in America: An Introduction* (New York: McGraw-Hill, 1983), p. 2.
2. Bruce Smith, *Rural Crime Control* (New York: Columbia University, 1933), p. 40.
3. *Ibid.*, pp. 42–44.
4. *Ibid.*
5. *Ibid.*
6. *Ibid.*, pp. 182–184.
7. *Ibid.*, pp. 188–189.
8. *Ibid.*, p. 192.
9. *Ibid.*, pp. 218–222.
10. *Ibid.*, pp. 245–246.
11. *Ibid.*
12. Craig Uchida, "The Development of American Police: An Historical Overview," in *Critical Issues in Policing: Contemporary Readings*, ed. Roger G. Dunham and Geoffrey P. Alpert (Prospect Heights, IL: Waveland Press, 1989), p. 14.
13. Charles Reith, *A New Study of Police History* (London: Oliver and Boyd, 1956).
14. Carl Klockars, *The Idea of Police* (Beverly Hills: Sage Publications, 1985).
15. *Ibid.*, pp. 45–46.
16. *Ibid.*, p. 46.
17. David R. Johnson, *American Law Enforcement History* (St. Louis: Forum Press, 1981), p. 4.
18. *Ibid.*, p. 5.
19. *Ibid.*
20. *Ibid.*, p. 6.
21. *Ibid.*, p. 1.
22. *Ibid.*, pp. 8–10.
23. *Ibid.*, p. 11.
24. *Ibid.*, p. 13.
25. David A. Jones, *History of Criminology: A Philosophical Perspective* (Westport, CT: Greenwood Press, 1986), p. 64.
26. Johnson, *American Law Enforcement History*, pp. 14–15.
27. *Ibid.*, pp. 17–18.
28. *Ibid.*, pp. 18–19.
29. Leon Radzinowicz, *A History of English Criminal Law and Its Administration from 1750*, vol. IV, *Grappling for Control* (London: Stevens & Son, 1968), p. 163.
30. Johnson, *American Law Enforcement*, p. 19.
31. *Ibid.*, pp. 19–20.
32. *Ibid.*, pp. 20–21.
33. A. C. Germann, Frank D. Day, and Robert R. J. Gallati, *Introduction to Law Enforcement and Criminal Justice* (Springfield, IL: Charles C. Thomas, 1962), p. 63.
34. Clive Emsley, *Policing and Its Context, 1750–1870* (New York: Schocken Books, 1983), p. 37.
35. Pamela D. Mayhall, *Police-Community Relations and the Administration of Justice*, 3d ed. (New York: John Wiley & Sons, 1985), p. 425.
36. Johnson, *American Law Enforcement*, p. 26.
37. *Ibid.*, pp. 26–27.
38. *Ibid.*, p. 27.

39. *Ibid.*
40. *Ibid.*, pp. 28–29.
41. *Ibid.*, pp. 30–31.
42. James F. Richardson, *Urban Policing in the United States* (London: Kennikat Press, 1974) pp. 47–48.
43. James F. Richardson, *The New York Police: Colonial Times to 1901* (New York: Oxford Press, 1970), p. 259.
44. James F. Richardson, *Urban Policing in the United States*, p. 48.
45. Richardson, *The New York Police*, pp. 195–201.
46. Richardson, *Urban Policing in the United States*, p. 51.
47. *Ibid.*
48. *Ibid.*, pp. 53–54.
49. *Ibid.*, pp. 55–56.
50. *Ibid.*, pp. 59–60.
51. Johnson, *American Law Enforcement*, p. 92.
52. *Ibid.*
53. *Ibid.*, pp. 96–97.
54. *Ibid.*, p. 98.
55. *Ibid.*
56. *Ibid.*, pp. 100–101.
57. Eric H. Monkkonen, *Police in Urban America, 1860–1920* (New York: Cambridge University Press, 1981), p. 158.
58. John E. Eck, *The Investigation of Burglary and Robbery* (Washington, D.C.: Police Executive Research Forum, 1984).
59. See George L. Kelling, "Juveniles and Police: The End of the Nightstick," in *From Children to Citizens, Vol. II: The Role of the Juvenile Court*, ed. Francis X. Hartmann (New York: Springer-Verlag, 1987).
60. Herman Goldstein, *Policing a Free Society* (Cambridge, MA: Ballinger, 1977).
61. August Vollmer, "Police Progress in the Past Twenty-five Years," *Journal of Criminal Law and Criminology*, 24 (1933): 161–175.
62. Alfred E. Parker, *Crime Fighter: August Vollmer* (New York: Macmillan, 1961).
63. Douthit, "August Vollmer," p. 102.
64. *Ibid.*
65. Paul Jacobs, *Prelude to Riot: A View of Urban America from the Bottom* (New York: 1966), pp. 13–60.
66. *Ibid.*
67. Samuel Walker, *A Critical History of Police Reform: The Emergence of Professionalism* (Lexington, MA: Lexington Books, 1977), p. 81.
68. *Ibid.*, pp. 80–83.
69. See Gene E. Carte and Elaine H. Carte, *Police Reform in the United States: The Era of August Vollmer, 1905–1932* (Berkeley, CA: University of California Press, 1975) for a chronology of Vollmer's career and a listing of his publications.
70. See Orlando Wilson, *Police Administration* (New York: McGraw-Hill, 1950).
71. Mark H. Moore and George L. Kelling, "'To Serve and Protect': Learning from Police History," *The Public Interest* 70 (Winter 1983): 49–65.
72. Johnson, *American Law Enforcement History*, pp. 119–120.
73. *Ibid.*, pp. 120–121.
74. Richardson, *Urban Police in the United States*, pp. 139–143.
75. Peter K. Manning, "The Researcher: An Alien in the Police World," in *The Ambivalent Force: Perspectives on the Police*, 2d ed. (Hinsdale, IL: The Dryden Press, 1976), pp. 103–121.
76. Herman Goldstein, *Problem-oriented Policing* (New York: McGraw-Hill, 1990), p. 9.
77. William G. Doerner, *Introduction to Law Enforcement: An Insider's View* (Englewood Cliffs, NJ: Prentice Hall, 1992), pp. 21–23.
78. Samuel Walker, "'Broken Windows' and Fractured History: The Use and Misuse of History in Recent Police Patrol Analysis." In *Classics in Policing* (Cincinnati, OH: Anderson, 1996), ed. Steven G. Brandl and David E. Barlow, pp. 97–110.
79. Thomas J. Baker, "Designing the Job to Motivate," *FBI Law Enforcement Bulletin* 45 (1976): 3–7.
80. Police Foundation, *The Newark Foot Patrol Experiment* (Washington, D.C.: Author, 1981).
81. *Ibid.*, p. 71.

CHAPTER 2

Policing Levels, Roles, and Functions

Key Terms and Concepts

Department of Justice
Department of the Treasury
FBI National Academy
Municipal police departments
National Crime Information
 Center (NCIC)

Sheriff's departments
State police
Styles and functions
 of policing
Uniform Crime Reports

Blessed are the peacemakers.—*Matthew 5:8*

INTRODUCTION

This chapter examines contemporary American policing at the federal, state, and local levels. Included is an overview of the agencies contained within the U.S. Department of Justice and the Department of the Treasury. Then we examine selected characteristics of state and local police agencies. The chapter concludes with an examination of the roles, functions, and styles of policing—areas that contain considerable conflict, debate, and disagreement.

In sum, this chapter demonstrates how the police operate in a democratic society and how our nation's police agencies interrelate to attempt to control crime at different jurisdictional levels—a very costly and labor-intensive undertaking in contemporary society.

The appendix contains information about obtaining a position in federal, state, and local law enforcement agencies. Another significant burgeoning level of policing—that which is performed by the "private" police, or the security industry—is examined in Chapter 12, "Issues and Trends."

FEDERAL LAW ENFORCEMENT AGENCIES

The U.S. government employs about 83,000 law enforcement agents who are authorized to carry firearms and to make arrests.[1] The three largest employers of federal officers are all in the Justice Department: the Immigration and Naturalization Service (INS), the Federal Bureau of Prisons, and the Federal Bureau of Investigation (FBI). Table 2–1 shows the number of sworn personnel that are employed by federal agencies having more than 500 full-time agents or officers.

Following is an analysis of the major law enforcement arms of the federal government, which are under the aegis of the Justice and Treasury Departments.

AGENCIES OF THE DEPARTMENT OF JUSTICE

The U.S. **Department of Justice** is headed by the attorney general, who is appointed by the president with Senate approval. The president also appoints the attorney general's assistants and the U.S. attorneys for each of the judicial districts. U.S. attorneys in each judicial district control and supervise all criminal prosecutions and represent the government in legal suits. These attorneys may appoint committees to investigate other governmental agencies or offices when questions of wrongdoing are raised or when possible violations of the laws of the United States are suspected or detected. The Department of Justice is the official legal arm of the government of the United States. Several law enforcement organizations that investigate violations of federal laws are branches of this department: the Federal Bureau of Investigation, the Drug Enforcement Administration, the Immigration and Naturalization Service, and the U.S. Marshals Service.

Federal Bureau of Investigation

The FBI has been one of the most colorful and even most glamorous law enforcement organizations. During Prohibition, the FBI was instrumental in capturing several notable gangsters, including John Dillinger, Charles "Pretty Boy" Floyd, and George "Machine Gun" Kelly. *The FBI* was a popular television program in

TABLE 2–1 Federal Agencies Employing 500 or More Full-time Officers with Authority to Carry Firearms and Make Arrests, June 1998

Agency	Number of Full-time Officers with Arrest and Firearm Authority
Immigration and Naturalization Service	16,552
Federal Bureau of Prisons	12,587
Federal Bureau of Investigation	11,285
U.S. Customs Service	10,539
U.S. Secret Service	3,587
U.S. Postal Inspection Service	3,490
Internal Revenue Service	3,361
Drug Enforcement Administration	3,305
U.S. Marshals Service	2,705
Administrative Office of the U.S. Courts	2,490
National Park Service[a]	2,197
Bureau of Alcohol, Tobacco and Firearms	1,723
U.S. Capitol Police	1,055
GSA — Federal Protective Service	900
U.S. Fish and Wildlife Service	831
U.S. Forest Service	601

Note: Table excludes employees based in U.S. territories or foreign countries.

[a]National Park Service total includes 1,524 Park Rangers commissioned as law enforcement officers and 673 U.S. Park Police officers.

Source: Brian A. Reaves and Timothy C. Hart, *Federal Law Enforcement Officers, 1998* (Washington, D.C.: U.S. Department of Justice, Bureau of Justice Statistics Bulletin, 2000), p. 3.

the 1970s, portraying the abilities of the bureau and the professional demeanor of its agents.

The FBI was created and funded through the Department of Justice Appropriation Act of 1908. The FBI was originally known as the Bureau of Investigation. With 35 agents, it had no specific duties other than the prosecution of crimes, focusing on bankruptcy frauds, anti-trust crimes, neutrality violations, and crimes on Native American reservations. Espionage and sabotage incidents during World War I, coupled with charges of political corruption that reached into the Department of Justice and the bureau itself, prompted angry demands for drastic changes.

A new era began for the FBI in 1924, with the appointment of J. Edgar Hoover as director (see box). Hoover was determined that the organization would become a career service in which appointments would be made strictly on personal qualifications and abilities and promotions would be based on merit. Special agents had to be college graduates with degrees in law or accounting. A rigorous

John Edgar Hoover (1895–1972)

(Courtesy FBI)

Born in Washington, D.C., J. Edgar Hoover received his LL.B. and LL.M. degrees from George Washington University in 1917. After admission to the bar, he began a career with the Justice Department and served as special assistant to the attorney general from 1919 to 1921. President Calvin Coolidge appointed Hoover director of the Federal Bureau of Investigation in 1924; he would serve in that capacity until his death in 1972. Hoover set out to remove politics from the FBI appointment process and to establish better training programs for new agents. Very early he recognized the importance of improved technical methods of police work and instituted such aids as a national fingerprint filing system; he also coordinated development of the Uniform Crime Reporting System. Hoover personally supervised the pursuit of such notorious criminals as "Ma" Barker, "Machine Gun" Kelly, Bonnie and Clyde, and John Dillinger. In the 1940s and 1950s, Hoover became noted for his anti-Communist views. Still, history has not been kind to Hoover's memory; a decline in his reputation began soon after his death, when the nation began to learn the full extent of his abuse of power, his persecution of Martin Luther King Jr. and other civil rights activists, of feminists, and of the early environmentalists. There are also strong suspicions that Hoover was a devout gambler (particularly betting on horse races) and that this compulsion was one of the reasons why Hoover maintained a less-than-aggressive posture with organized crime investigations for much of his career, even refusing to publicly acknowledge the existence of the Mafia. Yet Hoover enjoyed many accomplishments during his tenure as director of the FBI. The building housing the bureau's headquarters in Washington, D.C.—11 stories high and nearly 3 million square feet in size—bears his name.

course of training had to be completed, and agents had to be available for assignment wherever their services might be needed.

The bureau's Identification Division was created on July 1, 1924. In 1932, with a borrowed microscope and a few other pieces of scientific equipment, the bureau's laboratory opened. Eventually, the laboratory came to assist both federal and local investigations by examining and analyzing blood, hair, firearms, paint, handwriting, typewriters, and other types of evidence. Highly specialized techniques are now performed with DNA, explosives, hairs and fibers, tool marks, drugs, plastics, and bloodstains. All of these examinations are performed without expense to state and local police agencies.

In 1933, all of the bureau's functions, including those of the old Bureau of Prohibition, were consolidated and transferred to a Division of Investigation headed by a director of investigation. The Division of Investigation became the Federal Bureau of Investigation on March 22, 1935.

The FBI has jurisdiction to investigate nearly 300 types of crimes. Recently, its jurisdiction was expanded to include carjackings, "deadbeat dads," hate crimes, and health care fraud. Today the bureau also participates with local police in about 70 task forces nationwide that target fugitives and violent gangs. It also investigates civil matters and cases of national security that involve intelligence.

The J. Edgar Hoover FBI Building, Washington, D.C. (*Courtesy FBI*)

(*Courtesy FBI*)

Another feature of the FBI, inaugurated in 1950, was the "Ten Most Wanted Fugitives" list. Over the years, the list has contained such notable fugitives as Willie Sutton, James Earl Ray, and Ted Bundy. As of mid-2000, the bureau had caught about 460 Top Ten fugitives; the Internet has also helped to invigorate the program: The Ten Most Wanted Web page receives about 25 million hits per month.[2]

The FBI now has a budget in excess of $3.6 billion[3] and employs more than 11,000 full-time, sworn special agents[4] in its 15 divisions. The Identification Division assists local police departments in identifying criminal suspects, missing persons, and accident victims. The bureau also maintains an Instrumental Analysis Unit, where samples of various substances are stored. It also renders expert testimony and investigative and training assistance to local police agencies.

The FBI operates the **National Crime Information Center (NCIC)**, through which millions of records relating to stolen property and missing persons and

2000 CRIME CLOCK

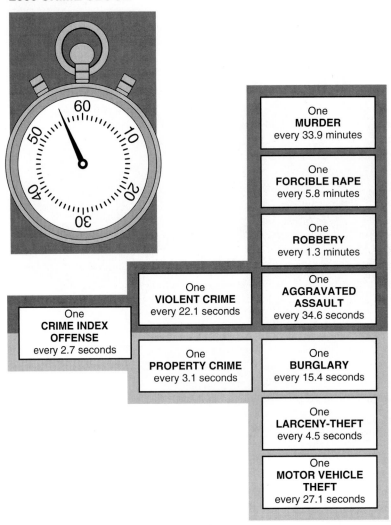

FIGURE 2–1 The UCR Crime Clock. (*Source:* U.S. Department of Justice, Federal Bureau of Investigation, Crime in the United States, 2000. Washington, D.C., GPO, 2001, p. 6.)

fugitives are instantaneously available to local, state, and federal authorities across the United States and Canada. (See the discussion of the NCIC 2000 system in Chapter 14.)

One of the FBI's several annual publications is the **Uniform Crime Reports** (UCR), which includes data reported from more than 15,000 state and local police agencies concerning 29 types of offenses: eight Part I (or "index") offenses (criminal homicide, forcible rape, robbery, aggravated assault, burglary, larceny theft (except motor vehicle theft), motor vehicle theft, and arson) and Part II offenses. The UCR also includes a so-called Crime Clock (Figure 2–1). The UCR data have several shortcomings. First, many crime victims do not report their offenses to the police; the local police, therefore, are unaware of these "hidden" offenses. Also, the entire reporting system is voluntary, and there is no penalty if the police do not report cases to the FBI. Furthermore, the reporting system is not uniform. Crimes may be reported incorrectly or inaccurately. At best, the UCRs have limitations and must be used cautiously.

Recently—especially after the Ruby Ridge (1992) and Waco (1993) debacles—the FBI has come under extreme criticism for several blunders, for which many people blame the recent rapid retirement of supervisory and regular special agents and the resultant high incidence of "the new leading the new":[5]

- July 1996: A homemade pipe bomb exploded during the Summer Olympic Games in Atlanta; a suspect was later exonerated and accused the FBI of trying to trick him into confessing.

- April 1997: The Justice Department's Inspector General report slams the once-renowned FBI crime lab for sloppy, inaccurate work and for providing testimony slanted in the prosecution's favor.

- December 1999: Wen Ho Lee, a former nuclear weapons scientist at Los Alamos National laboratory, is arrested and charged with 59 counts of illegally removing classified weapons data from the lab and selling secrets to China. In April 2000, a federal judge orders Lee's release. Lee pleads guilty to one felony count, and the judge apologizes to him for the government's conduct.

- February 2001: Longtime FBI agent Robert Hanssen is discovered to have been selling secrets to Moscow for 15 years before being arrested. The FBI is pummeled in the press for failing to police itself for spies.

- May 2001: The execution of convicted Oklahoma City bomber Timothy McVeigh is delayed over the FBI's failure to turn over to defense attorneys some 4,000 pages of related documents (McVeigh was executed in June 2001).[6]

- July 2001: FBI agents could not account for 184 missing laptop computers (some of which contained classified data) and nearly 450 handguns, rifles, and submachine guns.[7]

Notwithstanding these occurrences, it is still widely believed that no other law enforcement agency is better at apprehending offenders than the FBI.

(Courtesy DEA)

Drug Enforcement Administration

The beginning of what would eventually become the Drug Enforcement Administration (DEA) occurred with passage of the Harrison Narcotic Act, signed into law on December 17, 1914, by President Woodrow Wilson. The Harrison Act was primarily a tax law, but Section 8 made it unlawful for any "nonregistered" personnel to possess heroin, cocaine, opium, morphine, or any of their products. Drug enforcement began in 1915, and during their first year of activity, agents seized 44 pounds of heroin and achieved 106 convictions (mostly a result of illicit activities of physicians).

In the 1920s, federal narcotics agents focused on organized gangs of Chinese immigrants suspected of importing opium. In 1920, the Volstead Act (initiating Prohibition) was enacted; the Narcotics Division of the Prohibition Unit of the Revenue Bureau consisted of 170 agents working out of 17 offices around the country. New authority was granted to agents by the Narcotic Drugs Import and Export Act of 1922.

Today's DEA is also an outgrowth of the former Bureau of Narcotics, established in 1930 under the direct control of the Treasury Department. In 1968, the Bureau of Narcotics was transferred from Treasury to the Department of Justice and renamed the Bureau of Narcotics and Dangerous Drugs. This reorganization occurred, in part, because Bureau of Narcotics special agents were accused of selling some of the drugs they had confiscated.[8] In 1973, the DEA was established under a plan that combined the functions of several agencies. In January 1982, the DEA was given primary responsibility for drug and narcotics enforcement, sharing this jurisdiction with the FBI.[9]

The major responsibilities of the DEA, under the United States Code, include the following:

1. The development of an overall federal drug enforcement strategy, including programs, planning, and evaluation

2. Full investigation of and preparation for the prosecution of suspects for violations under all federal drug-trafficking laws
3. Full investigation of and preparation for the prosecution of suspects connected with illicit drugs seized at U.S. ports of entry and international borders
4. Conduct of all relations with drug enforcement officials of foreign governments
5. Full coordination and cooperation with state and local police officials on joint drug enforcement efforts
6. Regulation of the legal manufacture of drugs and other controlled substances

The DEA employs about 3,305 special agents and more than 400 intelligence specialists.[10] Overseas, the DEA maintains 71 offices in 50 foreign countries.[11]

Immigration and Naturalization Service

Congress was first given the power to establish a naturalization service under Article I, Section 8 of the U.S. Constitution. In 1798, the Alien Act gave the president the power to expel aliens who were considered a threat to national security. The Office of Commissioner of Immigration was created in 1864 under the Department of State, and in 1898, the Commissioner of the Bureau of Immigration was established. In 1903, this bureau was placed under the control of the Department of Commerce and Labor. By 1913, it was divided into the Bureau of Labor and the Bureau of Naturalization, both of which were placed under the Department of Labor. Later these two bureaus were combined into the Immigration and Naturalization Service. The INS became part of the Department of Justice in 1940.

Today more than 16,500 agents of the INS have full powers of arrest, search, and seizure under the U.S. Code, and the INS has numerous functions.[12] Its major function includes overseeing immigrants admitted for legal permanent residence, refugees and asylees, nonimmigrant arrivals (for example, tourists and students), aliens, aliens apprehended and expelled, and aliens inspected at ports of entry. Each year, the INS grants permanent legal residence to nearly 1 million persons. The INS also admits about 20 million nonimmigrants (mostly tourists) into the

(*Courtesy Border Patrol*)

country, apprehends deportable aliens, investigates people at ports of entry to the United States, and examines petitions for naturalization.[13]

Under the McCarran-Walter Act of 1952, the INS was charged with three responsibilities:

1. The reunification of families
2. The immigration of persons with needed labor skills
3. The protection of the domestic labor force

Perhaps the best-known agency of the INS is the Border Patrol, which was established in 1924 with 450 officers (discussed further in Chapter 8).[14]

United States Marshals Service

The United States Marshals Service (USMS) is one of the oldest federal law enforcement agencies, established under the Judiciary Act of 1789; George Washington appointed 13 marshals, one for each of the original 13 states. The USMS formally became a bureau in 1974. It now has about 2,700 officers with arrest powers.[15] Each district headquarters office is managed by a politically appointed U.S. marshal and a chief deputy U.S. marshal, who direct a staff of supervisors, investigators, deputy marshals, and administrative personnel. As in the Wild West, the backbone of the USMS are the deputy U.S. marshals, who pursue and arrest DEA fugitives (persons wanted for federal drug violations) and escaped federal prisoners and provide a secure environment for the federal courts and judges.

(Courtesy USMS)

Virtually every federal law enforcement initiative involves the USMS. Its almost 2,700 agents[16] produce prisoners for trial; protect the courts, judges, attorneys, and witnesses; track and arrest fugitives; and manage and dispose of seized drug assets.[17]

In 1971, the USMS created the Special Operations Group, or SOG, consisting of an elite, well-trained group of deputy marshals. The group provides support in priority or dangerous situations, such as movements of large groups of high-risk prisoners and at trials involving alleged drug traffickers or members of subversive groups.

Another important function of the USMS is the operation of the Witness Protection Program. Federal witnesses are at times threatened by defendants or their associates, depending on the nature of the case (e.g., they sometimes testify against an organized crime figure). If certain criteria are met, the USMS will provide complete changes of identity for witnesses and their families, including new Social Security numbers, residences, and employment. This protection program was established in 1970.

AGENCIES OF THE DEPARTMENT OF THE TREASURY

Several federal law enforcement agencies are under the direct control of the **Department of the Treasury** including the Bureau of Alcohol, Tobacco and Firearms; the Customs Service; the Internal Revenue Service; and the United States Secret Service.

Bureau of Alcohol, Tobacco and Firearms

The Bureau of Alcohol, Tobacco and Firearms (ATF) originated as a unit within the Internal Revenue Service (IRS) in 1862, when certain alcohol and tobacco tax statutes were created. In 1863, Congress authorized the hiring of three "detectives" to aid in the prevention of tax evasion and the detection and punishment of tax evaders. Originally called the Alcohol, Tobacco, Tax Unit, it eventually became the Alcohol, Tobacco and Firearms Division within the IRS. In 1972, it became the ATF, under the direct control of the Treasury Department.

Like the FBI and several of the other federal agencies, the ATF has a rich and colorful history, much of which concerns capturing bootleggers and disposing of illegal whiskey stills during Prohibition.[18] From 1920 to 1933, congressional legislation was in effect that made it illegal to manufacture, possess, or sell intoxicating liquors in the United States (with certain exceptions). Still, America was awash with liquor. History is replete with accounts of violations of Prohibition laws; much has been written and portrayed in movies of that era, when the moonshiners tried to outsmart and outrun the law.[19] Speakeasies (secret bars) proliferated across America to satisfy the American yearning for liquor. This era bolstered the popularity of such G-men as Eliot Ness, on whose 1920s and 1930s career a 1960s television program, *The Untouchables*, was based. Each year the ATF

(*Courtesy ATF*)

regulates the production of millions of barrels and gallons of beer, wine, and distilled spirits.[20]

The ATF's general mission is to reduce the illegal use of firearms and explosives. However, today the ATF's law enforcement functions extend to not only liquor, tobacco, and firearms trafficking, but also to arson, high explosives, and automatic weapons. The ATF has the charge of investigating the character and background of applicants for such weapons; the bureau also authorizes licenses for firearms dealers, importers and exporters, and ammunition manufacturers. The ATF also seeks to curtail arson-for-profit schemes.[21]

The problem of high explosives has become a major ATF concern as well. Easy-to-make bombs are beginning to proliferate among gangs and drug traffickers.[22] Virtually anyone can make or buy a bomb with easily obtained, low-cost ingredients. Criminals fearing paper trails and scrutiny can turn to the black market, creating a major problem of thefts of explosives from poorly guarded construction sites, quarries, explosives manufacturers, and even the military. Half a pound of C-4 plastic explosive can demolish an average police station or destroy a commercial airliner.[23] The use of bombs has also increased among gangs, as bombs allow gang members to intimidate and maim on a much larger scale.[24]

The ATF has about 1,723 special agents[25] and 850 inspectors and specialists.[26] Along with the FBI, the ATF received considerable criticism in the aftermath of the tragic raid on the Branch Davidian compound in Waco, Texas, on February 28, 1993. Two U.S. House of Representatives subcommittees conducted joint hearings on the debacle in 1993, and another congressional hearing was held in 1995, both of which attempted to determine what went wrong.

Customs Service

The U.S. Customs Service was created by President Washington under the Tariff Act of 1789, when Congress authorized it to collect duties on goods, wares, and merchandise.[27] The United States Code authorizes the Customs Service to conduct searches and inspections of all ships, aircraft, and vehicles entering U.S. borders. Customs officers may seize any illegal merchandise or wares, and agents are empowered to arrest suspects engaging in illegal activities. The history of the Customs Service includes its responsibilities for preventing the smuggling of war munitions to the Confederacy from the North during the Civil War, stopping the exportation and importation of alcoholic beverages during Prohibition, monitoring the actions of warships and commerce during the two world wars, and the inspecting of and providing security for packages from the Middle East during the Persian Gulf War.

Normally the Customs Service maintains a close liaison with other federal law enforcement agencies, especially in the area of international drug trafficking and smuggling. Customs officers enforce the federal laws at airports and in cargo areas, on piers, and in terminals. They perform duties on foot, on horseback, in helicopters, and by boat and airplane. They use specially trained dogs to locate contraband. And, in close cooperation with the INS and the ATF, they investigate illegal aliens and the illegal smuggling of firearms. The 1994 Violent Crime Control and Law Enforcement Act mandated monies for 50 new agents and inspectors who are dedicated to investigating the export of stolen vehicles. Customs has also become more actively involved in preventing child pornography materials from entering the United States.

(Courtesy U.S. Customs Service)

The Customs Service employs about 10,500 full-time, sworn agents.[28] In addition to making conventional arrests, the Customs Service also seizes vehicles, aircraft, and vessels, as well as financial assets.[29]

Internal Revenue Service

The Internal Revenue Service (IRS) has as its main function the monitoring and collection of federal income taxes from American individuals and businesses. Since 1919, the IRS has had a Criminal Investigation division (CI), employing 3,400 "accountants with a badge" who investigate possible criminal violations of income tax laws and recommend appropriate criminal prosecution whenever warranted.[30]

The first chief of the Special Intelligence Unit, Inspector Elmer I. Irey, gained notoriety by participating in investigations that included income tax evasion charges against Mafia kingpin Alphonse (Al) Capone and the kidnapping of Charles Lindbergh's baby in 1932.[31] Since then, the list of celebrated prosecuted CI "clients" has been impressive; it includes federal judges, politicians (including former Vice President Spiro T. Agnew and former Secretary of the

(Courtesy IRS)

Treasury Robert B. Anderson), musicians (such as Chuck Berry), athletes (such as Pete Rose and Darryl Strawberry), mafiosi (such as John Gotti), and entrepreneurs (such as Leona Helmsley). Indeed, today there is a much greater appreciation for what a financial investigator can do for almost any type of criminal investigation.[32]

The IRS's 3,361 agents[33] are armed; the United States Code authorizes them to execute search warrants, make arrests without warrants for internal revenue–related offenses, and make seizures of property relating to offenses of the tax laws. Agents engage in investigations of money laundering under Title 18 of the United States Code and investigate individuals who organize, direct, and finance high-level criminal enterprises for tax and currency violations.

CI enforces nearly all of the provisions of the Bank Secrecy Act, which requires financial institutions or individuals to report certain domestic and foreign currency transactions to the federal government. CI also enforces the wagering tax laws and conducts investigations concerning the pornography industry. Another important area of CI is the Questionable Refund Program, which proactively attempts to detect and stop claims for multiple fictitious tax refunds.

United States Secret Service

The United States Secret Service (USSS) originated in 1865 as the Secret Service Division, established primarily to capture and punish counterfeiters. In 1908, the division was transferred to the Department of Justice, and presidential security became a function of the office in 1917. In 1965, following the assassination of President Kennedy, the division was renamed the United States Secret Service and placed under the Treasury Department.

The USSS is instrumental in preventing presidential assassinations. USSS agents are part of the retinue of people often seen in the company of the president wherever he is in public and possibly exposed to danger. USSS agents have been seen overpowering such would-be presidential assassins as John Hinckley (who shot President Reagan in 1981) and Lynette "Squeaky" Fromme (who shot at President Ford in 1975). The agents are also responsible for making security arrangements for the president prior to his arrival at a new location.

The 3,600 Secret Service agents[34] are authorized to carry firearms and to make arrests for violations of any federal laws. The United States Code also authorizes them to investigate credit- and debit-card frauds or frauds relating to electronic fund transfers, such as those involving automatic teller machines.[35]

The FBI Academy and the Federal
Law Enforcement Training Center

Since the first class graduated from the **FBI National Academy** in 1935, thousands of local police managers from across the country have received training at the academy in Quantico, Virginia, which covers 385 acres and has 21 buildings. The FBI also provides extensive professional training to national supervisory-level

(Courtesy FLETC)

police officers at the National Academy. (More information is provided about local police training in Chapter 3.)

Consolidated federal law enforcement training at the Federal Law Enforcement Training Center (FLETC) was established in 1970. In 1975, the center was located in Glynco, Georgia, where it occupies a 1,500-acre campus with state-of-the-art classrooms from which it provides training for 70 U.S. federal law enforcement organizations.[36] FLETC trains about 26,000 students per year, including about 3,500 state and local officers and about 600 personnel from foreign countries.[37]

Central Intelligence Agency

The National Security Act of 1947 established the National Security Council, which in 1949 created a subordinate organization, the Central Intelligence Agency (CIA). Considered the most clandestine government service, the CIA participates in undercover and covert operations around the world. The formal functions of the CIA include the collection, production, and dissemination of foreign intelligence and counterintelligence information (including narcotics production and trafficking) and the transmission of information to other law enforcement agencies, including the FBI and the Customs Service, as part of a nationwide program to curtail the distribution of illegal narcotics. Although the intelligence-gathering functions of the CIA have overshadowed the agency's participation in narcotics-trafficking investigations, the CIA has been very effective in the latter function as well.

During a recent intensive recruitment campaign, the agency sought computer scientists, engineers, and fluent speakers of Chinese and Arabic. A more diverse workforce is also desired, as agents today are expected to infiltrate terrorist cells and drug cartels abroad.[38]

The CIA is an agency of neither the Department of Justice nor the Department of the Treasury, and it has no police function in the United States.

STATE AND LOCAL POLICE AGENCIES

GENERAL INFORMATION

Today, Sir Robert Peel would be amazed; there are now about 17,000 general-purpose local police departments and sheriff's offices in the United States:[39] about 420,000 sworn, full-time municipal police officers,[40] 175,000 sworn sheriff's personnel;[41] and 55,000 sworn state police.[42] (This is in addition to the 83,000 sworn, full-time federal law enforcement agents mentioned earlier.)

STATE POLICE

Sixty-nine percent of the full-time sworn personnel in the 49 **state police** agencies (Hawaii has no such agency) are uniformed officers who regularly respond to calls for service.[43] Most states have a highway patrol and a bureau of investigation or identification, many of which serve as the repository for state crime data—collecting, analyzing, and publishing state crime information. Agents of these state bureaus perform routine criminal investigation functions as well as specialized operations, such as investigations of organized crime or thefts of livestock brands. State investigators often assist local police agencies in investigative matters, such as performing background checks on police applicants.

Thirty-nine (or about 80 percent) of the nation's state police agencies authorize the use of one or more types of semi-automatic sidearms by their sworn officers; the 9mm was authorized by about three-fourths of those agencies. Less-than-lethal weapons were authorized for use by 45 state agencies (91 percent).[44]

MUNICIPAL POLICE AGENCIES

Municipal police departments have general police responsibilities, including traffic enforcement, accident investigation, patrol and first response to incidents, property and violent crime investigation, and death investigation.

Fourteen percent of local police departments require recruits to have completed at least some college. Field and classroom training requirements for new

Eighty percent of all sheriff's departments operate a jail. (*Courtesy Washoe County, Nevada, Sheriff's Department*)

recruits average about 1,000 hours combined. More than half (54 percent) of all departments serving 10,000 or more residents have either a full-time community policing unit or designated community policing personnel (community policing is examined in Chapter 6). Ninety-four percent of all local police departments authorize the use of semi-automatic sidearms. Pepper spray is authorized for use in 91 percent of all local agencies, and about half of all these agencies require officers to wear protective body armor while on duty. Nearly all local police departments are computerized; 29 percent use in-car computers.[45]

SHERIFF'S DEPARTMENTS

Fourteen percent of all **sheriff's departments** require recruits to have completed at least some college. Field and classroom training requirements for new recruits average about 800 hours combined. About one-third of all sheriff's departments serving 100,000 or more residents have a full-time community policing unit. Ninety-five percent of all sheriff's departments authorize the use of semi-automatic sidearms. Pepper spray is authorized for use in 87 percent of these agencies, and about 39 percent require deputies to wear protective body armor while on duty. Ninety-three percent of sheriff's departments are computerized; 29 percent use in-car computers. Most (about 80 percent) of the nation's sheriff's departments are responsible for operating a jail.[46]

More information is presented about women and minorities in local police and sheriff's departments in Chapter 12.

ROLES, FUNCTIONS, STYLES OF POLICING

DEFINING AND UNDERSTANDING THE ROLE

Why do the police exist? What are they supposed to do? Often these questions are given oversimplified answers, such as "They enforce the law," or "They 'serve and protect.'"[47]

But policing is much more complex. As Herman Goldstein put it, "Anyone attempting to construct a workable definition of the police role will typically come away with old images shattered and with a new-found appreciation for the intricacies of police work."[48] Even with all of the movies and television series depicting police in action, most Americans probably still do not have an accurate idea of what the police really do. This confusion is quite understandable, because the police are called upon to perform an almost countless number of tasks as part of their functions. Police are even used as prosecutors in some states, such as New Hampshire.

Who defines the police role? There are several groups and individuals who do so:

- Private citizens obviously influence the nature of the role by their contacts with the police, by memberships in community oriented policing and problem solving groups (discussed in Chapter 6), and through the election of public officials who set policy and appoint police administrators.
- Legislative bodies influence the role of the police by enacting statutes, both those that govern the police and those that the police use to govern others. In addition, legislative bodies determine budgets for the police.
- The courts have been actively policing the police by handing down decisions that regulate police conduct concerning the Bill of Rights.
- Executives such as city managers and prosecutors help to define the police role by determining, respectively, the types of cooperative agreements and evidence necessary for a prosecutable case.
- Police officers themselves define their roles by choosing to intervene in some incidents while ignoring others.[49]

Until relatively recently, few attempts were made to accurately describe the police role. In 1971, the American Bar Association's *Standards Relating to the Urban Police Function* identified eleven elements of the police role:

1. To identify criminals and criminal activity and, where appropriate, apprehend offenders and participate in court proceedings
2. To reduce the opportunities for the commission of crime through preventive patrol and other measures
3. To aid individuals who are in danger of physical harm
4. To protect constitutional guarantees
5. To facilitate the movement of peoples and vehicles
6. To assist those who cannot care for themselves
7. To resolve conflict
8. To identify problems that are potentially serious law enforcement or government problems
9. To create and maintain a feeling of security in the community
10. To promote and preserve civil order
11. To provide other services on an emergency basis[50]

One of the greatest obstacles to understanding the American police is the crime-fighter image. Many people believe that the role of the police is confined to law enforcement: the prevention and detection of crime and the apprehension of criminals. This is not an accurate contemporary view of police.[51] It does not describe what the police do on a daily basis. First, only about 20 percent of the police officer's typical day is devoted to fighting crime per se.[52] And, as Jerome Skolnick and David Bayley point out, the crimes that terrify Americans the most—robbery, rape, burglary, and homicide—are rarely encountered by police on patrol: "Only 'Dirty Harry' has his lunch disturbed by a bank robbery in progress. Patrol officers individually make few important arrests. The 'good collar' is a rare event. Cops spend most of their time passively patrolling and providing emergency services."[53]

The crime-fighter image persists, although it is extremely harmful to the public, police departments, and individual officers.[54] The public suffers from this image because it gives rise to unrealistic expectations about the ability of the police to catch criminals. Cops in the entertainment industry nearly always catch the bad guys. Also, this image harms individual officers, who believe that rewards and promotions are tied only to success in capturing criminals. Also, many individuals enter policing expecting it to be exciting and rewarding, as depicted on television and in the movies. Later they learn that much of their time is spent with boring, mundane tasks that are anything but glamorous. Much of the work is trivial, and paperwork is seldom stimulating.

ROLE CONFLICTS

Role conflicts may also develop with officers and their departments. A family disturbance is a good example. Assume that Jane Smith reports to the police that her husband, John, is assaulting her. Police must respond to the disturbance, and the law empowers them to intervene, to enforce the law, and to maintain order. For the combatants, it is a very trying experience, not only because of their family dysfunction but also because the police have been summoned to their home. The more experienced officers might view the domestic call as trivial and an inconvenience, leaving the scene as quickly as possible to go perform "real" police duties.

By the same token, the role of the police is often in the eye of the beholder. For example, in the same domestic argument just described, the matter might best seem to fit in the category of maintaining order. However, if the responding officer(s) are trained in crisis intervention or if they refer the couple to counseling, they arguably have provided a social service. On the other hand, if John is found to have assaulted Jane, it is likely a criminal matter. If the police make an arrest or even just assist Jane in swearing out a warrant, the matter becomes a law enforcement issue. The category to which this incident is assigned will vary greatly from agency to agency, officer to officer, and researcher to researcher, making it difficult to draw any solid conclusions about the police role.

Many local law enforcement agencies operate boot camp programs in their jail. (*Courtesy Washoe County, Nevada, Sheriff's Department*)

Still, it is important to be as explicit as possible about the police role. First, we can recruit and select competent police personnel only if we have a clear vision of what the police are supposed to accomplish. Second, evaluation for retention and promotion is useful only to the extent that we evaluate in terms of what the police are supposed to do. Third, budgetary decisions should be based on an accurate analysis of police roles. Fourth, efficiency and effectiveness in police organizations depend on accurate task description. And, finally, public cooperation with the police depends on developing reasonable expectations of the roles of the police and the public.[55]

The police should identify those crimes on which police resources should be concentrated, focusing on the crimes that generate the most public fear and economic loss. The chief executive should have written policies to ensure that the police mission, and the objectives used to achieve that mission, are maintained by the police department. In other words, it is not enough for the police to "maintain order" or "provide justice." A police department may use many methods to maintain order and provide justice. In China or South Africa, those methods would be far different from those generally employed in the United States. (See Chapter 12, "Comparative Perspectives.") But would "justice" result? In America, the police must maintain order without resorting to extralegal means or violating human rights.

POLICING FUNCTIONS AND STYLES

When broken down to its fundamental terms, police may be said to perform four basic functions: (1) enforcing the laws; (2) performing services (such as maintaining or assisting animal control units, reporting burned-out street lights or damaged or stolen traffic lights and signs, delivering death messages, checking the welfare of people in their homes, or delivering blood); (3) preventing crime (patrolling, providing the public with information on locks and lighting to reduce the opportunity for crime); and (4) protecting the innocent (by investigating crimes, police are systematically removing innocent people from consideration as crime suspects).

James Q. Wilson looked at the **functions of the police** differently, determining that the police perform two basic functions: maintaining order (peacekeeping) and enforcing the law. Maintaining order constitutes most of the activities of the police; as noted earlier, less than 20 percent of the calls answered by police are directly related to crime control or law enforcement. Much of an officer's time is spent with such service activities as traffic control, routine patrol, and others described earlier. Indeed, in some cases the police deliberately avoid enforcing the law in an attempt to maintain order. For example, if the police know of a busy street where many drivers speed, they may desist from setting up a speed trap during rush hours so as not to impede the flow of traffic and possibly cause accidents.

The law enforcement function means upholding the statutes, but this is not as simple and straightforward a function as it might seem. First, it might be stated that the police are really not very good at performing the law enforcement function for several reasons: The police have not traditionally been successful at preventing crime or providing long-term solutions to neighborhood disorder (although the relatively new community policing and problem solving concepts are addressing this shortcoming). Second, there are also several crimes—such as white-collar crime—with which the local police seldom deal. Third, the police, representing only about 2.3 officers per 1,000 population in the United States, cannot effectively control the public alone. Finally, the police are successful in solving only a fraction of the property and personal crimes that occur.[56]

Wilson also maintained that there are three distinctive **styles of policing**— the *watchman* style, the *legalistic* style, and the *service* style.[57]

- The *watchman style* involves the officer as a "neighbor." Here, officers act as if order maintenance (rather than law enforcement) is their primary function. The emphasis is upon using the law as a means of maintaining order rather than for regulating conduct through arrests. Police will ignore many common minor violations, such as traffic and juvenile offenses. These violations and so-called victimless crimes, such as gambling and prostitution, will be tolerated; they will often be handled informally. Thus, the individual officer has wide latitude concerning whether to enforce the letter or the spirit of the law; the emphasis is upon using the law to give people what they "deserve." It assumes that some people, such as juveniles, are occasionally going to "act up."

- The *legalistic style* casts the officer as a "soldier." This style takes a much harsher view of law violations. Here, police will issue large numbers of traffic citations, detain a high volume of juvenile offenders, and act vigorously against illicit activities. Large numbers of other kinds of arrests will be found as well. Chief administrators want high arrest and ticketing rates not only because violators should be punished, but also to reduce the opportunity for their officers to engage in corrupt behavior. This style of policing assumes that the purpose of the law is to punish.

• The *service style* views the officer as a "teacher." This style falls in between the watchman and legalistic styles. The police take seriously all requests for either law enforcement or order maintenance (unlike in the watchman-style department) but are less likely to respond by making an arrest or otherwise imposing formal sanctions. Police see their primary responsibility as protecting public order against the minor and occasional threats posed by unruly teenagers and "outsiders" (tramps, derelicts, out-of-town visitors). The citizenry expects its service-style officers to display the same qualities as its department store salespersons: They should be courteous, neat, and deferential. The police will frequently use informal sanctions instead of making arrests.

WHICH ROLE, FUNCTION, AND STYLE IS TYPICALLY EMPLOYED?

We have seen in the previous discussion that the role, function, and style of the police will differ by time and place. They will also be fluid within the agency, changing with the times, the political climate, and the problems of the day. Most police agencies do not determine which problems they address; rather, they respond to those problems that citizens believe are important and depend on the goals set by the community, the chief executive, and the individual officers. Sometimes roles, functions, and styles overlap, but most of the time they are distinct.

SUMMARY

This chapter described the major federal law enforcement agencies contained within the United States Departments of Justice and Treasury, as well as selected characteristics of state and local police agencies. Also examined were the roles, functions, and styles of the local police in this country. This chapter has perhaps demonstrated that after more than a century and a half after the adoption of the early British model of policing (discussed in Chapter 1), there is still widespread disagreement, conflict, and debate about what we truly want our police to do, represent, and become.

This chapter has also shown that a tremendous amount of specialization has evolved in today's law enforcement agencies. Policing has developed into a highly organized discipline with many branches and narrow fields of jurisdiction and responsibility.

ITEMS FOR REVIEW

1. What are the primary functions of the major federal law enforcement agencies?
2. How would you describe the "typical" local police and sheriff's departments?
3. What are the primary differences between federal, state, and local law enforcement agencies?

4. What are the police supposed to do in our society, and who decides their role?
5. Describe the police role as set forth by the American Bar Association in 1971 and why the crime-fighter image is the greatest obstacle in accepting this or any other single explanation of the police role.
6. Explain the primary functions and styles of policing.

NOTES

1. Brian A. Reaves and Timothy C. Hart, *Federal Law Enforcement Officers, 1998* (Washington, D.C.: U.S. Department of Justice, Bureau of Justice Statistics Bulletin, March 2000), p. 1.
2. Jeff Glasser, "In Demand for 50 Years: The FBI's 'Most Wanted' List—Good Publicity, and a History of Success," *U.S. News and World Report*, March 20, 2000, p. 60.
3. Chitra Ragavan, "FBI, Inc.: How the World's Premier Police Corporation Totally Hit the Skids," *U.S. News and World Report*, June 18, 2001, p. 14.
4. U.S. Department of Justice, Bureau of Justice Statistics Bulletin, "Federal Law Enforcement Officers—1998," March 2000, p. 2.
5. Ragavan, "FBI, Inc.," p. 16.
6. *Ibid.*, pp. 18–19.
7. Chitra Ragavan, "Gunning for Gumshoes: More Foul-ups at the FBI," *U.S. News and World Report*, July 30, 2001, p. 24.
8. Donald A. Torres, *Handbook of Federal Police and Investigative Agencies* (Westport, CT: Greenwood, 1985), pp. 126–127.
9. *Ibid.*, p. 126.
10. U.S. Department of Justice, Bureau of Justice Statistics Bulletin, "Federal Law Enforcement Officers, 1998," p. 2.
11. U.S. Department of Justice, Drug Enforcement Administration, "Factsheet: DEA Data," July 1995, pp. 1–2.
12. U.S. Department of Justice, "Federal Law Enforcement Officers, 1998," p. 2.
13. *1993 Statistical Yearbook of the Immigration and Naturalization Service* (Washington, D.C.: U.S. Government Printing Office, September 1994), p. 11.
14. *Ibid.*, p. 178.
15. U.S. Department of Justice, Bureau of Justice Statistics Bulletin, "Federal Law Enforcement Officers, 1998," p. 1.
16. U.S. Department of Justice, "Federal Law Enforcement Officers, 1998," p. 2.
17. U.S. Department of Justice, United States Marshals Office, *The FY 1993 Report to the U.S. Marshals* (Washington, D.C.: Author, 1994), pp. 188–189.
18. Torres, *Handbook of Federal Police and Investigative Agencies*, p. 63.
19. See, for example, Patrick G. O'Brien and Kenneth J. Peak, *Kansas Bootleggers* (Manhattan, KS: Sunflower University Press, 1991); Joseph E. Dabney, *Mountain Spirits* (New York: Charles Scribner's Sons, 1974); Jess Carr, *The Second Oldest Profession: An Informal History of Moonshining in America* (Englewood Cliffs, NJ: Prentice-Hall, 1972).
20. U.S. Department of the Treasury, Bureau of Alcohol, Tobacco, and Firearms, *Ready Reference 1994* (Washington, D.C.: U.S. Government Printing Office, 1994), p. 3.
21. U.S. Department of the Treasury, Bureau of Alcohol, Tobacco and Firearms, *1993 Explosives Incidents Reports* (Washington, D.C.: Author, 1994), p. 65.
22. *Ibid.*, p. 18.
23. *Ibid.*, pp. 29, 37.
24. James Popkin, "Bombs Over America", *U.S. News and World Report*, July 29, 1991, pp. 18–20.
25. U.S. Department of Justice, "Federal Law Enforcement Officers, 1998," p. 2.
26. U.S. Department of the Treasury, *Ready Reference 1994*, p. 2.
27. Torres, *Handbook of Federal Police and Investigative Agencies*, p. 393.
28. U.S. Department of Justice, "Federal Law Enforcement Officers, 1998," p. 2.
29. U.S. Department of the Treasury, U.S. Customs Service, *Office of Investigations Accomplishments: FY 1994*, no date, pp. 17–19.
30. U.S. Department of Justice, "Federal Law Enforcement Officers, 1998," p. 1.

31. Ludovic Henry Coverley Kennedy, *The Airman and the Carpenter: The Lindbergh Kidnapping and the Framing of Richard Hauptmann* (New York: Viking, 1985).
32. Don Vogel, quoted in Department of the Treasury, Internal Revenue Service, *CI Digest*, Pub. No. 1827, June 1994, p. 12.
33. U.S. Department of Justice, Bureau of Justice Statistics Bulletin, "Federal Law Enforcement Officers, 1998," p. 2.
34. *Ibid.*
35. U.S. Department of Justice, "Federal Law Enforcement Officers, 1998," p. 3.
36. Department of the Treasury, Federal Law Enforcement Training Center, *Catalog of Training Programs* (Washington, D.C.: Author, 1995), p. 1.
37. U.S. Department of Justice, "Federal Law Enforcement Officers, 1998," p. 5.
38. Bruce B. Auster, "Spies Wanted: James Bond Need Not Apply," *U.S. News and World Report*, June 15, 1998, p. 32.
39. U.S. Department of Justice, Bureau of Justice Statistics Bulletin, "Census of State and Local Law Enforcement Agencies, 1996," June 1998, p. 1.
40. U.S. Department of Justice, Bureau of Justice Statistics Executive Summary, "Local Police Departments, 1997," October 1999, p. 1.
41. U.S. Department of Justice, Bureau of Justice Statistics Executive Summary, "Sheriffs' Departments, 1997," October 1999, p. 1.
42. U.S. Department of Justice, "Census of State and Local Law Enforcement Agencies," p. 1.
43. U.S. Department of Justice, Bureau of Justice Statistics Bulletin, *Census of State and Local Law Enforcement Agencies, 1996* (Washington, D.C.: Office of Justice Programs, 1998), p. 11.
44. U.S. Department of Justice, "State and Local Police Departments, 1998," p. 12.
45. U.S. Department of Justice, "Local Police Departments, 1997," pp. 1–3.
46. U.S. Department of Justice, "Sheriff's Departments, 1997," pp. 1–3.
47. Samuel Walker, *The Police in America: An Introduction*, 2d ed. (New York: McGraw-Hill, 1992), p. 61.
48. Herman Goldstein, *Policing a Free Society* (Cambridge, MA: Ballinger Publishing Company, 1977), p. 21.
49. Steven M. Cox, *Police: Practices, Perspectives, Problems* (Boston, MA: Allyn and Bacon, 1996), pp. 18–19.
50. American Bar Association, *Standards Relating to the Urban Police Function* (Chicago: American Bar Association, 1973), p. 7.
51. *Ibid.*, p. 61.
52. See Albert Reiss, *The Police and the Public* (New Haven: Yale University Press, 1971), p. 96.
53. Jerome H. Skolnick and David H. Bayley, *The New Blue Line: Police Innovation in Six American Cities* (New York: The Free Press, 1986), p. 4.
54. Patrick V. Murphy and Thomas Plate, *Commissioner: A View from the Top of American Law Enforcement* (New York: Simon and Schuster, 1977). Also see Walker, *The Police in America*, pp. 55–56.
55. Cox, *Police: Practices, Perspectives, Problems*, pp. 18–19.
56. *Ibid.*
57. James Q. Wilson, *Varieties of Police Behavior* (Cambridge, MA: Harvard University Press, 1968), pp. 140–226.

CHAPTER 3

Police Subculture

The Making of a Cop

Key Terms and Concepts

Computer-based training
  (CBT)
Field training officer
Hurdle process
Police cynicism

Police subculture
Recruiting/academy training
Sixth sense
Traits of good officers
Working personality

I think the necessity of being *ready* increases. Look to it.—*Abraham Lincoln*

Make thy Model before thou buildest; and go not too far in it without due Preparation.—*Thomas Fuller*

Learn how to be a policeman, because that cannot be improvised.
—*Pope John XXIII*

INTRODUCTION

This chapter focuses on the **police subculture**. It describes how and where the officer's career begins and, to a large extent, how his or her occupational personality is formed. Studying the subculture of the police helps us to define the "cop's world" and the officer's role in it; this subculture shapes the officer's attitudes, values, and beliefs.

The idea of a police subculture was first proposed by William Westley in his 1950 study of the Gary, Indiana, police department, where he found, among many other things, a high degree of group cohesion, secrecy, and violence.[1] It is now widely accepted that the police develop traditions, skills, and attitudes that are unique to their occupation because of their duties and responsibilities.[2]

We begin at the threshold, looking at some of the methods, challenges, and problems connected with the recruitment of qualified individuals. Then we track the typical police applicant's progression through the various types of tests that may be employed—written, psychological, physical, oral, character, and medical.

Then formal police training at the academy is examined; this is where the initiation of the officer-to-be into the police subculture commences in earnest. Discussed in this chapter section are types of academies, their general curriculum, and some of the informal learning that takes place. We then look at post-academy training—the field training officer concept. Following that is a look at how officers adopt their working personality—the formal and informal rules, customs, and beliefs of the occupation. This portion of the chapter includes an assessment of the traits that make a good officer.

FIRST THINGS FIRST: RECRUITING QUALIFIED APPLICANTS

WANTED: THOSE WHO WALK ON WATER

Recruiting an adequate pool of applicants is an extremely important facet of the police hiring process. August Vollmer said that those persons to be selected as law enforcers should

> have the wisdom of Solomon, the courage of David, the patience of Job and leadership of Moses, the kindness of the Good Samaritan, the diplomacy of Lincoln, the tolerance of the Carpenter of Nazareth, and, finally, an intimate knowledge of every branch of the natural, biological and social sciences.[3]

Many people believe that the police officer has the most difficult job in America. Police officers are solitary workers, spending most of their time on the job unsupervised; also, people who are hired today will become the supervisors of the future. For all of these reasons, police agencies must attempt to attract the best individuals possible.

Police applicants typically come from lower-middle-class or working-class backgrounds;[4] they generally have a high school education and a history of employment. They also tend, at the application stage, to be enthusiastic, idealistic, generally lacking in knowledge about what police work is really like, and very different from the stereotype of the police officer as authoritarian, suspicious, and insensitive.[5]

Some studies indicate that police applicants are primarily motivated by the need for job security.[6] Other researchers have found that both males and females listed the same six factors—helping people, job security, fighting crime, job excitement and prestige, and a lifetime interest—as strong positive influences in their career choices.[7] Joel Lefkowitz[8] concluded that police candidates were lower than average in their desire to do autonomous work, and other studies have indicated that applicants tend to favor a more directive leadership style. Such findings are not unusual, given that most police agencies are highly structured and paramilitary in nature. Studies do *not* establish that police candidates fit the stereotypes of harsh, controlling people who wish to control others. An attribute that *is* sought, however, is that of assuming leadership, or the ability to take charge of situations. Some researchers even have found the typical police applicant to be very similar to the average college student.[9]

Bruce Carpenter and Susan Raza, using the Minnesota Multiphasic Personality Inventory (MMPI), found that police applicants differed from the general population in several important ways.[10] Police applicants, they learned, are somewhat more psychologically healthy, are generally less depressed and anxious, and are more assertive and interested in making and maintaining social contacts. Furthermore, few police aspirants have emotional difficulties, and they have a greater tendency to present a good impression of themselves than the general population. They are a more homogeneous group.

Female police applicants tend to be more assertive and nonconforming and to have a higher energy level than male applicants. They are also less likely to identify with traditional sex roles than male applicants. Older police applicants tend to be less satisfied, have more physical complaints, and are more likely to develop physical symptoms under stress than younger applicants. Applicants to large city police forces tend to be less likely to have physical complaints and a higher energy level than applicants to small or medium towns. (This is probably explained by the fact that applicants in large cities are significantly younger.[11]) Some departments are mandated to recruit special groups of people, such as women, African Americans, and Hispanics; gays have also been recruited in several cities.

What psychological qualities should be sought? According to psychologist Lawrence Wrightsman,[12] it is important that they be *incorruptible*, of high moral character. They should be *well adjusted*, able to carry out the hazardous and stressful tasks of policing without "cracking up," and thick-skinned enough to operate without defensiveness. They should be *people oriented*, with a genuine interest in people and a compassionate sense of their innate dignity. Applicants should also be *free of emotional reactions*, they should not be impulsive or overly aggressive, and they should exercise restraint. This is especially important given their active role in crime detection.

Finally, they need *logical skills* to assist in their investigative work. An interesting example of some of the logical skills needed for police work is provided by Al Seedman, former chief of detectives in the New York Police Department:

> In the woods just outside of town they found the skeleton of a man who'd been dead for three months or so. I asked whether this skeleton showed signs of any dental work. But the local cops said no, although the skeleton had crummy teeth. No dental work at all. Now, if he'd been wealthy, he could have afforded to have his teeth fixed. If he'd been poor, welfare would have paid. If he was a union member, their medical plan would have covered it. So this fellow was probably working at a low-paying non-unionized job, but making enough to keep off public assistance. Also, since he didn't match up to any family's missing-person report, he was probably single, living alone in an apartment or hotel. His landlord never reported him missing, either, so most likely he was also behind on his rent and the landlord probably figured he had just skipped. But even if he had escaped his landlord, he would never have escaped the tax man. The rest was simple. I told these cops to wait until the year is up. Then they can go to the IRS and get a printout of all single males making less than $10,000 a year but more than the welfare ceiling who paid withholding tax in the first three quarters but not in the fourth. Chances are the name of their skeleton will be on that printout.[13]

Other desirable traits of entry-level officers are discussed later.

RECRUITING PROBLEMS AND SUCCESSES

Whether one blames a generally strong national economy, higher educational requirements, or noncompetitive wages and benefits, police and sheriff's departments in virtually every region of the country agree that the generous pool of applicants from which they once sought qualified candidates is becoming increasingly shallow.[14] In fact, the National Association of Police Organizations (NAPO) recently named **recruitment** one of the top problems facing police agencies.[15]

The state of the economy has a strong influence over recruitment. The long-running boom in California's economy has caused problems for recruiters in that state's police agencies. With the unemployment rate running below 5 percent, police departments find it hard to compete with the private sector, despite offering relatively high pay, full benefits, and rich retirement plans. The Los Angeles Police Department saw its applications decline by about half in three years.[16] Chicago, which tested 25,000 applicants in 1993, had only 1,900 testees in the year 2000.

The requirement that applicants possess college credits is believed to make recruitment problems more acute. Several agencies now require two years of college credits and believe this standard reduces their applicant pool significantly. (The benefits of higher education for police officers are discussed in Chapter 12.) At the same time, a Census Bureau report put the average salary of a nonsupervisory

---

EXHIBIT 3–1

Speaking in Tongues

English-only Isn't Enough for Many Police Agencies

. . . According to preliminary findings from the U.S. Census Bureau, the Hispanic population grew nationwide by 39 percent between the years 1990 and 1999, but its growth in some Midwestern states has been significantly higher.

With Latinos drawn to the region in search of jobs in agriculture, meat packing and food distribution, preliminary census data reported in August and September found the percentage of Hispanics in Wisconsin rose by 50.4 percent from 1990 to 1999. Iowa experienced an 89-percent increase since the 1990 census. . . .

In Nebraska, the percentage of Hispanics increased by 108 percent, making them the state's single largest minority group. . . .

Despite the increases, Hispanics are still just a fraction of the overall population in many places, but the situation has led to a far more frequent need for Spanish-speaking officers and interpreters. Some law enforcement agencies . . . cannot recruit bilingual officers fast enough. . . .

According to Des Moines Police Chief William Moulder, Iowa officials have actively sought to make the state a magnet for immigration so as to meet labor force needs. In addition to Hispanics, there are also Chechnyans and Sudanese immigrants as well as Southeast Asians. . . . "We have a few officers who can speak Spanish fairly fluently, a few that speak the Asian dialects. I don't think anyone can speak Russian." . . .

"We tried to keep our interpreters up," said Mehlin. "Eight to 10 years ago, it wasn't a problem at all. We would have an occasional need for one but it wasn't difficult to find people. But over the last year or so, it has become more and more critical. I think it's because the interpreters we have, it has become tiresome to be called in the middle of the night."

Recruitment, already a challenge with the nation's continued economic upturn, is particularly difficult, said Mehlin. . . .

Source: Reprinted with permission from *Law Enforcement News*, October 31, 2000, p. 1. John Jay College of Criminal Justice (CUNY) 555 West 57th St., New York, NY 10019.

---

police officer at $34,700; this compares to an average $40,546 that a technical support worker can expect to make, and the average of $51,351 that managers or executives earn. According to NAPO, add to that the media attention given to corrupt officers, police shootings, and department scandals, and many people believe it is no wonder that people are shying away from police careers. To counter this declining interest in a police career, departments are accentuating the occupation's positive aspects: solid insurance, excellent retirement plans, long vacations, and opportunities to advance.[17]

Recruiting and retaining females in police service remain particularly problematic. Gender bias (reflected in the absence of women being hired and then promoted into policymaking roles) and sexual harassment concerns prevent many women from applying and cause many female officers to leave, and quickly: About 60 percent of female officers who leave their agency do so during their second and fifth years on the job.[18] (Problems of bringing women into policing—and retaining them—are discussed more fully in Chapter 12.)

But some agencies have successfully addressed the recruitment dilemma. For example, the New York State Police (NYSP) recently swore in its largest class in 30 years. Using an academic survey developed by a high-ranking trooper with a doctorate, the agency asked people what would make them consider joining the state police (their answer: job enrichment or more interesting work). The survey also learned that the two factors playing a key role in a person's decision to enter police work were the ability to help others and the opportunity to serve the community—two factors now pushed in the NYSP mission and values statements and in its recruitment drives. Also emphasized in recruitment literature and television announcements are how female troopers can balance family and career and the various specialized jobs—scuba diving, for example—that are available within the NYSP. The organization also enlisted the entire force as recruiters and recruited at nontraditional locations: women's road races and health clubs, for example.[19]

Other methods employed by police agencies in their attempts to develop a bigger pool of applicants include seeking applicants far from home (as examples, Los Angeles recruits in Chicago; Chicago recruits in Wisconsin), having application forms on the Web where they can be easily downloaded, lowering the minimum age from 22 to 21, and allowing some applicants to substitute work experience for college credits.[20]

TESTING: THE HURDLE PROCESS FOR NEW PERSONNEL

Once the recruit is allowed in the door, much work still remains to be done before he or she will be deemed worthy of the title police officer. The new recruit must successfully complete what is known as the **hurdle process**. Next we consider some of the kinds of tests that are employed to weed out undesirable candidates.

Not all of the types of tests shown in Figure 3–1 will be employed by the 17,000 police agencies in America; nor will these steps necessarily occur in the sequence shown. It is known, however, that nearly all local police agencies (99 percent) use criminal record checks, background investigations (98 percent), driving record checks (98 percent), and medical exams (97 percent) to screen

Recruits at pistol range. (*Courtesy R. Brand*)

applicants. Psychological (91 percent), aptitude (84 percent), and physical agility (78 percent) tests are also used.[21]

Also note that under affirmative action laws and court decisions, a burden rests with police administrators to demonstrate that the tests used are job related. The hiring sequence shown in Figure 3–1, called the Multiple Hurdle Procedure,[22] may take longer than three months to complete, depending on the number and type(s) of tests used and the ease of scheduling and performing them.

WRITTEN EXAMINATIONS: GENERAL KNOWLEDGE AND PSYCHOLOGICAL TESTS

Measures of general intelligence and reading skills are the best means a police agency can use for predicting who will do well in the police academy.[23] Of course, any such test must be reliable and valid. To achieve reliability and validity, many if not most police agencies purchase and use "canned" test instruments—those prepared by professional individuals or companies.

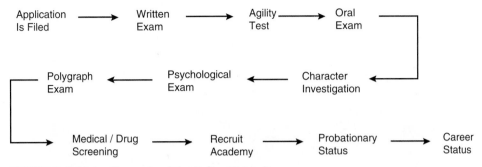

FIGURE 3–1 Major Elements of the Police Hiring Process.

Four types of written tests are used by larger police departments and state police agencies: cognitive tests (measuring aptitudes in verbal skills and mathematics, reasoning, and related perceptual abilities), personality-type tests (predominantly the MMPI), interest inventories (the Strong-Campbell, the Kuder, and the Minnesota Interest tests), and biographical data inventories.[24]

A study of deputy sheriffs found that candidates with written test scores above the 97th percentile were most apt to be successful.[25] However, a study of the Tucson, Arizona, police department determined that the IQ scores of officers who dropped out of the force were significantly higher than those of a norm group. The study concluded that one can be too bright to be a cop, unless an alternate career-development program can be developed to challenge and use highly intelligent people.[26] Of course, there is much more to police work than reading skills.

General intelligence tests are often administered and scored by the civil service or the central personnel office. Most frequently, those who fail the entrance examination do not make the minimum score (usually set at 70 percent) and will go on to other careers, although most jurisdictions allow for a retest after a specified period of time has elapsed. The names of those who pass will be forwarded to the police agency for any further in-house testing and screening.[27]

The other form of written examination for police applicants is the psychological screening test. Candidates must be carefully screened in order to exclude those who are emotionally unstable, overly aggressive, or suffering from some personality disorder. The two primary tests of suitability of police candidates are the MMPI and the California Personality Inventory (CPI).[28] There are two major concerns in using such tests to screen out applicants: stability and suitability. Stability is a major legal concern. If an officer commits a serious, harmful, and inappropriate act, the question of his or her stability will be raised. The police agency may be asked to provide documentation about why the officer was deemed stable at the time of employment. It has been found that 2 to 5 percent of the police applicant pool may be eliminated due to severe emotional or mental problems.[29]

Police recruits must demonstrate proficiency in using several types
of less-than-lethal weapons as well as in weaponless defense.
(*Courtesy NYPD*)

PHYSICAL AGILITY TEST

Entry-level physical examinations range from a minimally acceptable number of pushups to timed running and jumping tests, such as dragging weights, pushing cars, leaping over six-foot walls, walking on horizontal ladders, crawling through tunnels, and negotiating monkey bars. The problem is that very few of these activities are actually performed by police officers on the job.

Therefore, the challenge for police executives, and an area of lawsuit vulnerability, is finding and using a truly job-related physical agility examination. A need exists to determine the nature and extent of physical work performed by police officers, followed by the development of an instrument to measure applicants' ability to perform that work. One such test is based on the theory that police officers must perform three basic physical functions: getting to the problem (possibly needing to run, climb, vault, and so forth), resolving the problem (perhaps needing to fight or wrestle with an offender), and removing the problem (often requiring that the officer carry heavy persons or objects). To establish the testing protocol for a given jurisdiction, the officers fill out written forms concerning the kinds of physical work that they performed during their workday for one month. Information from the forms is then computer analyzed and, using these data, a physical agility test is developed that actually measures the recruit's ability to do the kinds of work performed by police officers in that specific locale.[30] If challenged in court, such tests seek to show that they test for the actual job requirements of that jurisdiction and do not discriminate on the basis of gender, race, height, age, and so on.

ORAL INTERVIEW

The oral interview is used by more than 90 percent of all police agencies as part of the selection process.[31] Candidates appear individually before one or more oral boards that are comprised of persons selected from the police agency and often the community. Candidates may also be asked to participate in a clinical interview with a psychologist; studies have indicated that the clinical interview complements the written psychological tests well.[32]

The purpose of the oral board is to assess aspects of the candidate that cannot be measured on other tests, such as appearance, the ability to communicate and reason (often using situational questions), and general poise and bearing. The interview is not normally well suited for judging character, dependability, initiative, or other such factors.

A primary advantage of the interview is that evaluators can ask applicants to explain how they would behave and use force in given situations. Any number of

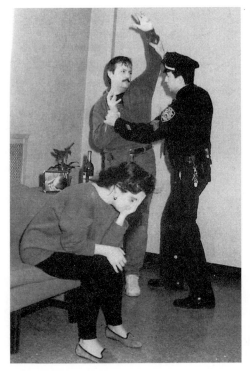

Role-playing scenarios are also effective for training recruits to handle difficult situations. (*Courtesy New York Police Department Photo Unit*)

possible scenarios exist; following are examples of the kind of situations that might be posed to police applicants to see how well they think on their feet, develop appropriate responses, and prioritize their actions.

1. You are dispatched to a neighborhood park to check out a young man who is acting strangely. Upon arrival, you see the youth standing near a group of children playing on a merry-go-round. He is holding a .22 caliber rifle. What is your next action?

2. You are in the men's locker room at the end of shift. You hear another male officer talking about a female officer's anatomy. What response would you have?

3. You are at home on a weekend, watching a football game. Your neighbor comes to your door and frantically claims that his door has been kicked in and that he believes someone is inside. What do you do? Suppose the neighbor claims that his daughter might be upstairs in his house. How would you proceed?

4. You are in a downtown area making an arrest. A crowd gathers and you begin to hear comments about "police harassment." Soon the crowd becomes more angry in nature. How do you react?

5. You and another officer are dispatched to a burglary call at an office building. While searching the scene, you observe the other officer remove an expensive fountain pen from a desktop and put it into his pocket. What would be your response?

CHARACTER INVESTIGATION

As indicated earlier, nearly all (98 percent) of local police departments use background checks or character investigations—probably the most important element of the selection process. If properly done, it can be one of the most time-consuming and costly elements as well.

Character is one of the most subjective yet most important factors an applicant brings to the job; it cannot normally be measured with data and interviews. It involves talking to the candidate's past and current friends, co-workers, teachers, neighbors, and employers, all of whom have personal knowledge of the candidate and so can provide crucial information. The applicant should be informed that references will be checked and through them the investigation will mushroom to others. No expense should be spared in talking with those people; if the job is done properly, the investigator will not only have a complete knowledge of the person's character but will also know where any bones may be buried in the applicant's background.

POLYGRAPH EXAMINATIONS

A survey of the nation's 626 largest police agencies found that 62 percent conducted polygraph examinations as part of their selection process. The agencies indicated that the primary benefits of polygraph screening are that applicants are more honest and that it produces higher-quality employees; 90 percent of these agencies automatically reject applicants who refuse to undergo polygraph testing.[33]

A survey of the benefits of polygraph examinations for police applicants by Richard Arther, director of the National Center of Lie Detection,[34] supported the need for the polygraph for police recruitment:

- An applicant for a police position in Lower Merion, Pennsylvania, came to that agency highly recommended by a police lieutenant and his employer at a home for blind retarded children. During the polygraph examination, the applicant admitted to at least 50 instances of sexually abusing the children.
- An applicant with the Wichita, Kansas, Police Department admitted to the polygraphist that he had been involved in many burglaries. Its detective division was able to clear eight of its unsolved burglaries as a result of this applicant's confession.
- A police officer in one California police department applied for employment in the Salinas, California, Police Department. He appeared to be a model police officer, was in excellent physical condition, and was familiar with state codes. His previous experience made him a potentially ideal candidate. However, during the polygraph exam, he admitted to having committed over a dozen burglaries while on duty and having used his patrol car to haul away the stolen property. He also admitted to planting stolen narcotics on innocent suspects in order to make arrests and having had sexual intercourse with girls as young as 16 in his patrol vehicle.
- An applicant for the San Diego, California, Sheriff's Department admitted to that agency's polygraphist that on weekends he would go from bar to bar pretending to be drunk. He would then seek out people to pick fights with, since he could only have an erection and orgasm while inflicting pain on others. In addition to these sadistic tendencies, he also admitted that he got rid of his frustrations by savagely beating "niggers, Chicanos, and long-haired pukes who cause all the trouble."

These are but a few examples of how the investment of time and money for polygraph examinations can spare the public and the police agency a tremendous amount of trouble and expense later on. It is doubtful that few if any of these behaviors would have surfaced during the course of an oral interview or a background investigation. Polygraph testing is discussed in greater detail in Chapter 7, in connection with criminal investigations.

MEDICAL TESTS/DRUG SCREENING

Someone once said that police medical examinations are often of the "hear thunder/see lightning" variety. It is also widely believed that policing is only for those young people who are in peak physical condition. Whether these statements are facetious or not, it is certainly true that policing is no place for the physically unfit. Such persons would not only be a hazard to themselves but also to their fellow officers. The job, with its stress, shift work, many hours of inactivity during patrol time, and other such contributing factors, can be physically debilitating enough for veteran officers, especially those who fail to exercise and diet properly. Police administrators certainly do not wish to bring people who are physically unfit to begin with into the police service. The FBI, for example, will not consider applicants whose weight exceeds the norm for their height and body type. Unfit personnel would also have lower energy levels and would probably give less attention to duty and take more sick days. Early retirements and disability often result, as well as increased operating expenses for replacing ill officers and hiring and training new permanent replacements.

More and more, police agencies, like the private sector, the military, and other sensitive government agencies, are compelling prospective employees to submit to a drug screen examination. Substance abuse remains a very real problem in the workplace, resulting in poor productivity, lowered agency morale, and increased accidents and injuries.

THE RECRUIT'S FORMAL ENTRY INTO POLICING: ACADEMY TRAINING

TYPES OF ACADEMIES

Today, with the concern for well-trained officers and civil liability, academy training is regarded as a vital part of the hiring process, or, as Clinton Terry put it, as "merely another filter through which the candidate must successfully pass."[35] The recruit academy is a major point in the career of the officer-to-be; for many police agencies, **academy training** provides the bulk of the formal training that the officer will acquire during an entire career. The academy also plays a significant role in shaping the officer's attitudes and is the beginning point for the occupational socialization of the officer.

With in-house police academies, the department is authorized by some certifying body to train its own officers; there are also state and regional academies. Some of these academies are operated by community colleges and universities, using instructors from the area. In a relatively new concept, civilians attend police academies at their own cost, hoping to gain employment as a free agent with a police agency after graduating and becoming formally certified. This pre-service model is becoming more and more popular; police administrators are realizing tremendous savings by not paying salaries, registration fees, living expenses, and other costs normally accrued while their employees attend academies.

The length of academy training varies by state; the norm is about 400 hours. Many metropolitan police agencies, some federal organizations (such as the FBI), and many state agencies operate their own basic training academies. Several federal agencies send their personnel to the Federal Law Enforcement Training Center in Glynco, Georgia.

THE CURRICULUM: STATUS AND ONGOING NEED FOR REVISION

Recruit training curricula also vary from state to state, but most are heavily weighted toward the technical aspects of police work. Typically, today's basic recruit academy provides training in the following general areas: examining the criminal justice system, the law, human values and problems, patrol and investigation procedures, police proficiency (including armed and unarmed defense, riot and prisoner control, physical conditioning, and driver training), and administration (including departmental rules and regulations, policies, and organization).[36] More specifically, the modern training academy environment will include presentations about diverse cultural groups, resolving disputes, victim and witness assistance, conducting field sobriety tests, and use of computers. Modern recruits can also expect to learn about proper radio procedures, recognizing and managing stress, courtroom demeanor, and preparing accurate reports. Training may also be provided on such diverse topics as organized crime, state alcoholic beverage control, and wildlife and game laws.

John J. Broderick[37] contended that there are essentially three types of police academies: the plebe system (or stress academy), the technical training model, and the college system. The stress academy has been described as "a cross between Paris Island marine boot camp and college."[38]

Stress academies are intended to emphasize physical, mental, and emotional activities that resocialize the recruit into a disciplined police officer. Typically, many recruits drop out or lose their self-esteem.[39] The technical training model, adopted by many departments, is like advanced military training in that it teaches useful operational skills and the use of equipment. The graduate of this program has technical competence. Officers who graduate are good at report writing, can cite many of the laws they are to enforce, know the radio code, have qualified with their weapons, and are able to testify in court. Critics of this model, however, contend that it requires only that recruits be receptive to the "pour it in with a funnel" approach of instructors and be able to recite what has been taught.

The college system often occurs in a college setting and stresses professionalism. The college system of academy training, recognizing that recruits are intelligent and capable of reason and decision making, has gotten a good deal of attention in recent years. Less emphasis is put on stress, discipline, and technical training, and more is placed on discussion and problem analysis.

Studies indicate that upon entry into the academy, recruits are normally confident and believe that their success is all but guaranteed. John Van Maanen wrote that this level of confidence is not always justified, however:

> The individual usually feels upon swearing allegiance to the department, city, state and nation that "he's finally made it." However, the department instantaneously and somewhat rudely informs him that until he has served his probationary period he may be severed from the membership rolls at any time without warning, explanation or appeal. It is perhaps ironic that in a period of a few minutes, a person's position vis-a-vis the organization can be altered so dramatically.[40]

An analysis of the curricula of basic training academies in 30 states by David Bradford and Joan Pynes revealed that little has changed since 1986. Less than 3 percent of basic training academy time is spent in the cognitive and decision-making domain; the remaining time is spent in task-oriented activities. The exception is the Commonwealth of Massachusetts, which recently revised its basic curriculum to have nearly all of its training hours in the cognitive domain.

According to the Massachusetts Criminal Justice Training Council, the objective of the cognitive training is to get the officer to

> understand how to speak to, reason with, and listen to people and learn to use communication skills to manage a wide range of problematic situations. Physical tactics and tools, though readily available, are secondary to the primary response of communication.[41]

Bradford and Pynes assert that community oriented policing and problem solving (COPPS, discussed in Chapter 6) requires the utilization of extensive cognitive and reasoning aptitudes and effective interpersonal skills and judgment. However, their examination of current basic training curricula found that problem solving and interpersonal skills development are low priorities of administrators and training directors.[42]

Bradford and Pynes concluded that there is a lack of congruence between the curriculum taught in many police academies and the job responsibilities of COPPS officers; given the rapid spread of the COPPS approach, as well as the availability of related textbooks, they asked, "Why, then, are training academy curricula not changing?"[43]

A NEW UNIFORM AND DEMEANOR

As the academy training begins, recruits adopt a new identity and a system of discipline in which they learn to take orders and not to question authority. They learn that loyalty to fellow officers, a professional demeanor and bearing, and respect for authority are highly valued in this occupation. The classroom teaches the recruit how to approach situations. Outside the classroom, as they hear and share with each other war stories discussed with academy staff, they informally transmit the proper attitudes to each other. Thus, the recruits begin to form a collective understanding of policing and how they are supposed to function and gradually develop a common language and demeanor. Many people also believe that the police develop a swagger, or a confident, authoritarian way of walking and presenting themselves. This is the beginning of the police officer's working personality.[44]

A uniform may also be worn for the first time during academy training; this is typically an awe-inspiring experience for the new recruits. The uniform sets them apart from society at large and certainly conveys a sense of authority and responsibility to the recruit and to the public. "Image is everything," according to a popular saying, and the choice of agency uniform can go a long way toward setting the image and tone of the department. Police uniforms come in various colors, styles, and fabrics. Some agencies even have their officers wearing blue jeans or shorts and T-shirts (for beach patrol).

The belt is one of the most important components of the patrol uniform; it is certainly one of the heaviest. It often exceeds 20 pounds when laden with weapon, cuffs, baton, radio, flashlight, extra ammunition, Mace, and so on. The uniform hat comes in several styles and is probably the piece of equipment that most readily identifies the officer and the department's image; each type of hat makes a certain statement to the public about the officer and the authoritarian image that is to be conveyed. The officer's badge also conveys a tremendous sense of authority. The most popular are customized shields, incorporating everything from the state motto and seal to symbols that, again, convey the agency's image and philosophy. When designing its badge, the department will want to consider the agency's tradition and history as well as that of the community.[45]

A SIXTH SENSE

The recruits are taught to nurture a **sixth sense:** suspicion. A suspicious nature is as important to the street officer as a fine touch is to a surgeon. The officer should not only be able to visually recognize, but also be able to physically *sense* when something is wrong or out of the ordinary. As a Chicago Police Department bulletin stated:

> Actions, dress, or location of a person often classify him as suspicious in the mind of a police officer. Men loitering near schools, public toilets, playgrounds and swimming pools may be sex perverts. Men loitering near . . . any business at closing time may be robbery suspects. Men or youths walking along looking into cars may be car thieves or looking for something to steal. Persons showing evidence of recent injury, or whose clothing is disheveled, may be victims or participants in an assault or strong arm robbery.[46]

Officers are trained to be observant, to develop an intimate knowledge of the territory and people, and to "notice the *normal* . . . only then can he decide what persons or cars under what circumstances warrant the appellation 'suspicious.'"[47] They must recognize when someone or something needs to be checked out. Sights such as the following will often warrant a field investigation:

- Persons who do not "belong" where they are observed
- Automobiles that do not "look right" (such as dirty cars with clean license plates or a vehicle with plates attached with wire or some other object)
- Businesses open at odd hours or not according to routine or custom
- Exaggerated unconcern over contact with the officer or being visibly "rattled" when near the officer
- Solicitors or peddlers in a residential neighborhood
- Lone male sitting in a car near a shopping center or near a school paying unusual attention to women or children
- Hitchhikers
- A person wearing a coat on a hot day[48]

The academy also teaches the neophyte officers that their major tools are their bodies; like mountain climbers, acrobats, or athletes, their bodies are an essential tool for the performance of their trade. The gun and nightstick initially fascinate the recruits, but until they are adequately trained in their use, officers using them would be more a menace than protector to society. Proper handling and safety measures are drilled into the recruits; the message is unequivocal that recruits will not be trusted with these potentially lethal weapons until they become proficient in their use. The new officers must be taught to measure their capacity to do the job, to carefully assess the physical assets of people they will confront on the street, and to determine whether or not someone can be subdued without assistance or the risk of injury if a physical altercation should develop.[49]

The officers are also told, however, that they cannot approach every situation with holster unsnapped or baton raised or twirling; they must demonstrate poise and not be eager to use force. It is constantly instilled in them that the days of the club-swinging cop are gone. Thus, knowing that the body is a tool, the recruits are taught how to unobtrusively position it, whether at a vehicle stop or while engaged in a discussion on the street, in order to gain a physical advantage should trouble arise. They are taught when to use force and when to relent, to always keep control of the situation and feel that they would emerge victorious should force be required. Thus, in addition to weapons training, they may be given some weaponless defense training, including some holds that can be applied to subjects to bring them into compliance.

They are taught some aspects of human nature and are encouraged not to be prejudicial in their actions or speech. They learn to deal with criminal suspects, offenders, victims, and witnesses and also learn to be suspicious of "eyewitness" accounts (for example, 25 "witnesses" claimed they helped carry the body of Abraham Lincoln from Ford Theater into the little house where he died; eight different persons said they held his head, and 84 people said they were in the room that night).[50]

Recruits often participate in hands-on training, practicing their new techniques in the field in simulated situations. Quite possibly the ultimate in hands-on training is the Hogan's Alley complex at the FBI Academy in Quantico, Virginia, which opened in 1987 and covers almost 35 acres. This facility combines training, office, and classroom space on one site, increasing training effectiveness. Hogan's Alley (the name given to many turn-of-the-century training facilities, apparently after an old comic strip about mischievous Irish kids) resembles a fully developed urban area. The set includes a business area and a residential street with townhouses and apartments. The use of Hollywood technology gives the illusion of depth and space. All furnishings—including a fleet of cars, furniture, desks, and even a pool table—are property forfeited by convicted criminals. Federal agents are trained in the practical skills of crime scene investigation and photography, surveillance techniques, the mechanics of arrest, and other investigative skills. Trainees use only deactivated weapons that fire blanks.

Other methods of police training that are in use include **computer-based training (CBT)**, electronic bulletin boards, satellite training and teleconferencing, on-line computer forums, and correspondence courses. With computer costs declining, CBT has been increasingly used and shown to be more effective. As CBT simulates real-life situations through the use of computer-modeled problems, it closely duplicates the way we think. One study found that police officers who learned about the exclusionary rule (discussed in Chapter 9) through CBT understood the material significantly better than the non-CBT control group.[51]

A segment of Hogan's Alley. (*Courtesy FBI*)

Virtual reality is another available (although very costly) form of police training. Here, trainees wear a head-mounted device that restricts their vision to two monitors and projects a computer-generated, three-dimensional illusion that engulfs the senses of sight, sound, and touch. Virtual reality may one day be commonly used for training police officers in such areas as pursuit driving, firearms training, critical incident management, and crime scene processing.

Finally, graduation day arrives and the academy experience becomes a rite of passage. Graduation also means new uniforms, associates, responsibilities, and a rise in pay and status. As Arthur Niederhoffer observed, for many officers academy graduation is a worthy substitute for a college education. But "the very next morning the graduate is rudely dumped into a strange precinct where he must prove himself."[52]

POST-ACADEMY FIELD TRAINING

THE FIELD TRAINING OFFICER CONCEPT

Once the recruits leave the academy, the process of their knowledge of and acceptance into the police subculture is not yet complete. Another very important part of this process is their being assigned to one or more veteran officers for initial field instruction and observation. This veteran will often be known as a **field training officer**, or FTO. The earliest formal field training program was apparently begun in the San Jose, California, Police Department in 1972.[53] This training program provides the recruits with an opportunity to make the transition from the academy to the streets while still under the protective arm of a veteran officer. Recruits are on probationary status, normally ranging from six months to one year; they understand that they may be immediately terminated if their over-all performance is unsatisfactory during that period.

Most FTO programs consist of several identifiable phases: an *introductory phase* (the recruit learns agency policies and local laws); *training and evaluation phases* (the recruit is introduced to more complicated tasks confronted by patrol officers); and a *final phase* (the FTO may act strictly as an observer and evaluator while the recruit performs all the functions of a patrol officer).[54] The National Institute of Justice (NIJ), surveying nearly 600 police agencies, found that 64 per-cent had a field training program and that such programs had reduced the num-ber of civil liability suits filed against their officers and against standardized train-ing.[55] The length of time a rookie is assigned to FTOs will vary. Normally a formal FTO program will require such close supervision for a range of one to 12 weeks.

While, unfortunately, some police officers end their formal training with the conclusion of academy and field training, most police officers receive in-service training throughout their careers. Most states require a minimal number of hours of in-service training for their police officers, and many departments will often exceed the minimum requirement. News items, court decisions, and other rele-vant information can be covered daily at roll call prior to the beginning of each shift. Short courses ranging from a few hours to several weeks are available for the in-service officers through several means.

With the ever-changing laws and methods of policing and the constant specter of liability, such ongoing training is essential; police officers must keep abreast of current changes in their field. As Geoffrey Alpert and Roger Dunham put it, "Whether an officer is overweight or out-of-shape, a poor shot, uses poor judgment, or is too socialized into the police subculture to provide good commu-nity policing, in-service training can be used to restore the officer's skills or to improve his attitude."[56]

An Assist from New Technologies

New technology in the training function includes software known as ADORE, for Automated Daily Observation Report and Evaluation. FTOs in several agencies now field testing the software find it saves them time because they do not have to hand-write reports that consist of voluminous forms for each recruit that are kept in a five-inch thick three-ring binder. ADORE, which can be accessed either through a laptop or a Palm Pilot, provides for computerized note taking while FTOs are watching trainees at work; it also reduces paperwork by allowing train-ers to easily compile numbers for evaluating performance in dozens of categories.

The software is believed to also be reducing FTO burnout that is often a part of the paper-intensive process.[57]

Another new form of technology for police training that is being tested involves pursuit simulation. The training simulator is thought to be an effective means to determine how and when a vehicle pursuit should be halted. In one scenario, trainees in a simulated pursuit swerve around a computerized image of a transit bus, a produce truck, a minivan, and a child on a skateboard before the chased vehicle enters a school zone, where the officer should end the hot pursuit. These simulated pursuits also allow supervisors to see how well trainees conduct themselves in accordance with their agency's pursuit policy, which is often several pages long.[58]

HAVING THE "RIGHT STUFF": A WORKING PERSONALITY

DEVELOPING AND USING A POLICE PERSONALITY

Since William Westley first wrote about the police subculture in 1950, the notion of a *police personality* has become a popular area of study. In 1966, Jerome Skolnick described what he termed the **working personality** of the police. Basically, Skolnick determined that the police role contained two important variables: danger and authority. Danger is a constant feature of police work. Police officers, constantly facing potential violence, are warned at the academy to be cautious. They are told many war stories of officers shot and killed at domestic disturbances or traffic stops. Consequently, they develop a "perceptual shorthand," Skolnick said, that they use to identify certain kinds of people as "symbolic assailants"—persons whose gesture, language, and attire the officer has come to recognize as potentially violent.

The police, as Skolnick stated, represent authority. But unlike doctors, ministers, and the like, they must *establish* their authority. The symbols of that authority—the gun, the badge, and the baton—assist them, but officers' behavior and confidence are more important in social situations. As William Westley said,

> [An officer] expects rage from the underprivileged and the criminal but understanding from the middle classes: the professionals, the merchants and the white-collar workers. They, however, define him as a servant, not as a colleague, and the rejection is hard to take.[59]

Thus, officers cannot even depend on their symbols and position of authority in dealing with the public; often, they are left confused when the public does not automatically observe and accept their authority.

Although a police subculture exists, the question is often raised whether there is a police personality. Considerable research has compared the personality characteristics of the police with those of the general public,[60] and a number of differences have been discovered. One study found the average officer to be more intelligent, assertive, dependable, straightforward, and conscientious when compared to others.[61] Other researchers who recently studied state traffic officers and deputy sheriffs using the MMPI and CPI scales reported that the officers scored high on the values of achievement, a strong work ethic, ambition, leadership potential, and being well organized.[62] Studies have also found conservatism and a high degree of cynicism among the police, although those traits are found to be present in much of the society at large. The late LAPD Chief William Parker asserted that police were "conservative, ultraconservative, and very right wing."[63]

Niederhoffer reported his classic study of **police cynicism** in 1967, using the New York Police Department as the site of a longitudinal study. He found that although a typical recruit began his or her career without a trace of cynicism, police cynicism spikes most heavily immediately after recruits leave the basic academy. This is probably because they confront the reality of the streets—the pain and criminality of society—and perhaps lose friends. Cynical veteran peers frequently reinforce the worst aspects of the job. Then, from about two to six years of service, the cynicism level continues to increase but at a slower rate. The recruit has possibly begun to adapt to the occupation and the people to be dealt with every day. At about mid-career (about 8 to 13 years of service), the cynicism level actually begins to decline, possibly because the officer has accepted the job and has been promoted, earns a decent salary and benefits, and realizes that he or she is about halfway to retirement. Toward the end of the career, the degree of cynicism levels off; for many officers, this is a period of coasting toward retirement.

Once a policeman becomes cynical, his or her view of humanity becomes distorted because the great majority of people they deal with are offenders. They see miscarriages of justice (for example, improper or lenient court decisions, perjury on the witness stand, plea bargaining) and observe fellow officers who do not live up to their code of ethics. Cynicism does have a protective feature, however. It can help to make the officer calloused, allowing him or her to observe things that would sicken or horrify the average citizen without becoming mentally debilitated.

John Broderick presented another view of the working personality of the police. He believed that there are actually four types of police personalities: enforcers, idealists, realists, and optimists.[64] *Enforcers* are officers who believe that the job of the police primarily consists of keeping their beats clean, making good arrests, and sometimes helping people. These officers have sympathy for vagrants, the elderly, the working poor, and others seen as basically good people. However, drug users, cop haters, and others frustrate the efforts of enforcers to make them "good." This makes the enforcers very unhappy. Thus, they have high job dissatisfaction and an attitude of resentment, feeling a lot of people have hostility toward them.

Idealists are officers who put high value on individual rights and due process. They also believe it is their duty to keep the peace, protect citizens from criminals, and generally preserve the social order. As a group with a high percentage of college graduates, their commitment to the job is lowest of the four groups, and they are less likely to recommend the job to a son or daughter.

Realists place a relatively low emphasis on both social order and individual rights. They seem less frustrated and seem to have found a way to come to terms with a difficult job. For them, the reality of the job consists of manila envelopes and properly completed forms. Realists see many problems in policing, especially with such situations as politicians receiving special privileges. Reality is not warm bodies to be dealt with but rather the paperwork that the bodies leave behind. They work well in the ordered, predictable environment of a police records room.

Optimists also place a relatively high value on individual rights. Like idealists, they see their job as people- instead of crime-oriented. They see policing as providing opportunities to help people; they find the television version of policing to be totally unrealistic and view the 80 percent of a typical officer's time in service activities as rewarding. Optimists have the lowest amount of job resentment, are indeed committed to the job, and would choose policing as a career all over again. They enjoy the mental challenge of problem solving.

WHAT TRAITS MAKE A GOOD COP?

It is not too difficult to identify a bad cop through his or her unethical or criminal behavior. It is probably more difficult for the average person to decide what are the **traits of good officers**. How can a quantitative measure assess the work of police? Is it possible to judge the quality of one's work? These are challenging questions for police supervisors.

A major obstacle to assessing police performance rests with the nature of police work generally and the variation in kinds of work performed between shifts. The police role varies according to whether the officer is assigned to the day shift, evening (swing) shift, or night (graveyard) shift. (See a description of how police work varies by shift in Chapter 5.)

Dennis Nowicki acknowledged that while certain characteristics form the foundation of a police officer—honesty, ethics, and moral character—no scientific formula can be used to create a highly effective police officer. However, he compiled 12 qualities that he believes are imperative for entry-level police officers:

1. *Enthusiasm*: believing in what he or she is doing and going about even routine duties with a certain vigor that is almost contagious
2. *Good communications skills*: highly developed talking and listening skills; ability to interact equally well with a wealthy person or someone lower on the socioeconomic ladder
3. *Good judgment*: wisdom and analytical ability to make good decisions based on an understanding of the problem
4. *Sense of humor*: ability to laugh and smile, to help officers cope with the regular exposure to human pain and suffering
5. *Creativity*: ability to use creative techniques by placing themselves in the mind of the criminal and legally accomplishing arrests
6. *Self-motivation*: making things happen, proactively solving difficult cases, creating their own luck

7. *Knowing the job and the system*: understanding the role of a police officer, the intricacies of the justice system, and what the administration requires; using both formal and informal channels to be effective
8. *Ego*: believing they are good officers; having self-confidence that enables them to solve difficult crimes
9. *Courage*: ability to meet physical and psychological challenges; thinking clearly during times of high stress; admitting when they are wrong; standing up for what is difficult and right
10. *Understanding discretion*: enforcing the spirit of the law, not the letter of the law; not being hard nosed, hard headed, or hard hearted; giving people a break and showing empathy
11. *Tenacity*: staying focused; seeing challenges, not obstacles; viewing failure not as a setback but as an experience
12. *A thirst for knowledge*: being aware of new laws and court decisions; always learning (from the classroom but also via informal discussions with other officers)[65]

SUMMARY

This chapter examined how ordinary citizens are recruited, tested, and trained for their role as police officers; it demonstrated how people are socialized into the cop's world—how they are brought into the police subculture and transformed psychologically, physically, and emotionally in order to become competent to perform this very challenging occupation.

It was established that a working personality develops in police officers and that both formal training and informal peer relations are helpful in teaching the novice officer how to act and how to think about certain elements of the job. It was also shown that policing can easily inculcate in police officers a sense of cynicism by virtue of their having to deal with the everyday pain and problems of our society.

Clearly the recruiting and training process that is conducted at the outset of their careers and makes them new members of the police subculture is one of the most significant periods of the officer's career.

ITEMS FOR REVIEW

1. Describe some of the problems confronting today's police recruiters as well as some of the unique measures being taken to obtain a viable applicant pool.
2. Explain in general the hiring process of new police officers. Include a listing of the kinds of tests that are commonly employed.
3. Relate in general terms the kinds of skills and knowledge that are imparted to police trainees during the academy phase. Also, describe the typical subjects that are found in a police academy curriculum and some of its strengths and weaknesses.
4. Discuss methods and purposes of the FTO concept.

5. Describe what is meant by the term "police personality," how it is developed, and how it functions.
6. Explain police cynicism and how it operates, per Niederhoffer.
7. Review some of the ideal traits of persons who make good police officers.

NOTES

1. William A. Westley, *Violence and the Police* (Cambridge, MA: The MIT Press, 1970).
2. Geoffrey P. Alpert and Roger G. Dunham, *Policing Urban America,* 2d ed. (Prospect Heights, IL: Waveland Press, 1992), p. 80.
3. V. A. Leonard and Harry W. More, *Police Organization and Management,* 3d ed. (Mineola, NY: Foundation Press, 1971), p. 128.
4. Joel Lefkowitz, "Industrial-organizational Psychology and the Police," *American Psychologist* (May 1977): 346–364.
5. R. B. Mills, "Use of Diagnostic Small Groups in Police Recruit Selection and Training," *Journal of Criminal Law, Criminology and Police Science* 60 (1969): 238–241; John Van Maanen, "Police Socialization: A Longitudinal Examination of Job Attitudes in an Urban Police Department," *Administrative Science Quarterly* 20 (1975): 207–228.
6. C. Gorer, "Modification of National Character: The Role of the Police in England," *Journal of Social Issues* 11 (1955): 24–32; Arthur Niederhoffer, *Behind the Shield: The Police in Urban Society* (New York: Anchor Books, 1967), p. 140.
7. M. Steven Meagher and Nancy A. Yentes, "Choosing a Career in Policing: A Comparison of Male and Female Perceptions," *Journal of Police Science and Administration* 14 (1986): 320–327.
8. Lefkowitz, "Industrial-organizational Psychology and the Police."
9. J. D. Matarazzo, B. V. Allen, G. Saslow, and A. N. Wiens, "Characteristics of Successful Policemen and Fireman Applicants," *Journal of Applied Psychology* 48 (1964): 123–133.
10. Bruce N. Carpenter and Susan M. Raza, "Personality Characteristics of Police Applicants: Comparisons Across Subgroups and with Other Populations," *Journal of Police Science and Administration* 15 (1987): 10–17.
11. Carpenter and Raza also compared police applicants with other similar occupational groups, finding that police applicants appear to be most like nuclear submariners and least like air force trainees and security guards.
12. Lawrence S. Wrightsman, *Psychology and the Legal System* (Monterey, CA: Brooks/Cole Publishing Company, 1987), pp. 85–86.
13. Al Seedman and P. Hellman, *Chief!* (New York: Arthur Fields, 1974), pp. 4–5.
14. Jennifer Nislow, "Is Anyone Out There?" *Law Enforcement News,* October 31, 1999, p. 1.
15. Nicole Ziegler Dizon, "Searching for Police," Associated Press, June 3, 2000.
16. Marc Lifsher, "State Strains to Recruit New Police," *Wall Street Journal,* November 10, 1999, p. CA1.
17. Dizon, "Searching for Police."
18. "Plenty of Talk, Not Much Action," *Law Enforcement News,* January 15/31, 1999, p. 1.
19. Hiring Problem? What Hiring Problem? NYSP Has Answers to Recruiting Slump," *Law Enforcement News,* November 30, 2000, p. 1.
20. "Police Chiefs Try Many Recruiting Strategies to Boost Applicant Pool," *Crime Control Digest,* February 2, 2001, pp. 1–2.
21. U.S. Department of Justice, Bureau of Justice Statistics Executive Summary, *Local Police Departments, 1997* (Washington, D.C.: Author, October 1999).
22. Alfred Stone and Stuart DeLuca, *Police Administration* (New York: John Wiley & Sons, 1985).
23. Hans Toch, *Psychology of Crime and Criminal Justice* (Prospect Heights, IL: Waveland Press, 1999), p. 44.
24. Philip Ash, Karen B. Slora, and Cynthia F. Britton, "Police Agency Officer Selection Practices," *Journal of Police Science and Administration* 17 (December 1990): 259–264.
25. S. H. Marsh, "Validating the Selection of Deputy Sheriffs," *Public Personnel Review* 23 (1962): 41–44.
26. William H. Thweatt, "A Vocational Counseling Approach to Police Selection" (Unpublished dissertation, University of Arizona).

27. W. Clinton Terry III, *Policing Society* (New York: John Wiley & Sons, 1985), p. 194.
28. George E. Hargrave, "Using the MMPI and CPI to Screen Law Enforcement Applicants: A Study of Reliability and Validity of Clinicians' Decisions," *Journal of Police Science and Administration* 13 (1985): 221–224.
29. Roger G. Dunham and Geoffrey P. Alpert, *Critical Issues in Policing: Contemporary Readings* (Prospect Heights, IL: Waveland Press, 1989), p. 80.
30. For a complete discussion of the Sparks Police Officers Physical Abilities Test (POPAT), see Ken Peak, Douglas Farenholtz, and George Coxey, "Physical Abilities Testing for Police Officers: A Flexible, Job Related Approach," *The Police Chief* (January 1992): 51–56.
31. Terry Eisenberg, D. A. Kent, and C. R. Wall, *Police Personnel Practices in State and Local Governments* (Gaithersburg, MD: International Association of Chiefs of Police, 1973), p. 15.
32. George E. Hargrave and Deirdre Hiatt, "Law Enforcement Selection with the Interview, MMPI, and CPI: A Study of Reliability and Validity," *Journal of Police Science and Administration* (1987): 110–117.
33. Frank Horvath, "Polygraphic Screening of Candidates for Police Work in Large Police Agencies in the United States: A Survey of Practices, Policies, and Evaluative Comments," *American Journal of Police* 12 (1993): 67–86.
34. In Charles R. Swanson, Leonard Territo, and Robert W. Taylor, *Police Administration,* 2d ed. (New York: Macmillan Publishing Company, 1988), pp. 202–203.
35. Terry, *Policing Society*, p. 196.
36. See, for example, David Bradford and Joan E. Pynes, "Police Academy Training: Why Hasn't It Kept Up With Practice?" *Police Quarterly* 2 (September 1999): 283–301; N. Marion, "Police Academy Training: Are We Teaching Recruits What They Need to Know?" *Policing: An International Journal of Police Strategies and Management* 21 (1998): 54–79; Richard F. Brand and Ken Peak, "Assessing Police Training Curriculums: 'Consumer Reports,'" *The Justice Professional* 9, no. 1 (Winter 1995): 45–58.
37. John J. Broderick, *Police in a Time of Change* (Prospect Heights, IL: Waveland Press, 1987), p. 215.
38. Quoted in John M. Violanti, "What Does High Stress Police Training Teach Recruits? An Analysis of Coping," *Journal of Criminal Justice* 21 (1993): 411–417.
39. *Ibid.*, p. 416.
40. John Van Maanen, "On the Making of Policemen," in *Thinking About Police*, ed. Carl Klockars (New York: McGraw-Hill, 1983), pp. 388–400.
41. Quoted in Bradford and Pynes, "Police Academy Training," p. 289.
42. *Ibid.*, pp. 292, 297.
43. *Ibid.*, p. 298.
44. Alpert and Dunham, *Policing Urban America*, p. 50.
45. Lois Pilant, "Enhancing the Patrol Image," *The Police Chief* (August 1992): 55–61.
46. Wrightsman, *Psychology and the Legal System*, p. 86.
47. Quoted in Jerome Skolnick, "A Sketch of the Policeman's Working Personality," in *The Police Community*, ed. Jack Goldsmith and Sharon S. Goldsmith (Pacific Palisades, CA: Palisades Publishers, 1974), p. 106.
48. Thomas F. Adams, "Field Interrogation," *Police* (March-April 1963): 1–8.
49. Jonathan Rubenstein, "Cop's Rules," in *Police Behavior: A Sociological Perspective*, ed. Richard J. Lundman (New York: Oxford University Press, 1980), pp. 68–78.
50. Bruce Catton, "Eyewitness Reports on the Assassination of Abraham Lincoln," in *Criminal Justice: Allies and Adversaries*, eds. John R. Snortum and Ilana Hader (Pacific Palisades, CA: Palisades Publishing, 1978), pp. 155–157.
51. Tom Wilkenson and John Chattin-McNichols, "The Effectiveness of Computer-Assisted Instruction for Police Officers," *Journal of Police Science and Administration* 13 (1985): 230–235.
52. Niederhoffer, *Behind the Shield*, p. 51.
53. Dunham and Alpert, *Critical Issues in Policing*, p. 112.
54. *Ibid.*, p. 111.
55. *Ibid.*, pp. 112–115.
56. Alpert and Dunham, *Policing Urban America*, p. 58.
57. "Field Trainers Have Reports Well in Hand," *Law Enforcement News*, November 15, 2000, p. 5.
58. "Pursuit Simulation Training Is No Ordinary Crash Course," *Law Enforcement News*, November 15, 2000, p. 6.
59. William Westley, *Violence and the Police* (Cambridge, MA: MIT Press, 1970), p. 56.

60. Elizabeth Burbeck and Adrian Furnham, "Police Officer Selection: A Critical Review of the Literature," *Journal of Police Science and Administration* 13 (1985): 58–69.
61. J. Matarazzo, B. Allen, G. Saslow, and A. Wiens, "Characteristics of Successful Policemen and Firemen Applicants," *Journal of Applied Psychology* 48 (1964): 123–133.
62. George E. Hargrave, Deirdre Hiatt, and Tim W. Gaffney, "A Comparison of MMPI and CPI Profiles for Traffic Officers," *Journal of Police Science and Administration* 14 (1986): 250–258.
63. Quoted in Seymour M. Lipset, "Why Cops Hate Liberals—and Vice Versa," *Atlantic Monthly* 223 (March 1969): 76.
64. John J. Broderick, *Police in a Time of Change,* 2d ed. (Prospect Heights, IL: Waveland, 1987), pp. 22–115.
65. Adapted from Dennis Nowicki, "12 Traits of Highly Effective Police Officers," *Law and Order,* October 1999, pp. 45–46.

CHAPTER 4

Organization and Administration

Key Terms and Concepts

Bureaucracy
Chain of command
Communication
Compstat
First-line supervisors
Middle managers

Mintzberg model of CEOs
Organizational structure
Organizations
Police chief/sheriff
Policies and procedures
Rules and regulations

Uneasy lies the head that wears the crown.—*Shakespeare,* Henry IV

Leadership is like the Abominable Snowman, whose footprints are everywhere but who is nowhere to be seen.—*Warren Bennis*

INTRODUCTION

Imagine yourself as a police chief or sheriff in San Diego County, California—an area that has embraced community policing (discussed in Chapter 6) since the 1970s and remains one of the safest in the country. Still, envision yourself confronting such disasters as a gunman's shooting spree at a McDonald's, killing 21 people; the group suicide of members of a doomsday cult, who take a poisonous mixture of applesauce, vodka, and barbiturates (39 members die); and a school-yard shooting where a 15-year-old boy opens fire, killing two students and wounding 13 other people.[1] These and many other kinds of critical incidents—many of which are related to San Diego County's problems with illegal immigration—would challenge even the most seasoned police chief executive.

This chapter examines this very challenging aspect of policing: its organization and administration. After first looking at organizations and bureaucracies generally, we discuss organizational communication, including that which occurs within police organizations. Next we examine police agencies as organizations, including their structure, command principles, and use of policies and procedures. Then we review contemporary chiefs of police and sheriffs, including the use of the assessment center process for selecting chiefs and their qualifications, selection, and tenure. Following that, we use a management model to help better understand the general functions of police administrators in their interpersonal, informational, and decision-maker roles, including a brief illustration of good police administration: the Compstat process in New York City. Then we briefly consider the roles and functions of middle managers and first-line supervisors. Finally, we address the tenuous historical relationship between police administrators and politicians.

Note that later chapters deal with additional police organization and administration issues; for example, disciplinary policies and practices are examined in Chapter 10, and labor relations (which include unionization and collective bargaining) is discussed in Chapter 12. Also see Chapter 10 concerning police accountability.

ORGANIZATIONS AND THE POLICE

WHAT ARE ORGANIZATIONS?

Organizations are entities of two or more people who cooperate to accomplish an objective. In that sense, undoubtedly the first organizations were primitive hunting parties. Organization and a high degree of cooperation were required to bring down large animals. Organizations were also used to build pyramids and other monuments.[2] Organization can succinctly be defined as arranging and utilizing resources of personnel and materiel in such a way as to provide orderly attainment of specified objectives.

Every organization is unique. Larry Gaines, Mittie Southerland, and John Angell provided an excellent analogy that helps us understand organizations:

> Organization corresponds to the bones which structure or give form to the body. Imagine that the fingers were a single mass of bone rather than four separate fingers and a thumb made up of bones. The mass of bones could not, because of its structure, play musical instruments, hold a pencil, or grip a baseball bat. A police department's organization is analogous. It must be structured properly if it is to be effective in fulfilling its many diverse goals. Organization may not be important in a police department consisting of three officers, but it is extremely important in [larger] cities.[3]

As Gaines, Southerland, and Angell also noted, the development of an organization should be done with careful evaluation or the agency may become unable to respond efficiently to community needs. For example, the implementation of too many specialized units (such as community relations, crime analysis, or media relations units) in a police agency may obligate too many personnel to these functions and result in too few people to do the general grassroots work of the organization. (As a rule of thumb, at least 55 percent of all sworn police personnel should be assigned to patrol.)[4]

Formal organizational structures, which assist in this endeavor by spelling out areas of responsibility, lines of communication, and the chain of command, are discussed later.

ORGANIZATIONS AS BUREAUCRACIES

Organizations can be **bureaucracies**. Large organizations will be complex, with much more specialization, a more hierarchical structure, and a greater degree of authoritarian style of command.[5] Bureaucracies share several traits: People perform many different tasks toward a common goal, specialized tasks are placed in separate compartments or bureaus with a hierarchical structure and division of labor between personnel, and there is a clear chain of command through which information flows upward and commands flow downward. To ensure consistency and uniformity, there are normally an abundance of written rules for the performance of duties. Career paths up the organization allow employees to progress in an orderly fashion.[6]

ORGANIZATIONAL COMMUNICATION

DEFINITION AND CHARACTERISTICS

Communication is one of the most important dynamics of an organization. Indeed, a major role of today's administrators and other leaders is that of communication. Managers of all types of organizations spend an overwhelming amount of their time engaged in the process of—and coping with problems in—communication.

Today we communicate via facsimile machines, video camcorders, cellular telephones, satellite dishes, and so forth. We converse orally, in written letters and memos, through our body language, via television and radio programs, and through newspapers and meetings. Even private thoughts—which take place four times faster than the spoken word—are communication. Every waking hour our minds are full of ideas and thoughts. Psychologists say that nearly 100,000 thoughts pass through our minds every day, conveyed by a multitude of media.[7] Communication becomes exceedingly important and sensitive in nature in a police organization because of the nature of information that is processed by officers—who often see people at their worst and when they are in their most embarrassing and compromising situations. To "communicate" what is known about these kinds of behaviors could be devastating to the parties concerned.

Studies have long shown that communication is the primary problem in administration, however, and lack of communication is the primary complaint of employees about their immediate supervisors.[8] Managers are in the communications business. It has been said that

> Of all skills needed to be an effective manager/leader/supervisor, skill in communicating is *the* most vital. In fact, more than 50 percent of a [criminal justice] manager's time is spent communicating. First-line supervisors usually spend about 15 percent of their time with superiors, 50 percent of their time with subordinates, and 35 percent with other managers and duties. These estimates emphasize the importance of communications in everyday . . . operations.[9]

Several elements constitute the communication process: encoding, transmission, the medium, reception, decoding, and feedback.[10] Following are brief descriptions of these elements:

> *Encoding*: To convey an experience or idea to someone, we translate, or encode, that experience into symbols. We use words or other verbal behaviors and gestures or other nonverbal behaviors to convey the experience or idea.
>
> *Transmission*: This element involves the translation of the encoded symbols into some behavior that another person can observe. The actual articulation (moving our lips, tongue, etc.) of the symbol into verbal or nonverbal observable behavior is transmission.
>
> *Medium*: Communication must be conveyed through some channel, or medium. Media for communication may be our sight, hearing, taste, touch, or smell. Some other media are the television, the telephone, paper and pencil, and the radio. The choice of the medium is very important as well. For example, a message that is tranmitted via formal letter from the chief executive officer will carry more weight than if the same message is conveyed via a secretary's memo.
>
> *Reception*: The stimuli, the verbal and nonverbal symbols, reach the senses of the receiver and are conveyed to the brain for interpretation.
>
> *Decoding*: The individual who receives the stimuli develops some meaning for the verbal and nonverbal symbols and decodes the stimuli. These symbols are translated into some concept or experience of the receiver.

Feedback: When the receiver decodes the transmitted symbols, he or she usually provides some response or feedback to the sender. If somone appears puzzled, we repeat the message or we encode the concept differently and transmit some different symbols to express the same concept differently and transmit some different symbols to express the same concept. Feedback acts as a guide or steering device and lets us know whether the receiver has interpreted our symbols as we intended.

COMMUNICATION WITHIN POLICE ORGANIZATIONS

Communication within a police organization may be downward, upward, or horizontal. There are five types of *downward* communication within such an organization:

1. Job instruction—communication relating to the performance of a certain task
2. Job rationale—communication relating a certain task to organizational tasks
3. Procedures and practice—communication about organization policies, procedures, rules, and regulations
4. Feedback—communication appraisal of how an individual performs the assigned task
5. Indoctrination—communication designed to motivate the employee[11]

Upward communication in a police organization may encounter several deterrents. First, the physical distance between superior and subordinate impedes upward communication. Communication is often difficult and infrequent when superiors are isolated so as to be seldom seen or spoken to. In large police organizations, the administration may be located in headquarters that are removed from the operations personnel. The complexity of the organization may also cause prolonged delays of communication. For example, if a patrol officer observes a problem that needs to be taken to the highest level, normally this information must first be taken to the sergeant, then to the lieutenant, the captain, the deputy chief or the chief, and so on. At each level, these higher individuals will reflect on the problem, put their own interpretation on it (possibly including how the problem might affect them professionally or even personally), and possibly even dilute or distort the problem. Thus, delays in communication are inherent in a bureaucracy.

Horizontal communication thrives in an organization when formal communication channels are not open.[12] The disadvantage with horizontal communication is that it is much easier and more natural to achieve than vertical communication and, therefore, it often replaces vertical channels. Horizontal channels are usually of an informal nature, including the grapevine, discussed later. The advantage is that horizontal communication is essential if the subsystems within a police organization are to function in an effective and coordinated manner. Horizontal communication among peers may also provide emotional and social bonds that build morale and feelings of teamwork among employees.

THE GRAPEVINE

Something "heard through the grapevine" is a rumor from an anonymous source. The expression "grapevine telegraph" is also sometimes used, referring to the speed with which rumors spread. Communication also includes rumors, and there is probably *no* type of organization in our society that has more scuttlebutt than police agencies. Departments even establish rumor control centers during major riots. Compounding the usual barriers to communication is the fact that policing is a 24-hour, 7-day occupation, so that rumors are easily carried from one shift to the next.

The grapevine's most effective characteristics are that it is fast; it operates mostly at the place of work; and it supplements regular, formal communication. On the positive side, it can be a tool management can use to get a feel for employees' attitudes, to spread useful information, and to help employees vent their frustrations. The grapevine, however, can also carry untruths and be malicious. Without a doubt, the grapevine is a force for adminstrators to reckon with on a daily basis.

WRITTEN COMMUNICATION

Confidence is generally placed in the written word within complex organizations. It establishes a permanent record, but transmitting information in this way does not necessarily ensure that the message will be clear to the receiver, despite the writer's best efforts. This may be due in large measure to shortcomings with the writer's skills. Nonetheless, police organizations rely heavily on written communication, as evidenced by the proliferation of written directives and reports found in most of these agencies.

In the same vein, written communication is also preferred as a medium for dealing with citizens or groups outside the police agency. This means of communication provides the greatest protection against the growing numbers of legal actions taken against agencies by activists, citizens, and interest groups. In recent years, electronic mail (e-mail) has also proliferated as a communication medium in criminal justice organizations. E-mail can provide an easy-to-use and almost instantaneous communication with anyone else possessing a personal computer—in upward, downward, or horizontal directions. For all its advantages, however, e-mail messages can lack security and be ambiguous, not only with respect to the meaning of content, but also with regard to what it represents. Are such messages, in fact, mail, to be given the full weight of an office letter or memo, or should they be treated more as offhanded comments?[13]

BARRIERS TO EFFECTIVE COMMUNICATION

In addition to the previously mentioned inaccurate nature of the grapevine and the preponderance of poor writing skills, several other potential barriers to effective communication exist. Some people, for example, are not good listeners. Unfortunately, listening is one of the most neglected and the least understood of the communication arts.[14] We allow other things to obstruct our communication, including time, inadequate or too great a volume of information, the tendency to say what we think others want to hear, failure to select the best word, prejudices, and strained sender-receiver relationships.[15] Also, subordinates do not always have the same "big picture" viewpoint that superiors possess and do not always communicate well with someone in a higher position who is perhaps more fluent and persuasive than they are.

POLICE AGENCIES AS ORGANIZATIONS

CHAIN OF COMMAND

The administration of most police organizations is based on the traditional, pyramidal, quasi-military organizational structure that contains the elements of an organization and a bureaucracy. First, these agencies are managed by being organized into a number of specialized units. Figure 4–1 shows the hierarchy of managers within the typical police organization and the inverse relationship between rank and numbers of personnel; in other words, as rank increases, the number of

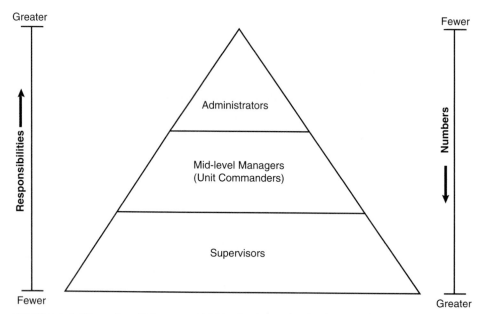

FIGURE 4–1 Hierarchy of Managers Within the Typical Police Organization. (*Source:* Larry K. Gaines, Mittie D. Southerland, and John E. Angell, *Police Administration* [New York: McGraw-Hill, Inc., 1991], p. 11. Used with permission of The McGraw-Hill Companies.)

persons in the hierarchical ranks decreases. Some larger agencies have additional ranks, such as corporal and major, but this can lead to a concern about becoming too top heavy. The rank hierarchy allows the organization to designate authority and responsibility at each level and to maintain a chain of command.

As shown in Figure 4–1, administrators (chiefs and assistant chiefs), middle managers (captains and lieutenants), and first-line supervisors (sergeants) exist to ensure that these units work together toward a common goal. Each unit working independently would lead to fragmentation, conflict, and competition and would subvert the goals and purposes of the entire organization. Second, police agencies consist of people who interact within the organization and with external groups, and they exist to serve the public.

Police departments are different from most other kinds of organizations for the simple reason that policing is significantly different from most other kinds of work. A special organizational structure has evolved to help carry out the complex responsibilities of policing. The highly decentralized nature and the varying size of American police departments, however, compel police agencies to vary in organization.

ORGANIZATIONAL STRUCTURE

Every police agency, no matter what its size, has an **organizational structure**, which is often prominently displayed for all to see in the agency's facility. Even a community with only a town marshal has an organizational structure, although the structure will be very horizontal, the marshal performing all of the functions displayed in the basic organizational chart for a small agency, shown in Figure 4–2.[16]

Operations (also known as line element) personnel are engaged in active police functions in the field. They may be subdivided into primary and secondary

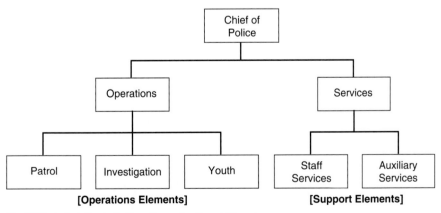

FIGURE 4–2 A Basic Police Organizational Structure.

operations elements. The patrol function—often called the backbone of policing—is the primary operational element because of its major responsibility for policing within the organization (the patrol function is examined in Chapter 5). In most small police agencies, patrol forces are responsible for all operational activities: providing routine patrols, conducting traffic and criminal investigations, making arrests, and functioning as generalists.[17]

The investigative and youth functions are the secondary operations elements (we discuss the investigative function thoroughly in Chapter 7 and the police role with juveniles in Chapter 9).

The support (or nonline) functions and activities can become quite numerous, especially in a large agency. These functions fall within two broad categories: staff (or administrative) services and auxiliary (or technical) services. The staff services usually involve personnel and include such matters as recruitment, training, promotion, planning and research, community relations, and public information services. Auxiliary services are the kinds of functions that a nonpolice or civilian person rarely sees. They include jail management, property and evidence, crime laboratory services, communications, and records and identification. Many career opportunities exist for persons interested in police-related work but who, for some reason, cannot or do not want to be a field officer.

Obviously the larger the agency, the greater the need for specialization and the more vertical the organizational chart will become. With greater specialization comes the need and opportunity for officers to be assigned to different tasks, often rotating from one assignment to another after a fixed interval. For example, in a medium-sized department serving a community of 100,000 or more, it would be possible for a police officer with ten years of police experience to have been a dog handler, a motorcycle officer, a detective, and/or a traffic officer while simultaneously holding a slot on the special weapons or hostage negotiation teams.

A basic organizational structure of the Portland, Oregon, Police Bureau (PPB) is shown in Figure 4–3. The City of Portland has a population of about 530,000, and the PPB has about 1,300 total personnel, 999 of whom are sworn. The police budget is about $121 million.[18] Some of the boxes shown in the structure are parochial in nature and demand further explanation: DPSST training is the Department of Public Safety Standards and Training, to which the PPB loans officers; PPA is the Portland Police Association, or the bureau's collective bargaining agent; the Management Services Division manages the bureau's facilities, liability, and loss control; the Sunshine Division includes personnel who work to provide food, clothing, and toys to needy families; the HAP liaisons and Safety Action Team include officers assigned to the Housing Authority of Portland; Douglas and Parkrose liaisons are officers working in school districts; APP liaison is a private association of businesses that fund one officer position for downtown; the Public Utility Commission has grant-funded positions for conducting traffic enforcement along interstate corridors; ROCN is the Regional Organized Crime and Narcotics task force; and WomenStrength is a program to teach women self-defense tactics.[19]

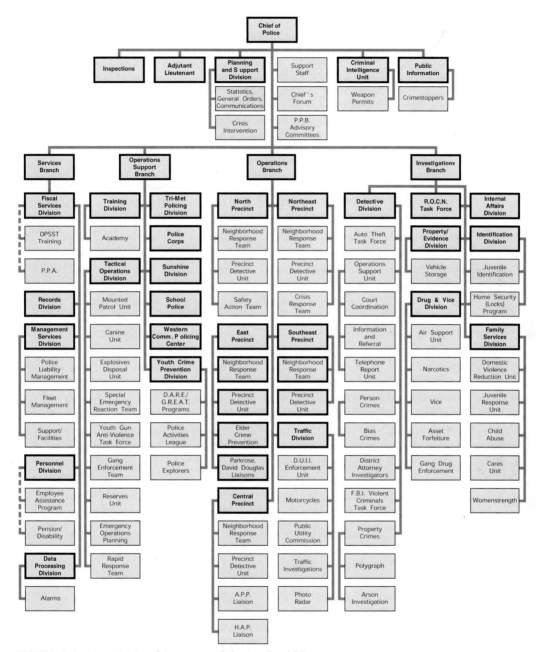

FIGURE 4–3 Organizational Structure of the Portland PD.

This organizational structure is designed to fulfill five functions: It apportions the workload among members and units according to a logical plan; it ensures that lines of authority and responsibility are as definite and direct as possible; it specifies a unity of command throughout, so there is no question about which orders should be followed; it places responsibility and authority, and if responsibility is delegated, the delegator is held responsible; and it coordinates the efforts of members so that all will work harmoniously to accomplish the mission.[20] In sum, this structure establishes the **chain of command** and determines lines of communication and responsibility.

In addition to these generally well-known and visible areas of specialization, there are other lesser known areas of policing, such as community crime prevention, child abuse, drug education, and missing children units.[21]

UNITY OF COMMAND AND SPAN OF CONTROL

A related principle is *unity of command*, an organizational principle dictating that every officer should report to one and only one superior (follow the chain of command) until that superior officer is relieved. Ambiguity about authority can and does occur in police organizations—who should handle calls, who is in charge at a crime scene, and so on. Nevertheless, the unity-of-command principle ensures that multiple and/or conflicting orders are not issued to the same officers by several supervisors. It is important that all officers know and follow the chain of command at critical incidents.

The term *span of control* refers to the number of persons one individual can effectively supervise. The limit is small; it is normally three to five at the top level of the organization and is often broader at the lower levels.[22] The tendency in modern police operations is to have supervisors spread too thinly.

ORGANIZATIONAL POLICIES AND PROCEDURES

It has been said that a well-written policy and procedure manual serves as the foundation of a professional law enforcement agency.[23] In policing, policies, procedures, rules, and regulations are also important for defining role expectations for officers. Police leaders rely on these directives to guide officers' behavior and performance. Because police agencies are intended to be service oriented in nature, they must work within well-defined, specific guidelines designed to ensure that all officers conform consistently to behavior that will enhance public protection.[24]

This tendency for organizations to promulgate rules, policies, and procedures has been caused by three contemporary developments. First is the *requirement for administrative due process* in employee disciplinary matters, encouraged by federal court rulings, police officer bill of rights legislation, and labor contracts. Another reason concerns *civil liability*. Lawsuits against local governments and their criminal justice agencies and administrators have become commonplace;

written guidelines by the agency prohibiting certain acts provide a hedge against successful civil litigation.[25] Finally, a third stimulus is the *accreditation movement*, particularly the trend toward police agencies pursuing the status and practical effects of becoming accredited. Agencies that are either pursuing accreditation or that have become accredited must follow policies and practice procedures.[26]

Policies are quite general and serve basically as guides to thinking rather than action. Policies reflect the purpose and philosophy of the organization and help interpret those elements to the officers. An example of a policy might be that all persons who are found to be driving while under the influence of drugs or alcohol shall be arrested or that all juveniles who are to be detained must be taken to a specified facility.

Procedures are more detailed than policies and provide the preferred methods for handling matters pertaining to investigation, patrol, booking, radio procedures, filing reports, roll call, use of force, arrest, sick leave, evidence handling, promotion, and many more job elements. Most police agencies are awash in procedures. Methods for accomplishing certain tasks are also found in myriad police agency general orders (such as when a new federal court decision is announced or a new state or local law takes effect on searches and seizures or the handling of domestic abuse cases or the use of force) and city or county administrative regulations.

Rules and regulations are specific guidelines that leave little or no latitude for individual discretion. Some examples are requirements that police officers not smoke in public, check the operation of their vehicle and equipment before going on patrol, not consume alcoholic beverages within a specified number of hours before going on duty, or arrive in court or at roll call early. Rules and regulations are not always popular, especially if they are perceived as unfair or unrelated to the job. Nonetheless, it is the supervisor's responsibility to ensure that officers perform these tasks with the same degree of professional demeanor as they do other job duties.

CONTEMPORARY POLICE CHIEFS AND SHERIFFS

Having analyzed police organizations, we now look at two of today's primary police chief executive officers, the chief of police (also known as superintendent, commissioner, or director) and the county sheriff. After looking at the qualifications and functions of persons holding these offices, we consider their roles in more detail with the Mintzberg model of chief executive officers.

The chief or sheriff in a 10-person agency is faced with many of the same problems and expectations as his or her big-city counterparts. The differences between managing large and small departments are a matter of scale. Executives of large departments are faced with a larger volume of many of the same problems that executives of small departments face. The leader of a small department must not only deal with all these managerial concerns, but in many cases must also perform the duties of a working officer.

Furthermore, the police manager's style must be flexible. The style used by a commander at the scene of a hostage situation would be much different from that used by a supervisor at a shoplifting scene. A less experienced employee will require a more authoritarian style of management than a more experienced employee. Management style is always contingent on the situation and the people being managed.[27]

CHIEFS OF POLICE: QUALIFICATIONS, SELECTION, TENURE

Qualifications for the position of **police chief** vary widely, depending on the size of the agency and the region of the country. Smaller agencies, especially those in rural areas, may not have any minimum educational requirement for the job. In large agencies, it is common to find a requirement for college education plus several years of progressively responsible police management experience.[28]

Although it is certainly cheaper to select a police chief from within the organization, the question is often raised about whether or not it is better to do so. That debate will probably continue well into the future, and there are obvious advantages and disadvantages to both practices. One study of police chiefs promoted from within and hired from outside in the West found significant differences in only one area: educational attainment. The outsiders were more highly educated, but the two groups did not differ in other areas, including background, attitudes, salary, tenure in current position or policing, or size of agency, community, and current budget.[29] Some states have made it nearly impossible for an outsider to come in. For example, California has mandated that the chief be a graduate of its full Peace Officers Standards and Training (POST) academy; New Jersey and New York also encourage "homegrown" chiefs.[30]

A recent survey by the Police Executive Research Forum of 358 police chiefs in larger jurisdictions (of 50,000 or more residents) found these chiefs were more educated than ever before (87 percent held a bachelor's degree and 47 percent had a master's degree) and were more likely to be chosen from outside the agencies they head. Even so, most chiefs spent less than five years in the position.[31]

To obtain the most capable people for executive positions in policing—and to avoid personnel, liability, and other kinds of problems that can arise from poor personnel choices—the *assessment center* method has surfaced as an elaborate yet efficacious means of hiring and promoting personnel. Although more costly and time-consuming than conventional testing methods (e.g., candidate's resumes are examined and oral interviews are held), they are well worth the extra investment. Monies invested at the early stages of a hiring or promotional process can save untold dollars and problems for many years to come. This process may include interviews; psychological tests; in-basket exercises; management tasks; group discussions; simulations of interviews with subordinates, the public, and news media; fact-finding exercises; oral presentation exercises; and written communications exercises.[32]

Individual and group role-playing provides a hands-on atmosphere during the selection process. For example, candidates may be required to perform in simulated police-community problems (such as having candidates conduct a "meeting" to hear concerns of local minority groups), react to a major incident (such as having candidates explain what they will do and in what order in a simulated shooting or riot situation), hold a news briefing, or participate in other such exercises. They may be given an in-basket situation in which they take the role of the new chief or captain who receives an abundance of paperwork, policies, and problems to be prioritized and dealt with in a prescribed amount of time. Writing abilities may also be evaluated by having candidates provide specific written products, such as a policy or a procedure.

During each exercise, several assessors analyze candidates' behavior and perform some type of evaluation. When the assessment center process ends, they submit their individual rating information to the person making the hiring or promotional decision. Typically selected because they have held the position for which candidates are now vying, assessors must not only know the types of problems and duties incumbent in the position but also should be keen observers of human behavior.

Job security of police chiefs ranges from full civil service protection in a small percentage of agencies to appointment and removal at the discretion of the mayor or city manager. There is a growing trend for a fixed term of office, such as a four- or five-year contract. Traditionally, however, the tenure of police chiefs has been low. A federal study found in the mid-1970s that the average length of office by chiefs of police was 5.4 years.[33] Another study by the Police Executive Research Forum (PERF) in the mid-1980s found the average to be practically unchanged, 5.5 years. That figure has not changed in more recent times.[34] Those who are appointed from within the agency tend to have longer tenure than those appointed from outside. This lack of job tenure has several negative consequences, including that long-range planning cannot be done, the possible negative effect of frequently having new policies and administrative styles, the inability of the short-term chief to develop a political power base and local influence, and the time and expense in hiring a new chief.

THE SHERIFF

Contemporary Nature and Functions

The stereotypical image of the county **sheriff**, derived from old movies (such as *Smoky and the Bandit*) and television programs (such as *The Dukes of Hazzard*) is often that of the corrupt character with a Southern drawl wearing a Stetson, sunglasses, and a five-pointed star. While it will be seen later that the sheriff's office, being largely elective, has the potential for problems, this is an overly simplistic and negative image of the American sheriff.

As discussed in Chapter 1, the position of sheriff has a long tradition. Sheriffs tend to be elected; thus, most candidates are aligned with a political party.

Therefore, it is possible that the only qualification a person brings to the office is the ability to get votes. (As an example, one elderly midwestern sheriff's only "qualification" for being newly elected was his popularity—he had owned and operated a beer hall in his home town practically all of his life.)

In some areas of the country, the sheriff's term of office is limited to two years, and the sheriff is prohibited from serving successive terms (thus, the office has been known to be "rotated" between the sheriff and undersheriff). In most counties, however, the sheriff has a four-year term of office and can be reelected. They enjoy no guarantee of tenure in office, although a federal study found that sheriffs (who average 6.7 years in office) had longer tenure in office than chiefs of police (who average 5.4 years).[35] This politicization of the office of sheriff can obviously result in high turnover rates of personnel who do not have civil service protection as well as a lack of long-range (strategic) planning.

Also, largely due to the political nature of the office, sheriffs tend to be older, be less likely to be promoted through the ranks of the agency, have less specialized training, and be less likely to be college graduates than police chiefs. Research has also found that sheriffs in small agencies have more difficulty with organizational problems (field activities, budget management), while sheriffs in large agencies find dealing with local officials and planning and evaluation to be more troublesome.

Because of the diversity of sheriff's offices throughout the country, it is difficult to describe a "typical" sheriff's department; those offices run the gamut from the traditional, highly political, limited-service office to the modern, fairly nonpolitical, full-service police organization.[36] It is possible, however, to list functions commonly associated with the sheriff's office:

1. Maintaining and operating the county correctional institutions
2. Serving civil processes (protective orders, liens, evictions, garnishments, and attachments) and other civil duties, such as extradition and transportation of prisoners
3. Collecting certain taxes and conducting real estate sales (usually for nonpayment of taxes) for the county
4. Performing routine order-maintenance duties by enforcing state statutes and county ordinances, arresting offenders, and performing traffic and criminal investigations
5. Serving as bailiff of the courts

Other general duties vary from one region to another.

Regional Role Differences

The status and operations of sheriff's offices have been studied in four regions of the country: the East, the South, the Midwest, and the West.[37] In most areas of the East, the sheriff has lost all law enforcement authority and has been reduced to functions such as court security and civil process serving. Even the traditional jail management function is being diminished with increasing use of independent jail management boards. The eastern sheriff's office is often quite small and has a tight budget. The office has become increasingly political, and legislation is regularly introduced to abolish the office altogether.

The sheriff in the South continues to be a strong law enforcement figure; larger agencies provide a full range of police services and have eliminated the need for state police forces. These sheriffs have maintained enough political clout to survive challenges to their existence and authority and have developed strong formal lobbying efforts and sheriff's associations. The political power of southern sheriffs has led to some abuse and corruption; nonetheless, these sheriff's offices tend to be efficient, effective, and secure, as was the historical model described in Chapter 1.

Midwestern sheriff's offices can be categorized into two types: one that is similar to the eastern model, and the other that is similar to the southern. Vast geographical distances, relatively low population density, and a general lack of demand for law enforcement services have created a unique situation. In many areas, staffing is small, with little specialization. These sheriffs provide services to unincorporated areas of the county, operate jails, and generally provide all the civil processes as found in the East. Midwestern sheriffs tend to have more professional status and more basic qualifications and receive a somewhat higher salary than those in the other two regions. The office is generally secure.

The western sheriff still carries many of the vestiges of the Wild West, with vast, wide-open territories of responsibility. In many respects, these sheriffs resemble the midwestern sheriffs, remaining as the chief law enforcement officer in the county.

THE POLICE CHIEF EXECUTIVE OFFICER: A MODEL

APPLYING THE MINTZBERG MODEL OF CEOS

What do contemporary police executives really do? Ronald Lynch described in simple terms their primary tasks:

> They listen, talk, write, confer, think, decide—about men, money, materials, methods, facilities—in order to plan, organize, direct, coordinate, and control their research service, production, public relations, employee relations, and all other activities so that they may more effectively serve the citizens to whom they are responsible.[38]

A police chief executive actually has many roles. Some chief executive officers (CEOs) openly endorse and subscribe to the philosophy of Henry Mintzberg,[39] who described a set of behaviors and tasks of CEOs in any organization. Following is an overview of the roles of the CEO—that is, the chief of police or sheriff—as adapted to policing, using the **Mintzberg model** and its interpersonal, informational, and decision-maker roles as an analytical framework.

The Interpersonal Role

First is the *interpersonal role*, which includes figurehead, leadership, and liaison duties. As a figurehead, the CEO performs various ceremonial functions. Examples would include riding in parades and attending other civic events; speaking before school and university classes and civic organizations; meeting with visiting officials and dignitaries; attending academy graduation and swearing-in ceremonies and some weddings and funerals; and visiting injured officers in the hospital. Like the mayor who cuts ribbons and kisses babies, the CEO performs these duties simply because of his or her title and position within the organization; the duties come with being a figurehead. While a chief or sheriff certainly cannot be expected to attend to every committee meeting or speaking engagement and all other events to which they are invited, they are certainly obligated from a professional standpoint to perform as many as they can.

The leadership function requires the CEO to motivate and coordinate workers while resolving different goals and needs within the department and the community. A chief or sheriff may have to urge the governing board to enact a code or ordinance that, whether popular or not, is in the best interest of the jurisdiction. For example, a police chief recently led a drive to pass an ordinance that prohibited parking by university students in residential neighborhoods surrounding the campus. This was a highly unpopular undertaking, but the chief pursued it because of the hardships suffered by the area residents. CEOs also provide leadership in such matters as bond issues (for more officers or new buildings) and should advise the governing body on the effects of proposed ordinances.

The role as liaison occurs when the CEO of a police organization interacts with other organizations and coordinates work flows. It is not uncommon for police executives from one geographical area—the police chief, the sheriff, the ranking officer of the local highway patrol office, the district attorney, the campus police chief, and so forth—to meet informally each month to discuss common problems and strategies. Also, the chief executives serve as liaisons between their agencies and others in forming regional police councils, narcotics units, crime labs, dispatching centers, and so forth. They also meet with representatives of the courts, the juvenile system, and other criminal justice agencies.

The Informational Role

The second major category under the Mintzberg model is the *informational role*. This role involves the CEO in monitoring/inspecting, disseminating information, and acting as spokesperson. In the monitoring/inspecting function, the CEO constantly looks at the workings of the department to ensure that things are operating smoothly (or as smoothly as a police agency can be expected to run). This function is often referred to as "roaming the ship," and many CEOs who isolated themselves from their personnel and the daily operations of the agency can speak from sad experience of the need to be alert and create a presence. Daily staff meetings are used by many police executives to discuss any information about the past 24 hours that might affect the department.

The dissemination tasks involve getting information to members of the department. This may include memoranda, special orders, general orders, and policies. The spokesperson function is related, but it is more focused on getting information to the news media. This is another very difficult task for the chief executive; news organizations are in a competitive field in which scoops and deadlines and the public's right to know the news are omnipotent. Still, the media must understand those occasions when a criminal investigation can be seriously affected by premature or overblown coverage. The police executive would do well to remember the power of the pen and not alienate the media; a wise person once said, "Never argue with someone who buys his ink by the barrel."

The Decision-Maker Role

Finally, as a decision maker, the CEO of a police organization serves as an entrepreneur, a disturbance handler, a resource alloca-tor, and a negotiator. As entrepreneur, the CEO must sell ideas to the governing board or the department. Ideas might include new computers or a new communications system, a policing strategy (such as community oriented policing and problem solving), or different work methods, all of which are intended to improve the organization. Sometimes there is a blending of roles, as when several police executives band together (functioning as liaisons) and go to the state attorney general and the legislature to lobby (in an entrepreneurial capacity) for new crime-fighting laws.

As a disturbance handler, the executive's tasks range from the minor (perhaps resolving trivial disputes between staff members) to the major (such as handling riots or muggings in a local park or cleaning up the city's downtown). Sometimes the intradepartmental disputes can reach major proportions, for example, when the patrol commander tells the street officer to arrest more public drunks, therefore causing a severe strain on the jail division commander's resources; this can cause enmity between the two commanders and force the chief executive to intervene.

The CEO as a resource allocator must be able to say "no" to subordinates. However, subordinates should not be faulted for trying to obtain more resources or improve their unit as best they can. The CEO must have a clear idea of the status of the budget and where priorities exist as well as listen to citizen complaints and act accordingly. For example, ongoing complaints of motorists speeding in a specific area will soon result in a shifting of patrol resources to that area or neighborhood.

As a negotiator, the police manager resolves employee grievances and sits as a member of the negotiating team for labor relations. A survey by PERF found that 7 out of 10 municipal police departments of more than 75 employees have some form of union representation.[40]

Collective bargaining puts the CEO in a difficult position. As a member of management, the CEO is often compelled to argue against salaries and benefits that would assist the rank and file. As mentioned earlier, however, as long as a limited supply of funds is available to the jurisdiction, managers will have to draw the line at some point and say "no" to subordinates. Again, the collective-bargaining unit and individual officers cannot be severely faulted for trying

PRACTITIONER'S PERSPECTIVE

Police Ingenuity and Entrepreneurship

Dennis D. Richards

Dennis D. Richards served as deputy, sergeant, and lieutenant with Los Angeles County Sheriff's Office (LACSO) from 1965 to 1986. He holds an associate's degree in police science from Pierre County Community College. He was a helicopter pilot with the LACSO, which first deployed helicopters for regular patrol. As a member of the LACSO administrative division, he was also head of the employee relations division (the sheriff's department was among the first to have a full-time unit as liaison between the administration and the labor union).

The thirty-odd miles of California shore loosely known as Malibu is arguably one of the most famous stretches of beach in the world. In the early 1970s, it was a composite of public recreation and expensive private homes that coexisted in a fragile and uneasy truce. The Malibu coast was also a part of the more than 240 square miles policed by the Malibu Station of the Los Angeles County Sheriff's Department. The public beaches generated great seasonal surges of transient population. Predictably, the police problems of the nearby expensive real estate were exacerbated by the summer influx. Patrol officers, driving conventional vehicles, made what reasonable observations they could from highways or parking lots and ventured onto the sand only in response to a call for service. Although it was generally agreed there was a need for increased police presence on the beach side, the question remained how best to provide that presence. I was asked to study the feasibility of routine beach patrol at Malibu and, if such an operation was practicable, suggest a proposal.

The first patrol vehicle under consideration was a three-wheel all-terrain motorcycle known as an ATC. The ATC, equipped with soft, oversized tires, was light and maneuverable enough to travel over both dry sand and the wet areas where tides rendered previous patrol activities marginal. However, it was not practicable for sand navigation. Furthermore, the ATC would be underpowered, unable to carry extra equipment in a secure manner, and would need significant renovation to carry and power a police radio.

We launched a search for a better patrol vehicle. Previous experiments with traditional four-wheel drive vehicles had been less than inspiring. Visibility of objects on the ground was poor, making it somewhat tricky for an officer to wend his way through a crowd of sunbathers without running over someone's arm or leg. The big four-wheelers were comparatively heavy and tended to bog down. Moreover, the beachgoers found these vehicles offensive because of their military, tanklike appearance. I then discovered that the lifeguards were experimenting with a dune buggy and were extremely enthusiastic regarding the buggy's performance. I tested it and was convinced that a dune buggy was the best mode of patrol, but the department's executives still had to be convinced of its efficacy.

I borrowed the lifeguard buggy, positioned it on a scenic stretch of beach, and photographed it with a uniformed deputy standing on either side. A department staff artist used this photograph as a basis for conceptual drawings, changing the bright lifeguard yellow to black and white and overlaying the department's radio car logo on the buggy's blunt shovel nose. The artist also attached a traditional light bar and siren to the crash bar and, in various drawings, dressed the deputies in standard patrol uniform and presented them in a softer image wearing shorts and baseball caps.

Armed with these visual aids, we made our presentation. The concept captured the administration's imagination and the dune buggy beach patrol was approved. The patrol was launched with considerable attention from the media. The sheriff personally chauffeured individual members of the press along Malibu Beach in a patrol buggy, and a grand time was had by all. More importantly, the stubby little cars were an instant hit with the public. Even the most die-hard libertarians found them more amusing than threatening. Beachgoers frequently flagged them down to be photographed with the vehicle. The positive public relations achieved by this operation was one of its most valuable aspects.

to improve salaries, benefits, and working conditions, but sometimes these associations go outside the boundary of reasonableness and reach an impasse or deadlock in contract negotiations with management. These situations can become uncomfortable and even disastrous, leading to work stoppages, work speedups, work slowdowns, or other such tactics (discussed more in Chapter 12).

An Example of Mintzberg in Action: NYPD's Compstat

A relatively new approach to crime analysis and prevention that employs several aspects of the Mintzberg model for chief executive officers—certainly including the interpersonal, informational, and decision-maker roles and several of their subcategories—is New York City's Compstat process.

Because of the massive size and bureaucracy of the New York Police Department (NYPD), New York City is probably the last city in which one would expect to find a major innovation in police administration and operations—one that has been described as having the "potential to become the dominant mode of policing in America."[41] But that situation changed with the election of Mayor Rudolph Giuliani in November 1993 and his appointment of Police Commissioner William Bratton in January 1994.

Before 1993, the city's daily crime statistics were almost useless for crime analysis purposes: At least two months' time was required for commanders to even obtain the figures. Meanwhile, crime was at a very high level and the NYPD, by its own admission, had gotten out of the habit of being held accountable for the crime problem. Bratton established three clear-cut goals for NYPD: reduce crime significantly, reduce the fear of crime, and work on improving the quality of life.[42] What quickly evolved in NYPD was a computer file to compare statistics—hence the acronym **Compstat**.

Computerization changed everything in NYPD, enabling the organization to move from being a reactive entity to being one that is proactive and goal specific. A four-step process became the essence of Compstat: collecting information in a timely and accurate manner, using effective tactics to respond to problems, deploying personnel and resources rapidly, and following up and assessing relentlessly. With these four steps, Compstat enables the NYPD to pinpoint and analyze crime patterns almost instantly and to respond in the most appropriate manner. A side benefit is that with Compstat, the organization is able to identify emerging leaders from deep within the ranks.[43]

Best known for its semi-weekly, high-stress 7:00 A.M. brainstorming sessions at police headquarters, where field commanders are grilled about crime in their areas, the results of Compstat have been impressive: From 1993 to 1996, New York City experienced the most dramatic decline in crime in the nation. As an example, according to FBI statistics, New York accounted for 61 percent of the nationwide decline in felonies during the first six months of 1995. The most unanticipated result of Compstat was that the decline in crime occurred in every police precinct in the city, debunking the theory that police can only displace crime from one area to another.[44] Because of its successes, by late 1996, the National Institute of Justice was attempting to replicate Compstat in two other jurisdictions.

According to one observer, "Compstat and the re-engineered NYPD are pointing the way to a new perspective on community policing, and demonstrating that a community orientation is not incompatible with aggressive, focused law enforcement."[45]

MIDDLE MANAGERS: CAPTAINS AND LIEUTENANTS

Few police administration books discuss the middle-management members of a police department: captains and lieutenants. This is unfortunate, because they are too numerous and too powerful within police organizations to ignore. Opinions vary about these midmanagement personnel, however, as will be seen later.

Normally, captains and lieutenants are commissioned officers, with the position of captain second in rank to the executive managers. Captains have authority over all officers of the agency below the chief or sheriff and are responsible only to them. Lieutenants are in charge of sergeants and all officers and report to captains. Captains and lieutenants may perform the following functions:[46]

- Inspecting assigned operations
- Reviewing and making recommendations on reports
- Helping develop plans
- Preparing work schedules
- Overseeing records and equipment
- Overseeing recovered or confiscated property
- Enforcing all laws and orders

Too often, **middle managers** become glorified paper pushers, especially in the present climate of myriad reports, budgets, grants, and so on. The agency should take a hard look at what managerial services are essential and whether or not lieutenants are needed to perform such services. Recently, some communities, such as Kansas City, Missouri, eliminated the rank of lieutenant, finding that this move had no negative consequences and some positive effects.[47]

Obviously, when a multilayered bureaucracy is created, a feudal kingdom and several fiefdoms will occupy the building. As Richard Holden observed, however, "If feudalism was so practical, it would not have died out in the Middle Ages."[48] It is important to remember that the two crucial elements to organizational effectiveness are (1) top administrators and (2) operational personnel. Middle management can pose a threat to the agency by acting as a barrier between these two primary elements. As has often been noted, the Roman Catholic Church serves many millions and employs many thousands of people with only five levels in the hierarchy.[49] Research has shown an inverse relationship between the size of the hierarchy in an organization and its effectiveness.[50] Normally, the closer the administrator is to the operations, the more effective the agency.

THE FIRST-LINE SUPERVISOR

A COMPLEX ROLE

At some point during the career of a patrol officer who has acquired the minimal years of experience, he or she has the opportunity to test for promotion to **first-line supervisor**, or sergeant. Competition for this position is normally quite keen in most departments. To compete well, officers are often advised to rotate into different agency assignments before testing for the sergeant's position to gain exposure to a variety of police functions and supervisors. The promotional system, then, favors not only those officers who are skilled at test-taking, but also those persons with experience outside the patrol division.[51]

Getting that first promotion.
(Courtesy Washoe County, Nevada,
Sheriff's Department)

The supervisor's role, put simply, is to get his or her subordinates to do their very best. This task involves a host of actions, including communicating, motivating, mediating, mentoring, leading, team building, training, developing, appraising, counseling, and disciplining. As such, no other rank in the police hierarchy will exert more direct influence over the working environment, morale, and performance of employees.

Adding to the complexity of the supervisor's role is the fact that the supervisor is generally in his or her first leadership position. A new supervisor must learn how to exercise command and be responsible for the behavior of several other employees. Long-standing relationships are put under stress when one party suddenly has official authority over former equals. Expectations of leniency or preferential treatment may have to be dealt with.

The supervisor is caught in the middle, working with rank-and-file employees—labor—on the one hand and middle or upper management on the other. While it is management's job to squeeze as much productivity out of workers as possible, labor's motivation often seems to be to avoid work as much as possible. Supervisors find themselves right in the middle of this contest. Their subordinates expect them to be understanding, to protect them from management's unreasonable expectations and arbitrary decisions, and to represent their interests. Management, though, expects supervisors to keep employees in line and to represent management's and the overall organization's interests.

For all of these reasons, given their role and functions, supervisors will in large measure determine the overall success of the police organization.

POLICE AND POLITICS

Another long-standing problem permeating police administration is that of political influence, which ranges from major policy and budgetary decisions to the overzealous city manager or city council member who appears unexpectedly at night to "help" officers at a burglary in progress and barely avoids being shot.

POLITICAL EXPLOITATION OF THE POLICE

Politics is the art or science of government. One definition of policing is very similar to that of politics: "the internal control and regulation of a political unit [such as a state or community] through exercise of governmental powers."[52] Both words derive from the Greek terms for "citizen" and "citizenship," and "police" also comes from the Greek work for city—"polis."

Historically, police departments in the United States have been political bodies, extensions of the municipal political authority.[53] Because of the close relationship between police departments and the political leadership of the community, abuse of political power has often occurred. From the beginning of this century, when the journalist Lincoln Steffens exposed corruption in American cities, to more recent times, when police scandals have rocked departments in New York City, Chicago, and Miami, politics has been shown to be entwined in the relationships that often bind criminals and police officers. Partisan politics has often been the cause of police corruption.[54]

Even in the nineteenth century, police forces were not autonomous; political figures outside the departments began to make key decisions regarding promotions, assignments, and disciplinary matters. For their own part, police officers often sought promotions, soft jobs, or particular beat locations. Some officers wanted a beat close to their homes. Corrupt officers wanted beats with many saloons and brothels to raise their incomes. In the late nineteenth century, promotion depended upon political influence and money, and some officers had to borrow the means for career advancement. Impartial police administration was the last thing politicians in large cities wanted. Their interests were best served by a policing system that was easily manipulated.[55]

POLICE EXECUTIVE RELATIONS AND EXPECTATIONS

The chief of police is generally considered to be one of the most influential and prestigious persons in local government. However, much of the power of the office has eroded because of an attrition rate, the increased power of local personnel departments, and the influence of police unions. Furthermore, mayors, city managers and administrators, members of the agency, citizens, special interest groups, and the media all have differing expectations of the role of chief of police, often conflicting in nature.

The mayor or city manager is likely to believe that the chief of police should be an enlightened administrator whose responsibility is to promote departmental efficiency, reduce crime, improve service, and so on. Other mayors and managers will appreciate the chief who simply "keeps the lid on" and manages to keep the morale high and the number of citizens' complaints low.[56] However, the mayor or city manager may also properly expect the chief to be part of the city management team, communicate city management's policies to police personnel, establish agency goals and objectives, select and effectively manage people, and be a responsible steward of the budget.[57]

The relationship between the police chief and mayor is difficult to articulate. However, Richard Brzeczek contends that several points are indispensable in the relationship. First, the mayor is boss. Indeed, the mayor possesses the legal or political power to fire or force the police chief out of office almost at whim. Second, the mayor has the responsibility for ensuring the public that the police are doing the best they can with available resources. Third, the police executive, if chosen on merit, has considerable knowledge about the problems of the community and a wide array of possible solutions—expertise that will serve city hall well. The mayor should come to rely upon the chief's pragmatism and take-charge approach as well. Finally, the mayor must give the chief the authority to run his department day to day; without this autonomy, perhaps guided by the mayor's input, the chief's authority will be eroded.[58]

Members of the police agency may have different expectations of the chief executive. They may be less concerned with cost effectiveness and more concerned with good salaries, benefits, and equipment. The officers expect the chief to be their advocate, backing them up when necessary and representing the agency well in dealings with judges and prosecutors who may be indifferent or hostile to their interests. Citizens, for their part, expect the chief of police to provide efficient and cost-effective police services while keeping crime and tax rates down and eliminating corruption and illegal use of force.

Special interest groups expect the chief to advocate desirable policy positions; for example, the Mothers Against Drunk Driving (MADD) group would insist upon strong anti-DUI measures by the police. Finally, the media expect the chief to fully cooperate with their efforts to obtain fast and complete crime information.

SUMMARY

This chapter has discussed police agencies as organizations and bureaucracies and the roles and functions of police chief executive officers, middle managers, and first-line supervisors. A management model was employed to clarify the general roles and functions of police administrators; included was a brief illustration of good police administration—the Compstat process in New York City—which is believed by many observers to be the dominant mode of policing in the future. Also addressed was the problem of politics and the police administrator.

The difficulties of leadership in policing were implied, as was the fact that all of the elements and issues of administration revolve around one very important component of the workplace: people. To be an effective administrator in this labor-intensive field, one should learn all that can be learned about human resources in addition to learning all one can about police methods.

ITEMS FOR REVIEW

1. Diagram and explain the elements of a basic, simplistic organizational structure of a police agency.
2. Describe the basic elements of the communications process.
3. Explain the processes of upward, downward, horizontal, and grapevine communication, applying these processes to a police organization.
4. Discuss the roles of the police executive, using the Mintzberg model of chief executive officers.
5. Contrast the roles and functions of contemporary chiefs of police and county sheriffs.
6. Discuss roles and functions of middle-level managers and first-line supervisors.
7. Explain how New York City's Compstat process functions and its potential promise for the future in other jurisdictions.
8. Describe the historical relationship and major problems between the police and politicians.
9. Review the advantages and possible component parts of an assessment for hiring and/or promoting police chief executives.

NOTES

1. Ben Fox, "Shootings Haunt San Diego Sheriff," Associated Press, March 11, 2001.
2. David A. Tansik and James F. Elliott, *Managing Police Organizations* (Monterey, CA: Duxbury, 1981), p. 1.
3. Larry K. Gaines, Mittie D. Southerland, and John E. Angell, *Police Administration* (New York: McGraw-Hill, 1991), p. 9.
4. *Ibid.*
5. Samuel Walker, *The Police in America: An Introduction,* 2d ed. (New York: McGraw-Hill, 1992), pp. 356–359.
6. *Ibid.,* p. 77.
7. *Ibid.,* p. 86.
8. *Interpersonal Communication: A Guide for Staff Development* (Athens, GA: Institute of Government, University of Georgia, August 1974), p. 15.
9. Wayne W. Bennett and Karen Hess, *Management and Supervision in Law Enforcement,* 2d ed. (St. Paul, MN: West, 1996), p. 85.
10. See R. C. Huseman, quoted in *Ibid.,* pp. 21–27. Material for this section was also drawn from Charles R. Swanson, Leonard Territo, and Robert W. Taylor, *Police Administration: Structures, Processes, and Behavior,* 5th ed. (Upper Saddle River, NJ: Prentice Hall, 2000).
11. D. Katz and R. L. Kahn, *The Social Psychology of Organizations* (New York: John Wiley & Sons, 1966), p. 239. As cited in P. V. Lewis, *Organizational Communication: The Essence of Effective Management* (Columbus, OH: Grid, 1975), p. 36.
12. See R. K. Allen, *Organizational Management Through Communication* (New York: Harper & Row, 1977), pp. 77–79.
13. Alex Markels, "Managers Aren't Always Able to Get the Right Message Across With E-Mail," *The Wall Street Journal,* August 6, 1996, p. 2.
14. Robert L. Montgomery, "Are You a Good Listener?" *Nation's Business* (October 1981): 65–68.
15. Wayne W. Bennett and Karen Hess, *Management and Supervision in Law Enforcement,* p. 101.
16. See George D. Eastman and Esther M. Eastman, eds., *Municipal Police Administration,* 7th ed. (Washington, D.C.: International City Management Association, 1971), p. 17.
17. *Ibid.,* p. 18.
18. Personal communication, Marsha R. Palmer, Planning and Support Division, City of Portland, Oregon, Bureau of Police, July 20, 2001.
19. *Ibid.,* June 4, 2001.

20. President's Commission on Law Enforcement and Administration of Justice, *Task Force Report: The Police* (Washington, D.C.: Government Printing Office, 1967), p. 46.
21. U.S. Department of Justice, Bureau of Justice Statistics, *Police Departments in Large Cities, 1987,* Special Report NCJ-119220 (Washington, D.C.: Author 1989), p. 5, Table 10.
22. M.D. Iannone and Nathan F. Iannone, *Supervision of Police Personnel,* 6th ed. (Upper Saddle River, NJ: Prentice Hall, 2000).
23. Michael Carpenter, "Put It In Writing: The Policy Policy Manual," *FBI Law Enforcement Bulletin,* October 2000, p. 1.
24. Robert Sheehan and Gary W. Cordner, *Introduction to Police Administration,* 2d ed. (Cincinnati, OH: Anderson, 1989), pp. 446–447.
25. Swanson, Territo, and Taylor, *Police Administration,* p. 248.
26. Stephen W. Mastrofski, "Police Agency Accreditation: The Prospects of Reform," *American Journal of Police* 5:3 (1986): 45–81.
27. Gaines, Southerland, and Angell, *Police Administration,* pp. 10–11.
28. *Ibid.,* p. 42.
29. Janice Penegor and Ken Peak, "Police Chief Acquisitions: A Comparison of Internal and External Selections," *American Journal of Police* 11, 1 (1992): 17–32.
30. Richard B. Weinblatt, "The Shifting Landscape of Chiefs' Jobs," *Law and Order,* October 1999, p. 50.
31. "Survey Says Big-city Chiefs are Better-educated Outsiders," *Law Enforcement News,* April 30, 1998, p. 7.
32. R. J. Filer, "Assessment Centers in Police Selection," in *Proceedings of the National Working Conference on the Selection of Law Enforcement Officers,* ed. C. D. Spielberger and H. C. Spaulding (Tampa, FL: University of South Florida, March 1977), p. 103.
33. National Advisory Commission on Criminal Justice Standards and Goals, *Police Chief Executive,* p. 7.
34. Weinblatt, "The Shifting Landscape of Chiefs' Jobs," p. 51.
35. National Advisory Commission on Criminal Justice Standards and Goals, *Police Chief Executive* (Washington, D.C.: Government Printing Office, 1976), p. 7.
36. Clemens Bartollas, Stuart J. Miller, and Paul B. Wice, *Participants in American Criminal Justice: The Promise and the Performance* (Englewood Cliffs, NJ: Prentice Hall, 1983), pp. 51–52.
37. *Ibid.,* pp. 53–59.
38. Ronald G. Lynch, *The Police Manager: Professional Leadership Skills,* 3d ed. (New York: Random House, 1986), p. 1.
39. Henry Mintzberg, "The Manager's Job: Folklore and Fact," *Harvard Business Review* 53 (July–August 1975): 49–61.
40. Witham, "The American Law Enforcement Chief Executive," p. xii.
41. Peter C. Dodenhoff, "LEN Salutes Its 1996 People of the Year, the NYPD and Its Compstat Process," *Law Enforcement News,* December 31, 1996, pp. 1, 4–5.
42. *Ibid.,* p. 4.
43. *Ibid.*
44. Eli B. Silverman, "Mapping Change: How the New York City Police Department Re-engineered Itself to Drive Down Crime," *Law Enforcement News,* December 15, 1996, p. 10.
45. Peter C. Dodenhoff, "LEN Salutes Its 1996 People of the Year, the NYPD and Its Compstat Process," p. 4.
46. Bennett and Hess, *Management and Supervision in Law Enforcement,* pp. 44–45.
47. Richard N. Holden, *Modern Police Management* (Englewood Cliffs, NJ: Prentice Hall, 1986), pp. 294–295.
48. *Ibid.,* p. 295.
49. *Ibid.,* p. 117.
50. Thomas J. Peters and Robert H. Waterman Jr., *In Search of Excellence* (New York: Warner Books, 1982), pp. 306–317.
51. John Van Maanen, "Making Rank: Becoming an American Police Sergeant," in *Critical Issues in Policing: Contemporary Readings,* ed. Roger G. Dunham and Geoffrey P. Alpert (Prospect Heights, IL: Waveland Press, 1989), pp. 146–161.
52. *Webster's Ninth New Collegiate Dictionary* (Springfield, MA: Merriam-Webster, 1983), p. 910.

53. Richard Brzeczek, "Chief-Mayor Relations: The View from the Chief's Chair," in *Police Leadership in America: Crisis and Opportunity*, ed. William A. Geller (New York: Praeger, 1985), pp. 48–55.
54. George F. Cole and Christopher Smith, *The American System of Criminal Justice*, 9th ed. (Belmont, CA: West/Wadsworth, 2001), p. 237.
55. James F. Richardson, *Urban Police in the United States* (Port Washington, NY: Kennikat Press Corp., 1974), pp. 55–58.
56. Bartollas, Miller, and Wice, *Participants in American Criminal Justice*, p. 35.
57. *Ibid.*, pp. 39–40.
58. *Ibid.*, pp. 49–50.

CHAPTER 5

On Patrol

The Beat, Deployment, and Discretion

Key Terms and Concepts

Beat culture

Deployment

Discretionary use of authority

Shift assignment

Tennesse v. Garner

Traffic control

For some must watch while some must sleep; so runs the world away.—*Shakespeare,* Hamlet

Get action; do things; be sane; don't fritter away your time; create, act, take a place wherever you are and be somebody; get action.—*Theodore Roosevelt*

You can observe a lot by watching.—*Yogi Berra*

INTRODUCTION

The patrol function has long been widely viewed as the backbone of policing. Since the founding of professional policing by Sir Robert Peel (discussed in Chapter 1), patrol has been considered the most important and visible part of police work. Indeed, this chapter serves as a prologue to many police activities described in later chapters, all of which branch off from the patrol function. Because patrol duties normally involve 60 percent to 70 percent of a police agency's workforce, this task has been a topic of considerable interest and analysis.

In addition to the patrol function, the work of policing revolves around the discretionary use of authority. Officers possess a wide range of options as they go about the business of patrolling, including whether or not to stop, whether or not to arrest, whether or not to use force, whether or not to shoot, and so on. From the relatively innocuous traffic stop to the use of lethal force, many choices—a number of which are major in nature—are involved.

This chapter begins by describing the culture of the beat: the purposes and nature of patrol, patrol work as a function of shift and beat assignment, and the kinds of dangers that are inherent in this aspect of police work. Next is an overview of what research has revealed concerning the patrol function. Then several methods are presented for determining the optimal size of the patrol force. Following that is an examination of a major related aspect of policing: discretionary use of police authority. An exercise is presented in the use of police discretion, followed by a review of the factors and political considerations that can enter into an officer's decision-making process and the advantages and disadvantages of such discretionary authority. We conclude the chapter with an examination of another function that is closely related to patrol: traffic.

Note that two closely related topics that are at the heart of the patrol function are discussed in later chapters: community oriented policing and problem solving, which is examined in Chapter 6, and the tools and high technology used by patrol officers in the performance of their duties, including less-than-lethal weapons, which are discussed in Chapter 14.

PATROL AS WORK: CULTURE OF THE BEAT

Purposes and Nature of Patrol

In Chapter 2, we discussed the role and functions of the police, including the four basic tasks of policing: preventing crime, enforcing the laws, protecting the innocent, and providing services. In this chapter section, we expand that discussion by looking at the **beat culture**, or some of the methods and problems that are connected with the patrol function.

Police work has certainly changed since 1910, when Leonard Fuld observed that "the policeman's life is a lazy life in as much as his time is spent doing nothing."[1] Today there are myriad duties for the patrol officer to perform, and danger is a constant adversary (danger as an element of patrol is discussed later).

When not handling calls for service, today's officers frequently engage in problem-solving activities (discussed in Chapter 6) and in random, preventive patrol, hoping to deter crime with a police presence. There are several forms of preventive patrol, including automobile, foot, bicycle, horse, motorcycle, marine, helicopter, and even snowmobile patrols. During all of these, the officer is alert for activities and people who look out of the ordinary. The traditional method of deployment of patrol officers should take into account the where and when of crime, attempting to distribute available personnel at the places and the times of day and days of week when trouble and crime seem to occur with greatest frequency. Unfortunately, many departments still deploy their patrol officers in their jurisdictions by using convenient beat dividers such as streets or water (such as lakes or rivers) instead of analyzing when and where crime and other disturbances are taking place. (We discuss methods for determining patrol deployment later.)

Patrol officers also attempt to effect good relations with the people on their beat, realizing that they cannot apprehend criminals or even maintain a quiet sector without public assistance. In many ways, the success of the entire police agency is dependent upon the skill and work of the patrol officer. For example, upon arriving at a crime scene, they must protect and collect evidence, treat and interview victims, locate and interview suspects and witnesses, and make such important discretionary decisions as whether or not to arrest and even perhaps whether or not to use their weapons.

The American Bar Association offered the following major purposes of police patrol:

1. To deter crime by maintaining a visible police presence
2. To maintain public order
3. To enable the police department to respond quickly to law violations or other emergencies
4. To identify and apprehend law violators
5. To aid individuals and care for those who cannot help themselves
6. To facilitate the movement of traffic and people
7. To create a sense of security in the community[2]

Officers must become very knowledgeable about their beats: They must be familiar with such details as where the doors and windows of building are, where the alleys are, where smaller businesses are located, and how the residential areas they patrol are laid out. Officers must learn what is *normal* on their beats and thus be able to discern people or things that are abnormal; in short, they should develop a kind of sixth sense that is grounded on suspicion—an awareness of something bad, wrong, harmful, without solid evidence; this is often termed "JDLR"—things "just don't look right."

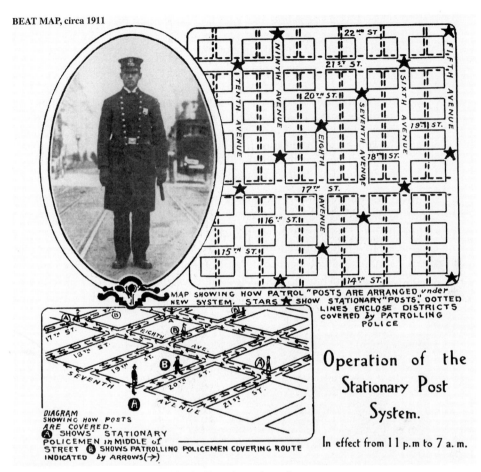

BEAT MAP, circa 1911

MAP SHOWING HOW PATROL "POSTS ARE ARRANGED *under* NEW SYSTEM. STARS ★ SHOW STATIONARY "POSTS." DOTTED LINES ENCLOSE DISTRICTS COVERED *by* PATROLLING POLICE

Operation of the Stationary Post System.

In effect from 11 p.m to 7 a.m.

DIAGRAM SHOWING HOW POSTS ARE COVERED. Ⓐ SHOWS STATIONARY POLICEMEN *in* MIDDLE *of* STREET Ⓑ SHOWS PATROLLING POLICEMEN COVERING ROUTE INDICATED *by* ARROWS(→)

Studying and creating beat configurations have long been a part of policing, as shown by this 1911 NYPD beat map. (*Courtesy NYPD Photo Unit*)

Patrol officers may also develop certain informal rules pertaining to their beats. For example, they may adopt the belief that "after midnight, these alleys belong to me." In other words, an officer may take the position that any person who is observed in "his" or "her" alley after midnight must be checked out—especially if that person is wearing dark clothing or is acting in a furtive or surreptitious manner.

Several authors have described, often in colorful but realistic terms, the kinds of situations encountered by officers on patrol. For example, as W. Clinton Terry III put it:

Patrol officers respond to calls about overflowing sewers, reports of attempted suicides, domestic disputes, fights between neighbors, barking dogs and quarrelsome cats, reports of people banging their heads against brick walls until they are bloody, requests to check people out who have seemingly passed out in public parks, requests for more police protection from elderly ladies afraid of entering their residence, and requests for information and general assistance of every sort.[3]

A former police chief described patrol duties as follows:

Cops on the street hurry from call to call, bound to their crackling radios, which offer no relief—especially on summer weekend nights. That is the time when the [city] throbs with noise, booze, violence, drugs, illness, blaring TVs, and human misery. The cops jump from crisis to crisis, rarely having time to do more than tamp one down sufficiently and leave for the next. Gaps of boredom and inactivity fill the interims, although there aren't many of these in the hot months.[4]

Indeed, the patrolling officer will encounter all manner of things while engaged in routine patrol—things they discover as well as problems phoned in by citizens. They are given "attempt to locate" calls (usually involving missing persons, ranging from juveniles who have not returned home on time to elderly people who have wandered away from nursing homes); "attempt to contact" and "be on the lookout" calls (such as an out-of-town individual who needs police to try to locate someone in order to deliver a message or for some other related reason); and

An officer in a beach patrol uniform. (*Courtesy Long Beach, California, Police Department*)

"check the welfare of" calls (a person has not been seen or heard from for some time, and relatives are concerned about that person's welfare).

One type of call is said to have broken the back of many police agencies: 911 emergency calls. Several hundred thousand of those calls are made per day across the nation—90 percent of them are for nonemergencies. Departments must find ways to free patrol officers from what has been called the "tyranny of 911": nonstop calls that send officers bouncing from one nonemergency call for service to the next. (Such an occasion occurred when the author, while standing in the communications center of a midwestern police agency, observed a 911 call answered by the dispatcher. The caller was reporting a goat standing on the front porch.) Indeed, the range of "emergencies" 911 callers report boggles the mind: Some people call because they want to know when the National Football League game begins that day, some people want to know the weather report, and some want help to exorcise the alien who entered their kitchen through the refrigerator's electrical cord. These calls also leave officers little time for community oriented policing and problem solving activities.[5]

Table 5–1 describes the complex nature of a patrol officer's work. Table 5–2 shows the difference in patrol allocation in selected large police departments. The table shows departments' diverse attempts to meet various needs (such as those created by streets, freeways, downtown areas, parks, and lakes) with different patrol staffing levels and types of patrol. For example, Detroit and Los Angeles staff over 60 percent of their patrol units with two officers, while Atlanta and Baltimore have no two-officer units. Furthermore, Atlanta, Chicago, and Houston commit over 25 percent of their officers to patrol on a 24-hour basis, while the remaining cities commit less than 20 percent of their officers to this task. There are also vast differences with respect to how agencies use different types of patrol. For example, while half of the agencies assign 80 percent to 90 percent of their patrol resources to automobiles, Los Angeles, New York, and Seattle assign 60 percent or less to automobiles.

FILLING OCCASIONAL HOURS OF BOREDOM

As indicated earlier, contrary to the image that is portrayed on television, some (or even much) of the time officers devote to patrolling consists of gaps of boredom and inactivity. During those periods of time (probably more so on the graveyard shift, when even late-night people and party-goers submit to fatigue and go home to sleep), patrol officers can engage in a variety of activities to pass the time:

1. They create "private places" for themselves—fire stations, hospitals, and other places where they can wash up, have a cup of coffee, make a phone call, or simply relax for a few moments.
2. They can engage in police-related activities, such as completing reports, checking license plates of vehicles that are parked at motels (to locate stolen vehicles or wanted persons), or meeting with other officers. Other more relaxing activities might include exercising in the station house workout room.

TABLE 5–1 Police Patrol: A Job Description

This behavioral analysis of a patrol officer's job provides one of the few empirical descriptions of the complex and varied demands of patrol work. Based on extensive field observations, the findings are reported as a list of the attributes that are required for successful performance in the field. Although completed over two decades ago, the findings appear to conform well to patrol activities of today. The researchers concluded that a patrol officer must do the following:

1. Endure long periods of monotony in routine patrol yet react quickly (almost instantaneously) and effectively to problem situations observed on the street or to orders issued by the radio dispatcher.
2. Gain knowledge of the patrol area, not only of its physical characteristics but also of its normal routine of events and the usual behavior patterns of its residents.
3. Exhibit initiative, problem-solving capacity, effective judgment, and imagination in coping with the numerous complex situations he or she is called upon to face, e.g., a family disturbance, a potential suicide, a robbery in progress, an accident, or a disaster.
4. Make prompt and effective decisions, sometimes in life-and-death situations, and be able to size up a situation quickly and take appropriate action.
5. Demonstrate mature judgment, as in deciding whether an arrest is warranted by the circumstances or a warning is sufficient, or in facing a situation where the use of force may be needed.
6. Demonstrate critical awareness in discerning signs of out-of-the-ordinary conditions or circumstances that indicate trouble or a crime in progress.
7. Exhibit a number of complex psychomotor skills, such as driving a vehicle in normal and emergency situations, firing a weapon accurately under extremely varied conditions, maintaining agility, endurance, and strength, and showing facility in self-defense and apprehension, as in taking a person into custody with a minimum of force.
8. Adequately perform the communication and record-keeping functions of the job, including oral reports, formal case reports, and departmental and court forms.
9. Have the facility to act effectively in extremely divergent interpersonal situations. A police officer constantly confronts persons who are acting in violation of the law, ranging from curfew violators to felons. He or she is constantly confronted by people who are in trouble or who are victims of crimes. At the same time, the officer must relate to the people on the beat—businessmen, residents, school officials, visitors. His or her interpersonal relations must range up and down a continuum defined by friendliness and persuasion on one end and by firmness and force at the other.
10. Endure verbal and physical abuse from citizens and offenders (as when placing a person under arrest or facing day-in and day-out race prejudice) while using only necessary force in the performance of his function.
11. Exhibit a professional, self-assured presence and a self-confident manner in his conduct when dealing with offenders, the public, and the courts.
12. Be capable of restoring equilibrium to social groups, e.g., restoring order in a family fight, in a disagreement between neighbors, or in a clash between rival youth groups.
13. Be skillful in questioning suspected offenders, victims, and witnesses of crimes.
14. Take charge of situations, e.g., a crime or accident scene, yet not unduly alienate participants or bystanders.
15. Be flexible enough to work under loose supervision in most day-to-day patrol activities and also under direct supervision in situations where large numbers of officers are required.
16. Tolerate stress in a multitude of forms, such as meeting the violent behavior of a mob, coping with the pressures of a high-speed chase or a weapon being fired, or dealing with a woman bearing a child.
17. Exhibit personal courage in the face of dangerous situations that may result in serious injury or death.
18. Maintain objectivity while dealing with a host of special-interest groups, ranging from relatives of offenders to members of the press.
19. Maintain a balanced perspective in the face of constant exposure to the worst side of human nature.
20. Exhibit a high level of personal integrity and ethical conduct, e.g., refrain from accepting bribes or "favors" and provide impartial law enforcement.

Source: Adapted from M. E. Baehr, J. E. Furcon, and E. C. Froemel. 1968. *Psychological Assessment of Patrolman Qualifications in Relation to Field Performance.* Washington, D.C.: Department of Justice, pp. 11–3 to 11–5.

TABLE 5–2 Patrol Allocation in Selected Large Police Departments, 1997

Department	Departments Using Patrol Type and Percentage of All Patrol Units Accounted For						Percentage of Officers on Patrol per 24 Hours	Percentage of Units with Two Officers
	Auto	Motorcycle	Foot	Bicycle	Horse	Marine		
Atlanta	R 77%	R 1%	R 14%	R 3%	R 1%	M 0%	31%	0%
Baltimore	R 96	R 1	R 2	S 0	R 1	R 0	16	0
Chicago	R 82	R 1	R 15	R 1	R 0	R 0	26	52
Detroit	R 84	R 1	R 7	R 3	R 1	R 4	16	68
Houston	R 95	R 1	R 1	R 1	R 1	R 0	27	11
Los Angeles	R 64	R 6	R 5	R 15	S 0	S 0	19	63
New York City	R 54	R 1	R 39	R 5	R 1	R 0	19	57
Seattle	R 42	R 14	R 8	R 16	R 4	R 3	16	31

Note: The codes for patrol use are as follows:
R = Patrol type is used on a routine basis
S = Patrol type is used for special events only
M = Patrol type is not used

Source: Adapted from B. A. Reaves and A. L. Goldberg. 1999. *Law Enforcement Management and Administrative Statistics, 1997.* Washington, D.C.: Department of Justice, pp. 71–80.

3. Even while engaged in routine patrol, the officer is often encouraged during recruit training to make good use of this slack time by engaging in "what if" mental exercises: "What if an armed robbery occurred at _____ (location)? How would I get there most rapidly? What would I do after arrival? Where would I find available cover?" Of course, officers can concoct any number of scenarios and types of calls for service to keep themselves mentally honed and ready to respond in the most efficacious manner.

4. An often overlooked part of policing is that patrol officers must also spend a lot of time—especially during the earlier part of their careers—memorizing many things: the "Ten Code," the numbering systems of streets and highways within their jurisdiction. (Indeed, new recruits can and do "wash out" during the field training phase of their careers because of their inability to read an in-car map of the city, thus being unable to arrive at their destination in a prompt manner.)

Patrol Work as a Function of Shift Assignment

Although the following analysis does not apply to all jurisdictions, it may be surprising to persons outside of policing to learn that the nature of patrol work is very closely related to the officer's particular **shift assignment**. Witness the following general descriptions of the nature of work on each of the three shifts (with three eight-hour shifts per day).

The *day shift* (approximately 8 A.M. to 4 P.M.) probably has the greatest contact with citizens. Officers may start their days by watching school crossings and unsnarling traffic jams. Speeding and traffic accidents are more common as people hurry to work in the morning. They also participate in school and civic presentations and other such programs. Most errands and nonpolice duties assigned to the police are performed by day shift officers, such as unlocking parking lots, escorting people, delivering agendas to city council members, transporting evidence to court, and seeing that maintenance is performed on patrol vehicles. Day shift officers are more likely to be summoned to such major crimes as armed robberies and bomb threats. This shift often has lull periods, as most people are at work or in school. Usually officers with more seniority work the day shift.

The *swing* or *evening shift* (4 P.M. to 12 A.M.) comes on duty in time to untangle evening traffic jams and respond to a variety of complaints from the public. Youths are out of school, shops are beginning to close, and, as darkness falls, officers must begin checking commercial doors and windows on their beat; new officers are amazed at the frequency with which businesspeople leave their buildings unsecured. Warm weather brings increased drinking and partying along with noise complaints. Domestic disturbances begin to occur and the action at bars and nightclubs is beginning to pick up; soon fights will break out. Many major events, such as athletic events and concerts, occur in the evenings, so crowd and traffic control are often performed. Toward the end of the shift, fast food and other businesses begin complaining about loitering and trash on the premises from gatherings of teenagers. Arrests are much more frequent than during the day shift, and officers must attempt to take one last look at the businesses on the beat before ending their shift to ensure that none have been burglarized during the evening and night hours. That done, arrest and incident reports often must be completed before officers may leave the station house for home.

The *night shift* (12 A.M. to 8 A.M.), known throughout history as the graveyard shift, is an entirely different world. Because of its adverse effects on the officer's sleeping and eating habits, this shift is usually worked by newer officers (who also, because of low seniority, work most weekends and holidays), but only long enough for the officer to build enough seniority to transfer to another shift. Few officers actually like the shift enough to want to devote a large number of years of their career working it. (Note that many agencies also have shift rotation, transferring their officers from one shift to another at fixed intervals.)

Officers on this shift come on duty fresh, ready for action in those states where bars and taverns legally must close at midnight. From about midnight to 3 A.M., the night shift is quite busy. Traffic is relatively heavy for several hours, and then it normally drops off to a trickle. The "night people" begin to come out— those people who sleep in the daytime and prowl at night, including the burglars. The nightly cat-and-mouse game begins between the cops and robbers. Night shift officers come to know who these people are and what vehicles they drive, what crimes they prefer, what their habits are, and where they hang out. Night shift officers spend much of the night patrolling alleys and businesses, working their spotlight as they seek signs of suspicious activity, open doors and windows in businesses, and unlawful entry. They also watch the residential areas, performing courtesy checks of homes in general or with greater scrutiny when people are away on vacation and have asked the police for a periodic check of their property.

Such patrol work is inevitably eerie in nature. Officers typically work alone under cover of darkness, often without hope of rapid backup units, although where possible, greater attention is given to providing backup to night shift officers, even during traffic stops. The police never know who or what awaits them around the next dark corner. The protective shroud of darkness given to the offenders makes the night shift officers more wary. After midnight, graveyard officers view alleys as theirs alone; anyone violating the peace of "their" alleys— especially one of the known "night people" or anyone wearing dark clothing or engaging in other suspicious activity—should be prepared to explain their actions and presence. Such persons may also be compelled to engage in a stop-and-frisk (pat down) search.

Once the alleys and buildings have been checked, the officers begin rechecking them, avoiding any routine pattern that burglars may discern. Some burglars can tell which beats are "open"—that is, wide open for burglarizing—by observing which patrol vehicles are parked at the station or at restaurants; therefore, officers should vary their patrol routine each night. At 2 or 3 in the morning, boredom can set in. Some officers welcome this change of pace, while others loathe it and look for ways to fight the monotony of the "dog watch." For them, the occasional high-speed chase may bring a welcomed adrenaline rush, as do crimes in progress. Other means of staying alert include meeting and chatting with other officers who are also bored with patrol and stopping for coffee. But these officers must be mentally prepared for action; they know that while this is normally a quiet shift after the initial activity, when something does occur on the night shift, it is often a major incident or crime.

INFLUENCES OF ONE'S ASSIGNED BEAT

Just as the work of the patrol officer is influenced by his or her shift assignment, the nature of that work is determined by the beat assignment. Each beat has its own personality, which may be quite different from other contiguous beats in its structure and demographic character, as seen in the following hypothetical examples:

Beat A contains a university with many large crowds that attend athletic and concert events; it also contains a number of taverns and bars where students congregate, resulting in an occasional need for police presence. A large hospital is located in this sector. Residents here are predominantly middle class. A large number of shopping malls and retail businesses occupy the area. The crime rate is quite low here, as are the numbers of calls for service. The university commands a considerable amount of officer overtime for major events as well as general officer attention for parking problems. During university homecoming week and other major events, officer(s) in this beat will be going from call to call, while officers assigned to other beats may find themselves completely bored. One portion of the beat contains several bars that attract working-class individuals and generate several calls for service each week due to fights, traffic problems, and so forth.

Beat B is almost totally residential in nature and is composed of the "old money" people of the community: upper- and upper-middle classes who "encourage" routine patrols by the police. Some of the community's banks, retail businesses, and industrial complexes are also located in this area. Most people have their homes wired for security, either to a private security firm or to the local police department. The crime rate and calls for service are relatively low in this beat, but there is a large amount of territory to patrol and a major thoroughfare runs on the beat's perimeter, generating some serious traffic accidents.

Beat C is composed primarily of blue-collar, working-class people. It generates low to medium numbers of calls for service relative to the other beats, and much of its geographical area is consumed by a small airport and a large public park and baseball diamond/golf course complex.

Beat D is the worst in the city in terms of quality of life, residents' income levels, and police problems. Though smaller in size than the other beats, it generates very high numbers of calls for service. It contains a large number of residents on the margins of the economy, lower-income housing complexes, taverns, barely surviving retail businesses, older mobile home parks and motels, and a major railroad switching yard. Officers are constantly driving from call to call, especially during summer weekend nights. At night, officers who are engaged in calls for service—even traffic stops—are given backup by fellow officers to the extent possible.

Of course, even the normal ebb and flow of beat activity will be greatly altered when a critical incident occurs; for example, an act of nature (e.g., a tornado, earthquake, or fire) or major event (a bank robbery or after-school kidnapping) can wreak havoc on a beat that is normally the most placid in nature.

Several other related, fundamental elements of beats—"cops' rules"—are part of the beat culture:

1. Don't get involved in another officer's sector; this means that each officer is accountable for his or her territory. Other officers are to "butt out" unless asked to come to a beat to assist, and each officer must live with the consequences of decisions made that pertain to the beat.
2. Don't leave work for the next duty shift; this means that such practical matters as putting gas in the patrol car and taking all necessary complaints should be accomplished prior to leaving the station house.
3. Hold up your end of the work: don't slack off.[6]

WHERE DANGER LURKS: THE HAZARDS OF BEAT PATROL

It is well known that police officers have a very dangerous job. Although several other occupations outrank that of policing in terms of the rate of employees who are killed each year, officers face an omnipresent peril; they never know if a citizen who is about to be confronted will be armed, high on drugs or alcohol, or even wanting to engage in a recent phenomenon known as "suicide by cop." During the first year of the new millennium, 151 police officers lost their lives in the line of duty; the annual average in the 1990s was 153, compared to 187 during the 1980s and 222 during the 1970s.[7] This is still a dangerous nation, notwithstanding the recent, highly publicized decreases of violent crimes during the past several years. An adverse omen is the fact that the number of murders committed in the nation's 30 largest cities seems to have leveled off, after dropping 37 percent from 1991 and 1999.[8] We discuss the dangers of police work at greater length in Chapter 12, when we examine police stress.

Jerome Skolnick and David Bayley described the beat's dangers as officers prepare for their tour of duty:

> Policing in the United States is very much like going to war. Three times a day in countless locker rooms, large men and a growing number of women carefully arm and armor themselves for the day's events. They begin by strapping on flak jackets. Then they pick up a wide, heavy, black leather belt and hang around it the tools of their trade: gun, mace, handcuffs, bullets. When it is fully loaded they swing the belt around their hips with the same practiced motion of the gunfighter in Western movies, slugging it down and buckling it in front. Inspecting themselves in a full-length mirror, officers thread their night sticks into a metal ring on the side of their belt.[9]

As John Crank stated, "This is not a picture of American youth dressing for public servitude. These are warriors going to battle, the New Centurions, as Wambaugh calls them. In their dress and demeanor lies the future of American policing."[10]

Danger is a constant companion of officers on patrol. (*Courtesy NYPD Photo Unit*)

As Crank also observed, police recognize many citizens for what they are: "Dangerous, unpredictable, violent, savagely cunning . . . in a world of capable and talented reptilian, mammalian, and pescian predators."[11]

This depiction of people confronted by officers on the beat may seem overly contrived, exaggerated, or brusque. Most patrol officers with any length of service, however, can attest to the fact that there are certain members of our society who, as one officer put it, are "irretrievable predators that just get off . . . on people's pain and on people's crying and begging and pleading. They don't have any sense of morality, they don't have any sense of right and wrong."[12] Therefore, most patrol officers, while arresting people and generally maintaining order, will during their careers have their lives verbally threatened by such individuals; they will take the great majority of such threats with a grain of salt. Occasionally, however, the "irretrievable predator" who possesses no sense of morality will issue such a threat, which the officer will (and must) take quite seriously; this is a very disconcerting part of the job.

The degree of importance that is attached to patrol officers providing backup to one another—especially during the hours of darkness—cannot be overstated, as shown in the following statement by Anthony Bouza:

> The sense of "us vs. them" that develops between cops and the outside world forges a bond between cops whose strength is fabled. It is widened by the dependence cops have on each other for safety and backup. The response to help is a cop's life-line. An "assist police officer" is every cop's first priority. The ultimate betrayal is for one cop to fail to back up another.[13]

In this same vein, patrol officers quickly come to know whom they can count on when everything "hits the fan"—which officers will race to assist another officer at a barroom brawl, a felony in progress, and so on—and who will not.

STUDIES OF THE PATROL FUNCTION

Because of the amount of resources devoted to the patrol function and a desire to make patrolling more productive and pleasant for the officers, several studies have been conducted that have uncovered several deficiencies and exposed several myths about preventive patrol. These studies also help us understand how the professional policing model put up walls between the public and the police, who became viewed by many people as an occupying force.[14] As the Police Foundation said, "Isolated in their rolling fortresses, police seem[ed] unable to communicate with the citizens they presumably served."[15]

The best-known study of patrol efficiency was in Kansas City, Missouri, in 1973, by George Kelling and a research team at the Police Foundation. The researchers divided the city into 15 beats that were then categorized into five groups of three matched beats each. Each group consisted of neighborhoods that were similar in terms of population, crime characteristics, and calls for police services. Patrolling techniques used in the three beats varied; there was no preventive patrolling in one (police only responded to calls for service), there was increased patrol activity in another (two or three times the usual amount of patrolling), and there was the usual level of service in the third beat. Citizens were interviewed and crime rates measured during the year the experiment was conducted. This experiment challenged several traditional assumptions of routine police patrol. The deterrent effect of policing was not weakened by the elimination of routine patrolling. Citizens' fear of crime and their attitudes toward the police also were not affected, nor was the ability of the police to respond to calls.

This experiment—known as the Kansas City Preventive Patrol Experiment, depicted in Figure 5–1—indicated that the old sacrosanct patrol methods were subject to question. As one of the study's authors stated, it "show[ed] that the traditional assumptions of 'Give me more cars and more money and we'll get there faster and fight crime' is probably not a very viable argument."[16]

In the mid-1970s, it was suggested that the performance of patrol officers would improve by using job redesign based on motivators rather than by attempting to change the individual officer selected for the job (by such means as increasing educational levels).[17] This suggestion later evolved into a concept known as *team policing,* which differed from conventional patrol in several areas. Officers were divided into small teams that were assigned permanently to small geographic areas or neighborhoods. Officers were to be generalists, trained to investigate crimes and attend to all of the problems in their area. Communication and coordination between team members and the community were to be maximized; team involvement in administrative decision making was emphasized as well.

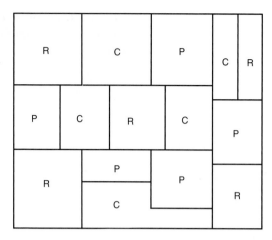

P = proactive
C = control
R = reactive

FIGURE 5–1 Schematic Representation of 15-Beat
Area, Kansas City Preventive Patrol
Experiment. (*Source:* George L. Kelling et al.,
*The Kansas City Preventive Patrol Experiment:
Final Report.* [Washington, D.C.: Police
Foundation, 1974], p. 9. Used with
permission.)

This concept, later abandoned by many departments, apparently because of its strain on resources, was the beginning of the 1980s movement to return to community oriented policing.

Two more attempts to increase patrol productivity, generally referred to as *directed patrol,* occurred in 1975. The New Haven, Connecticut, Police Department used computer data of crime locations and times to set up deterrent runs—D-runs—to instruct officers how to patrol. For example, the officer may be told to patrol around a certain block slowly, park, walk, get back in the car, and cruise down another street. A D-run took up to an hour, with each officer doing two or three of them per shift. Support for patrol officers was generally low; further, the program did not reduce crime but rather displaced it. After a year, it quietly died.[18] Wilmington, Delaware, instituted a split-force program, whereby three-fourths of the 250 patrol officers were assigned to a basic patrol unit to answer prioritized calls. The remainder of the officers were assigned to the structured unit; they were deployed in high-crime areas, usually in plain clothes, to perform surveillances, stakeouts, and other tactical assignments. An evaluation of the project found that police productivity increased 20 percent while crime decreased 18 percent in the program's first year.[19]

In the late 1970s, a renewed interest in foot patrol—a response to Peel's view that police officers should walk the beat—compelled the Police Foundation to evaluate the effectiveness of foot patrol in selected New Jersey cities between 1977 and 1979. It was found that, for the most part, crime levels were not affected by foot patrol, but it did have a significant effect on the attitudes of area residents. Specifically, residents felt safer, thinking that the severity of crimes in their neighborhoods had diminished. Furthermore, evaluations of the Flint, Michigan, Neighborhood Foot Patrol Program in 1985 found that foot officers had a higher level of job satisfaction[20] and felt safer on the job than motor officers.[21]

Other studies have illuminated patrol function as well. First, a long-standing assumption was that as police response time increased, the ability to arrest perpetrators proportionately decreased. Thus, conventional wisdom held, more police were needed on patrol in order to get to the crime scene more quickly and catch the criminals. In 1977, a study examined police response time in Kansas City, Missouri, finding that response time was unrelated to the probability of making an arrest or locating a witness. Furthermore, neither dispatch nor travel time was strongly associated with citizen satisfaction. The time it takes to report a crime, the study found, is the major determining factor of whether an on-scene arrest takes place and whether witnesses are located.[22] It is also known that two-person patrol cars are no more effective than one-person cars in reducing crime or catching criminals. Furthermore, injuries to police officers are not more likely to occur in one-person cars. Finally, most officers on patrol do not stumble across felony crimes in progress.[23]

While these studies should not be viewed as conclusive—different results could be obtained in different communities—they do demonstrate that old police methods should be viewed very cautiously. Many police executives had to rethink the sacred cows of beliefs about patrol functions.

PATROL OFFICER DEPLOYMENT

As indicated earlier, the largest, most costly, and most visible function in a police agency is that of patrol. Yet patrol **deployment** may well be that area of policing that receives the least amount of planning or analysis. An important part of police leadership lies in how to best allocate resources, especially in a time when new resources are more difficult to obtain and the police are being asked to do more with less.

Few police agencies pay regular attention to evaluating and adjusting patrol plans to meet service demands. Instead, patrol is often the first division in which a police administration seeks to reduce personnel in order to enhance specialized units or to create new programs. This practice often leaves the patrol division in need of personnel and often results in morale problems and unnecessary delays in responding to calls for service (CFS).

Patrol planning enables managers to properly assess service demands so that resources may be appropriately allocated across shifts and proportionate to workload; such planning also keeps supervisors and managers apprised of how resources are being utilized in order to make informed decisions about departmental operations and develop future plans. A patrol plan should be based on an analysis of data about the tasks that officers perform during their shifts of duty. These plans should also be flexible and constantly reviewed to meet the changing needs and goals of an organization.

Next we look at some methods for determining how many officers are required to do the job.

DETERMINING PATROL FORCE SIZE

The collection and analysis of data are the foundation of proper patrol deployment. Three crude methods can be used by police departments to determine resource needs:

Intuitive. This is basically educated guesswork based on the experience and judgment of police managers. It is probably the most commonly used method for small agencies in which the numbers of incidents and officers available are so few that more analytical analysis may not be necessary to determine when and where officers should be deployed.

Workload. This method requires comprehensive information, including standards of expected performance, community expectations, and the prioritization of police activities. Although rarely used by an entire police agency, it is most often used for determining resource needs for patrol or specific programs, such as crime prevention.

Comparative. Most commonly used by police agencies, this method is based on a comparison of agencies by numbers of officers per 1,000 residents. Data are available in the *Uniform Crime Reports* (UCR), published annually by the Federal Bureau of Investigation. For example, police administrators can compare the UCR's national average with their own numbers of full-time, sworn police officers per 1,000 residents.[24]

Another method for determining allocation needs is to set an objective related to the amount of time an agency wants officers to be committed to CFS and available for other functions. No established guideline exists, but agencies often set an objective that would restrict officers' time that is committed to CFS at 30 percent to 40 percent of total time available per shift. M. J. Levine and J. T. McEwen[25] provided the following guide for agencies when determining allocation needs using this formula:

Step 1.   Set an objective for patrol performance (e.g., 30 percent to 40 percent committed to CFS).

Step 2.   Select a time period to be analyzed.

Step 3.   Determine CFS workload for this time period.

Step 4.   Calculate the number of units needed based on the workload and the selected objective.

Step 5.   Calculate the number of on-duty officers needed per shift.

Step 6.   Multiply by the relief factor to obtain the total number of officers needed.

As an example, and using these steps, Table 5–3 shows the basic data for calculating the number of patrol officers needed in a city's patrol force. Assume that after discussion of how busy patrol units should be, and given that the agency is engaged in a COPPS strategy, the following objective is determined: "There should be sufficient units on duty so that the average unit utilization on CFS will

TABLE 5–3 An Example of Data for Determining Patrol Force Size

	MIDNIGHTS	DAYS	EVENINGS
1. Workload Data			
Calls for service	1,027	1,614	2,059
Average time (minutes)	32 min.	28	33
Assists	225	273	463
Average time (minutes)	22 min.	20	18
Traffic accidents	109	129	150
Average time (minutes)	63 min.	58	60
2. Hours of work for entire 4-week period	745	969	1,421
Average hours of work per shift	26.6	34.6	50.8
3. Units needed for 30%	11	14	21
4. Number of 1-officer units	8	10	15
Number of 2-officer units	4	4	6
5. Number of officers needed per shift	14	18	27
6. Total number of officers needed (relief factor = 2.2)	31	40	59

Source: U.S. Department of Justice, National Institute of Justice, *Patrol Deployment* (Washington, D.C.: U.S. Government Printing Office, 1985), p. 34.

not exceed 30 percent." Assume further that a mix of 70 percent one-officer and 30 percent two-officer units will be established for each shift. The data were collected during a four-week (28-day) period.

The first section of Table 5–3 shows the *total* number of calls for service, assists, and traffic accidents by shift for the four weeks and the average times for these activities for each shift. With these activities and average times, the total amount of work for the patrol force amounts to about 745 hours for the 12 A.M. to 8 A.M. shift; 969 hours for the 8 A.M. to 4 P.M. shift; and 1,421 hours for the 4 P.M. to 12 A.M. shift. Since a 28-day period was being studied, the average work *per shift* amounts to 26.6 hours, 34.6 hours, and 50.8 hours, respectively. (As an example, for midnights, $1{,}027 \times 32 = 32{,}864$; $225 \times 22 = 4{,}950$; $109 \times 63 = 6{,}867$, for a grand total of 44,681 minutes, or 745 hours, of work; 745 hours of work divided by 28 shifts = 26.6 hours of average work per shift.)

To calculate the number of patrol units needed to meet the desired objective—average unit utilization on CFS will not exceed 30 percent—we use the following formula:

$$\frac{\text{Average Hours of Work Per Shift}}{(\text{Shift Length}) \, (\text{Unit Utilization})} = \text{Number of Units Needed}$$

Again, using the midnight shift as an example, the calculation would be as follows:

$$\frac{26.6 \text{ hours}}{(8 \text{ hours}) \, (30\%)} = 11.8 \text{ units}$$

The answer must be rounded to 12 units, since fractions of units are not possible. Similar calculations for the day and evening shifts give results of 15 units and 21 units, respectively. Table 5–3 shows the officers needed for these shifts under the decision of a 70 percent/30 percent split between one-officer/two-officer units.

The final line in Table 5–3 multiplies the number of officers needed by the department's relief factor of 2.2 (to cover officers' absences due to days off, sick leave, vacations, training, etc.) to give a total of 35 officers for the midnight shift, 42 officers for the day shift, and 59 officers for the evening shift. A total of 136 officers would be required to meet the objective of an average 30 percent unit utilization. If an objective other than unit utilization had been selected, the same steps would have been followed to determine the number of units needed, but the calculations would have been different.

The selection of a 30 percent unit utilization objective is subject to criticism; remember that there is no universal rule to guide the choice of a percentage. A department should consider the big picture of patrol resource allocation; certainly, the existence of a COPPS philosophy should be weighed into this decision.

Finally, the use of a relief factor of 2.2 in our example and in Table 5–3 is also subject to debate. Another commonly accepted relief factor for determining the number of police officers that are needed to staff a shift annually is 1.66. Indeed, a very common calculation—one that uses the 1.66 relief factor—is that five officers are required to fill a position for an entire year: 3 (shifts) × 1.66 (officers) = 4.97, or 5 officers.

DISCRETIONARY USE OF POLICE AUTHORITY

The Link Between Patrol and Discretion

George Kelling, co-author (with James Q. Wilson) of the well-known "Broken Windows" article, described a Newark, New Jersey, street cop with whom he spent many hours walking a beat:

> As he saw his job, he was to keep an eye on strangers, and make certain that the disreputable regulars observed some informal but widely understood rules. Drunks and addicts could sit on the stoops, but could not lie down. People could drink on side streets, but not at the main intersection. Bottles had to be in paper bags. Talking to, bothering or begging from people waiting at the bus stop was strictly forbidden. Persons who broke the informal rules, especially [the latter], were arrested for vagrancy. Noisy teenagers were told to keep quiet.[26]

This quote points out the inextricable link between the patrol function and **discretionary use of police authority**. We cannot discuss one without including the other. To begin our discussion of police discretion, a hypothetical exercise is presented as food for thought.

AN EXERCISE IN DISCRETION

Imagine this scenario. You are a graveyard shift police officer alone on duty in a town of 3,000 people. It is about 1 A.M. and it is quiet; nothing is moving except for the local wildlife. You are parked on a two-lane highway that runs adjacent to your community, enjoying the solitude and peacefulness while indulging in your favorite summertime drink. Eventually a pair of headlights appear in the distance and approach you on the highway. The vehicle, a pickup truck, passes you by, and as it does, you notice that the vehicle begins to weave perceptibly, crossing the center line of the highway several times. Your training indicates that alcohol may be a factor and thus spurred to action, you pursue the pickup for about a half-mile and then pull it over with your red lights and siren.

The driver exits his truck, and you notice a slight stagger in his gait and then detect the telltale odor of alcohol on his breath. He is wearing a grey denim work shirt, with a patch reading "Joe's Heating and Cooling" on the backside, and "Bob" on the front. You ask Bob how much he has had to drink, and the answer (one that is given to police officers probably 90 percent of the time) is: "A couple of beers." You conduct a field sobriety test, and Bob's performance is poor. Your experience and training indicate that his blood alcohol content is approximately .15 (.10 being legally intoxicated).

Your mission is to list all of the possible legitimate and lawful measures that you as a police officer can employ to deal with this police matter. Do not concern yourself with trying to determine which options may be better or worse at this point; however, do discuss some obvious ramifications and advantages and disadvantages of each. A hint—there are at least a half dozen options available to you. Some possible answers and further discussion are provided at the end of the chapter.

ATTEMPTS TO DEFINE DISCRETION

Scholarly knowledge about the way police make decisions is limited. What is known, however, is that when police observe something of a "JDLR" nature, two important decisions must be made: (1) whether to intervene in the situation and (2) how to intervene. The kinds, number, and possible combinations of interventions are virtually limitless. What kinds of decisions are available for an officer who makes a routine traffic stop (as in the hypothetical situation just described)? David Bayley and Egon Bittner[27] observed long ago that officers have as many as 10 actions to select from at the initial stop (for example, order the driver out of the car), seven strategies appropriate during the stop (such as a roadside sobriety test), and 11 exit strategies (for instance, releasing the driver with a warning), representing a total of 770 different combinations of actions that might be taken!

Criminal law has two sides—the formality and the reality. The formality is found in the statute books and opinions of appellate courts. The reality is found in the practices of enforcement officers. In some circumstances, the choice of action to be taken is relatively easy, such as arresting a bank robbery suspect. In other situations, such as quelling a dispute between neighbors, the choice is more difficult. Drinking in the park is a crime according to the ordinance, but quietly drinking at a family picnic without disturbing others is not a crime according to the reality of the law, because officers uniformly refuse to enforce the ordinance in such circumstances. When the formality and the reality differ, the reality prevails.[28]

These examples demonstrate why the use of discretion is one of the major challenges facing U.S. police today. Our system tends to treat people as individuals. One person who commits a robbery is not the same as another person who commits a robbery. Our system takes into account *why* and *how* (one's intent, or *mens rea*) a person committed a crime. Under our judicial process, when A shoots B, a variety of possible outcomes can occur. The most important decisions take place on the streets, day or night, generally without the opportunity for the officer to consult with others or to carefully consider all the facts.

DETERMINANTS OF DISCRETIONARY ACTIONS

Ours is supposed to be a government of laws and not of people. That axiom is simply a myth, at least in the manner by which the law is applied. Official discretion pervades all levels and most agencies of government. The discretionary power of the police is awesome. Kenneth Culp Davis, an authority on police discretion, wrote, "The police are among the most important policy makers of our entire society. And they make far more discretionary determinations in individual cases than does any other class of administrators; I know of no close second."[29]

What determines whether the officer will take a stern approach (enforcing the letter of the law with an arrest) or will be lenient (issuing a verbal warning or some other outcome short of arrest)? Several variables enter into the officer's decision. First, the *officer's attitude* is a consideration. Police, being human, bring to work either a happy or unhappy disposition. If, on that same day, the officer received an IRS notice saying back taxes were owed; had a nasty spat with a spouse; was severely bitten on the leg by the family dog when leaving for work; and was stranded on the way to work because of car trouble, he or she may be more inclined to exercise an arrest in the scenario presented earlier. Conversely, if on that day the officer received an IRS check for overpayment of taxes; had a particularly amorous evening with his or her significant other; was licked affectionately by the dog when leaving for work; and can now afford to purchase a new car, he or she is more likely to give Bob a break. Personal views toward specific types of crimes also filter in; perhaps the officer in our scenario is especially fed up with the lives taken by intoxicated drivers (likely outcome: arrest) but is not especially outraged about kids who drag race on the highway (likely outcome: verbal or written warning).

Another major consideration in the officer's choice among various options is the *citizen's attitude*. If the offender is rude and condescending, denies having done anything wrong, and/or uses some of the standard cliches that are almost guaranteed to rankle the police ("You don't know who I am" or "I'll have your job" or "I know the chief of police/mayor" or "As a taxpayer, I pay your salary"), the probable outcome is obvious. On the other hand, the person who is honest with the officer, avoids attempts at intimidation and sarcasm, and does not try to "beat the rap" will normally fare better.

Several studies have also found that not only a citizen's demeanor but also his or her social class, sex, age, and race influence the decisions made by police patrol officers.[30] This may be discrimination on the part of the officers and points out that the police—like other citizens—are subject to stereotypes and biases that will affect their behavior.

PROS, CONS, AND POLITICS OF DISCRETIONARY AUTHORITY

Several ironies are connected with the way in which the police apply discretion. First is the inverse relationship between the officers' rank and the amount of discretion that is available. In other words, as the rank of the officer increases, the amount of discretion that he or she can employ normally decreases. The street officer makes discretionary decisions all the time—decisions about whether or not to arrest, shoot, and so forth. But the chief of police, who does very little actual police work, may be very constrained by department, union, affirmative action, or governing board guidelines and policies. Furthermore, the chief of police knows there are neither the resources nor the desire to enforce all the laws that are broken.

The police executive knows that the exercise of discretion is an essential part of police work. However, he or she cannot broadcast that fact to the rank and file or to the public. There is a myth of full enforcement of the laws—that all of the laws should be, and are, enforced all of the time and with impartiality. It is a delicate matter to tell police officers which laws they will enforce and which they will not. And when the chief is asked at a civic club luncheon which laws are not enforced, it is normally not prudent nor politically wise to list the offenses for which the police look the other way. Police agencies are also pressured to support the status quo, and there are legal concerns as well. For example, releasing drunk drivers cannot be the official policy of the agency.

Finally, the issue of police discretion is shrouded in controversy. There are several arguments both for and against discretion. Advantages include that it allows the officer to treat different situations in accordance with humanitarian and practical goals. Take, for example, an officer who attempts to pull over a speeding motorist, only to learn that the car is enroute to the hospital with a woman who is about to deliver a baby. While the agitated driver is endangering everyone in the vehicle as well as other motorists on the roadway, discretion allows the officer to be compassionate and empathetic, giving the car a safe escort to the hospital rather than issuing a citation for speeding. In short, discretionary use of authority allows the police to employ a philosophy of justice tempered with mercy.

Conversely, discretion smacks of partiality. In the scenario discussed earlier, Bob may be arrested for driving under the influence on a Tuesday, but in an identical situation, Jack is allowed by the same officer to sleep it off on Friday. Also, critics of discretion in policing can argue that such wide latitude in decision making may serve as a breeding ground for police corruption; an officer, for example, may be "bought" to exercise his or her discretion and overlook a traffic violation or even more serious offenses. And as Lawrence Sherman observed, another problem is that the police do not know the consequences of their discretionary decisions. The police mission, Sherman wrote, is defined as answering calls and being available to answer more calls; therefore, police managers have created only an input information system. Sherman contrasted the police to artisans and navigators, who receive feedback on the effects of their decisions. The police, however, have failed to create a feedback information system that tells them what happens after they leave a call or even after they make an arrest. Thus, police lack knowledge about the effects of their discretionary actions on suspects, victims, witnesses, and potential criminals.[31]

A large number of people in the nation would probably support greater control over police discretion. Incidents of abuse of discretion certainly lend weight to their argument. Control means guidelines, and even the National Advisory Commission on Criminal Justice Standards and Goals (1973) recommended a greater reliance on guidelines and policy over police discretion.[32] Certain aspects of policing will never be completely free of discretion, however; to a large extent, the work of a police officer is unsupervised and unsupervisable. And as the police strive to achieve professionalism, they will remember that discretion is a key element of a profession.

Police discretion is also part of the American political process.[33] As Kenneth Culp Davis observed, a major contributing factor to police discretion is that state legislative commands are ambiguous. Legislatures speak with three voices: They enact state statutes that seemingly require full enforcement of the laws; they provide only enough resources for limited enforcement of them; and, finally, they consent to such limited enforcement.[34] Some observers have even questioned the legality and morality of police discretion.[35] And, finally, it might be added, the statute books are often treated as society's "trash bins." A particular behavior is viewed negatively and a law is passed against it. The police are stuck with the dilemma of having to enforce it or ignoring what may be unpopular laws.

There are other aspects of politics in police discretion. For example, several state and local governments restricted police use of deadly force long before the Supreme Court, in *Tennessee v. Garner* (1985),[36] did away with the common-law "fleeing felon" doctrine.

A RELATED FUNCTION: TRAFFIC

O. W. Wilson reportedly said that "the police traffic function overshadows every other function." That may be an overstatement, but a strong link exists between the patrol function and **traffic control**. Traffic stops account for about half (52 per-cent) of the contact Americans have with the police. Nearly 20 million Americans will be subjected to such stops in a given year; more than half of them (54 per-cent) will be cited.[37] Therefore, the importance of a seemingly trivial traffic stop cannot be overstated, because the manner in which the officer conducts the stop may in large measure determine the citizen's view of the police for many years to come.

POLICING TODAY'S MOTORIZED SOCIETY

Police endeavor to reduce traffic deaths and injuries through enforcement of motor vehicle laws. Indeed, many citizens have had their only contact with a police officer by virtue of a traffic-related matter.

There are, of course, various levels of traffic enforcement policy. Some departments are relatively lenient, while others have ticket quotas and pressure

A modern-day patrol vehicle can be equipped with a computer, mobile digital terminal, videorecorder, and other state-of-the-art systems.
(*Courtesy Ron Glensor, Reno, Nevada, Police Department*)

officers to have a "ticket blizzard" to bring in revenues. These policies normally do not help police-community relations. Even with departmental policies, traffic enforcement carries wide discretion for patrol officers, who must decide whether or not to cite individuals.[38] It is an area in which many citizens think the police should practice discretion that is skewed toward leniency. Citizen perceptions of the police mission (or "Why aren't you out catching bank robbers?") and the perceived lack of seriousness of the offense often contribute to the citizen's unhappiness when cited for a traffic offense. As a result, traffic stops are major sources of friction between police officers and citizens.[39]

Notwithstanding the weight that many police agencies put on traffic enforcement, studies have found that crackdowns on speeding have no impact on fatality rates[40] and little influence on traffic violations or accident rates.[41] In fact, it has been shown that paying a small traffic fine and receiving a warning citation from an officer had a greater effect on future driving than appearance in a tough traffic court.[42]

Despite citizen disgruntlement with traffic enforcement, traffic stops and citations generally remain an integral part of police work. Police administrators find such work to be easily verifiable evidence that their officers are working.[43] Traffic enforcement has even gone high tech with the advent of the traffic camera, already nicknamed "the photocop." Traffic cameras, either mounted on a mobile tripod or permanently fixed on a pole, emit a narrow beam of radar that triggers a flash camera when the targeted vehicle is exceeding the speed limit by a certain amount, usually 10 miles per hour.

Police in Berkeley, California, have applied a new twist to the traffic function. Drivers who are "caught" driving safely and courteously are stopped and issued coupons good for movies or free nonalcoholic beverages at a local cafe. This Good Driver Recognition Program began with officers' donations and now receives city funding.[44]

TRAFFIC ACCIDENT INVESTIGATION

Patrol officers have long been required to engage in traffic accident investigation (TAI). In this era of accountability and litigation, and considering the vast damage done to persons and property each year as a result of traffic accidents, it is essential that officers understand this process of investigation and cite the guilty party—not only from a law-enforcement standpoint, but also in the event that the matter is taken to civil trial. Until officers receive formal training in this complex field, they are in a very precarious position.

In addition to basic TAI training normally provided at the police academy, several good in-service courses are provided, as well as Northwestern University's renowned traffic accident investigation courses. The process of analyzing road and damage evidence, estimating speeds, reconstructing what occurred and why, issuing citations properly, drawing a diagram of the scene, and explaining

The traffic function is a major part of the police role. (*Courtesy Washoe County, Nevada, Sheriff's Department*)

what happened in court is too serious to be left to untrained officers. The public demands skilled accident investigations.

IN PURSUIT OF THE "PHANTOM DRIVER"

One of the traffic-related areas in which the police enjoy wide public support is in their efforts to identify, apprehend, and convict the hit-and-run (or so-called phantom) driver. No one thinks highly of people who collide with another vehicle or person and leave the scene. Often these drivers are intoxicated. This matter is more of a criminal investigation than an accident investigation for the police. In some states, the killing of a human being while DUI is a felony. Physical evidence and witness statements must be collected in the same fashion as in a conventional criminal investigation; paint samples and automobile parts left at the scene are sent to crime laboratories for examination. The problem for the police is that unless the driver of the vehicle is identified—either by physical evidence, an eyewitness, or a confession—the case can be lost. If the phantom vehicle is located, the owner can simply tell the police his or her vehicle was stolen or on loan. Thus, the police often must resort to psychology to get a confession by convincing the suspect that incriminating evidence exists.

SUMMARY

This chapter has examined several elements of the backbone of policing: patrol. Included are the influences of an officer's shift and beat, some hazards of the job, how to determine the optimal number patrol officers for a jurisdiction, discretionary authority that patrol officers possess, and the traffic function.

Possible solutions for addressing the exercise on discretion presented in this chapter are offered next.

POSSIBLE SOLUTIONS TO THE DISCRETION EXERCISE

Following are some options available to the officer in the exercise at the beginning of the section on police discretion on page 135. Clearly, some options are not as good as others, and none are specifically recommended over others in all cases. Rather, the point is that officers in America's 17,000 police agencies can—and do—exercise daily all of the following alternatives in this kind of situation. Note that in each situation where Bob is separated from his vehicle, you should consider impounding it so it will not pose a safety hazard on the highway or be burglarized during the night.

1. Arrest the driver: This is obviously the formal, hard-line approach, certainly within the officer's power in order to remove Bob from the highway. Bob would be incarcerated for several hours, subjected to a blood alcohol test, and compelled to stay in jail until bond is posted. You, as the officer, would need to impound his truck as well. In addition to towing and impound fees, other possible repercussions might include an increase in his insurance premiums and he may lose his driver's license (as well as his job). His expenses for fines and attorney's fees may be considerable. He may also be compelled to attend a driving or DUI school.
2. Issue a verbal warning and release: Not a good choice unless the driver is clearly able to drive responsibly and the officer is certain that he will not injure others.
3. Issue a written warning and release: Again, this can be a poor choice, especially if he leaves and injures or kills others on the highway. If he does so, the officer would certainly be morally if not legally responsible for those injuries.
4. Book Bob for protective custody: Let him sleep it off in the local jail.
5. Let Bob ride around in the patrol car until he is able to drive: This option is probably employed more often than the public realizes, especially in smaller jurisdictions where most residents are known to the police. This option helps to break monotony on patrol.
6. Tell Bob to sleep it off in his truck. Remove his vehicle keys to ensure he doesn't leave; then give him the keys and release him when he is able to drive.
7. Call his employer or a friend or relative to come and get him.

What is the common thread of these options? Note that for all options except 2 and 3, the officer is getting the problem (Bob) off the highway. Again, the actual option that is selected will depend on several variables, including the officer's attitude on that day in general, his or her personal/agency/community view toward the offense (here, DUI), and Bob's attitude.

This exercise demonstrates the nature of police discretion. After engaging in such an exercise, you should better understand that several means are often available to police officers—both formal (arrest) and informal—for dealing with problems. Some offenses (for example, driving under the influence, serious crimes, or citizen calls for service for noise complaints) allow the officer little discretion, but most situations (especially order-maintenance tasks, such as disorderly conduct, trespassing, and public intoxication) allow for considerable latitude as to whether or not to arrest the offender.

ITEMS FOR REVIEW

1. Explain how the patrol function is affected by the officer's shift assignment and the nature of the beat to which he or she is assigned.
2. Describe some of the hazards that are inherent in beat patrol.
3. Review some of the major studies of the patrol function.
4. Define police discretion, some of its advantages and disadvantages, and factors that enter into the officer's decision-making process.
5. Describe the importance of the traffic function in patrol work.
6. Explain how a police administrator or supervisor may best determine the size of the patrol force.

NOTES

1. John A. Webster, "Patrol Tasks," in *Policing Society*, ed. W. Clinton Terry III (New York: John Wiley & Sons, 1985), pp. 263–313.
2. American Bar Association, *Standards Relating to Urban Police Function* (New York: Institute of Judicial Administration, 1974), Standard 2.2.
3. W. Clinton Terry III, *Policing Society: An Occupational View* (New York: Wiley, 1985), pp. 259–260.
4. Anthony V. Bouza, *The Police Mystique: An Insider's Look at Cops, Crime, and the Criminal Justice System* (New York: Plenum, 1990), p. 27.
5. U.S. Department of Justice, Office of Community Oriented Policing Services, "COPS Facts: 3-1-1 National Non-Emergency Number," October/November 1996, p. 1.
6. Elizabeth Reuss-Ianni, *Two Cultures of Policing: Street Cops and Management Cops* (New Brunswick, NJ: Transaction Books, 1983).
7. "More Police Killed This Year Than Last," Associated Press, December 29, 2000.
8. Kit R. Roane, "Deadly Numbers: Cops Fear New Surge," *U.S. News and World Report*, February 26, 2001, p. 29.
9. Jerome H. Skolnick and David H. Bayley, *The New Blue Line: Police Innovation in Six American Cities* (New York: The Free Press, 1986), pp. 141–142.
10. John P. Crank, *Understanding Police Culture* (Cincinnati: Anderson, 1998), p. 83.
11. *Ibid.*, p. 254.
12. Quoted in Mark Baker, *Cops: Their Lives in Their Own Words* (New York: Pocket Books, 1985), p. 298.
13. Bouza, *The Police Mystique*, p. 74.
14. Joel Samaha, *Criminal Justice,* 2d ed. (St. Paul, MN: West Publishing Company, 1991), pp. 163–164.
15. The Police Foundation, *The Newark Foot Patrol Experiment* (Washington, D.C.: Author, 1981), p. 9.

16. Kevin Krajick, "Does Patrol Prevent Crime?" *Police Magazine* 1 (September 1978): 4–16.
17. T. J. Baker, "Designing the Job to Motivate," *FBI Law Enforcement Bulletin* 45 (1976): 3–7.
18. Krajick, "Does Patrol Prevent Crime?" p. 10.
19. *Ibid.*, pp. 11–13.
20. Robert C. Trojanowicz and Dennis W. Banas, *Job Satisfaction: A Comparison of Foot Patrol Versus Motor Patrol Officers* (East Lansing: Michigan State University, 1985).
21. Robert C. Trojanowicz and Dennis W. Banas, *Perceptions of Safety: A Comparison of Foot Patrol Versus Motor Patrol Officers* (East Lansing: Michigan State University, 1985).
22. *Ibid.*, p. 235.
23. Skolnick and Bayley, *The New Blue Line*, p. 4.
24. Roy R. Roberg and Jack Kuykendall, *Police Organization and Management: Behavior, Theory, and Processes* (Pacific Grove, CA: Brooks/Cole, 1990), p. 284.
25. M. J. Levine and J. T. McEwen, *Patrol Deployment* (Washington, D.C.: Department of Justice, 1985), p. 35.
26. James Q. Wilson and George L. Kelling, "'Broken Windows': The Police and Neighborhood Safety," *Atlantic Monthly*, March 1982, pp. 28–29.
27. David H. Bayley and Egon Bittner, "Learning the Skills of Policing," in *Critical Issues in Policing: Contemporary Readings*, ed. Roger G. Dunham and Geoffrey P. Alpert (Prospect Heights, IL: Waveland, 1989), pp. 87–110.
28. Kenneth Culp Davis, *Police Discretion* (St. Paul, MN: West, 1975), p. 73.
29. Kenneth Culp Davis, *Discretionary Justice* (Urbana: University of Illinois Press, 1969), p. 222.
30. Richard J. Lundman, "Routine Police Arrest Practices: A Commonweal Perspective," *Social Problems* 22 (1974): 127–141; Donald Petersen, "Informal Norms and Police Practices: The Traffic Quota System," *Sociology and Social Research* 55 (1971): 354–361.
31. Lawrence W. Sherman, "Experiments in Police Discretion: Scientific Boon or Dangerous Knowledge?" *Law and Contemporary Problems* 47 (1984): 61–82.
32. National Advisory Commission on Criminal Justice Standards and Goals, *Police* (Washington, D.C.: GPO, 1973), pp. 21–33.
33. For a thorough discussion, see Gregory Howard Williams, "The Politics of Police Discretion," in *Discretion, Justice and Democracy: A Public Policy Perspective*, ed. Carl F. Pinkele and William C. Louthau (Ames, IA: Iowa State University, 1985), pp. 19–30.
34. *Ibid.*, p. 220.
35. See James F. Doyle, "Police Discretion, Legality, and Morality," in *Police Ethics: Hard Choices in Law Enforcement*, ed. William C. Heffernan and Timothy Stroup (New York: John Jay Press, 1985), pp. 47–68.
36. *Tennessee v. Garner*, 475 U.S. 1 (1985).
37. Patrick A. Langan, Lawrence A. Greenfield, Steven K. Smith, Matthew R. Durose, and David J. Levin, *Contacts Between Police and the Public: Findings from the 1999 National Survey* (Washington, D.C.: U.S. Department of Justice, Bureau of Justice Statistics, February 2001), p. 2.
38. Richardson, *Urban Police in the United States*, pp. 111–112.
39. *Ibid.*, p. 55; see also Jonathan Rubenstein, *City Police* (New York: Ballantine Books, 1973), pp. 153–155.
40. Donald Campbell and Harold L. Ross, "The Connecticut Crackdown on Speeding," *Law and Society Review* 3 (1968): 33–53.
41. John A. Gardiner, "Police Enforcement of Traffic Laws: A Comparative Analysis," in *City Politics and Public Policy*, ed. J. Wilson (New York: John Wiley & Sons, 1968), pp. 171–185.
42. Harold Ross, "Folk Crime Revisited," *Criminology* 11 (1973): 71–85.
43. Richard J. Lundman, "Working Traffic Violations," in *Policing Society: An Occupational View*, ed. W. Clinton Terry III (New York: John Wiley & Sons, 1985), pp. 327–333.
44. "The Right to Remain Foamy," *Newsweek*, July 17, 2000, p. 8.

CHAPTER 6

Community Oriented Policing and Problem Solving

Key Terms and Concepts

Community oriented policing
and problem solving
Crime prevention through
environmental design

Evaluation
Implementation
Repeat victimization
S.A.R.A.

No problem can be solved by the same consciousness that created it. We must learn to see the world anew.—*Albert Einstein*

If people are informed, they will do the right thing. It's when they are not informed that they become hostages to prejudice.—*Charlayne Hunter-Gault*

None of us know all the potentialities that slumber in the spirit of the people, or all the ways in which people can surprise us when there is the right interplay of events.—*Vaclav Havel*

INTRODUCTION

This is a uniquely challenging time to be entering police work. As mentioned in Chapter 1, a strategy that runs counter to the professional model of policing is spreading across the country: community oriented policing and problem solving. This chapter examines the rationale for, and methods of, that strategy, which represents a return to the philosophy and practices of policing of the early nineteenth century.

For the past several decades, the dominant police strategy emphasized motorized patrol, rapid response time, and retrospective investigation of crimes. Those strategies have some merit for police operations, but they were not designed to address root community problems. They were instead designed to detect crime and apprehend criminals, hence, the image of the "crime-fighter" cop. The conventional wisdom for policing now holds that the police cannot unilaterally attack America's burgeoning crime, drug, and gang problems that beset our society and drain our federal, state, and local resources. Communities must police themselves. We also understand that it is time for new police methods and measures of effectiveness.

This chapter begins by examining community oriented policing (COP), and then it reviews a more recent development, which extends COP by using the community to address crime and disorder: problem-oriented policing (POP). Included in this section is an overview of the S.A.R.A. problem-solving process. Following that, we look at how these two interrelated and complementary concepts work to engage the community in crime fighting through what has been termed community oriented policing and problem solving (COPPS). We review how COPPS should be implemented and evaluated and how it relates to two elements of crime prevention: environmental design and repeat victimization. The chapter concludes with two case studies of problem-solving efforts by police in Tulsa, Oklahoma, and San Diego, California.

COMMUNITY POLICING

BASIC PRINCIPLES

There is a growing awareness that the community can and *must* play a vital role in problem solving and crime fighting. A fundamental aspect of community oriented policing has always been that the public must be engaged in the fight against crime and disorder. And as we noted in Chapter 1, Robert Peel emphasized in the 1820s in his principles of policing that the police and community should work together.[1]

In the early 1980s, the notion of community policing emerged as the dominant model for thinking about policing. It was designed to reunite the police with the community.[2] No single program describes community policing. Community

Citizen input is crucial to the police for crime detection and prevention.
(*Courtesy Washoe County, Nevada, Sheriff's Department*)

policing has been applied in various forms by police agencies in the United States and abroad and differs according to the community needs, politics, and resources available.

COP is much more than a police-community relations program; it attempts to address crime control through a working partnership with the community. Community institutions such as families, schools, and neighborhood and merchants' associations are seen as key partners with the police in creating safer, more secure communities. The views of community members have greater status under community policing.[3]

COP is a long-term process that involves fundamental institutional change. This concept redefines the role of the officer on the street, from that of crime fighter to that of problem solver and neighborhood ombudsman. It forces a cultural transformation of the entire department, including a decentralized organizational structure and changes in recruiting, training, awards systems, evaluation, promotions, and so forth. Furthermore, this philosophy asks officers to break away from the binds of incident-driven policing and to seek proactive and creative resolution to the problems of crime and disorder.

The major points at which COP departs from traditional policing are shown in Table 6–1.

TABLE 6–1 Traditional vs. Community Policing: Questions and Answers

QUESTION	TRADITIONAL	COMMUNITY POLICING
Who are the police?	A government agency principally responsible for law enforcement.	Police are the public and the public are the police: The police officers are those who are paid to give full-time attention to the duties of every citizen.
What is the relationship of the police force to other public service departments?	Priorities often conflict.	The police are one department among many responsible for improving the quality of life.
What is the role of the police?	Focusing on solving crimes.	A broader problem-solving approach.
How is police efficiency measured?	By detection and arrest rates.	By the absence of crime and disorder.
What are the highest priorities?	Crimes that are high value (e.g., bank robberies) and those involving violence.	Whatever problems disturb the community most.
What, specifically, do police deal with?	Incidents.	Citizen's problems and concerns.
What determines the effectiveness of police?	Response times.	Public cooperation.
What view do police take of service calls?	Deal with them only if there is no real police work to do.	Vital function and great opportunity.
What is police professionalism?	Swift effective response to serious crime.	Keeping close to the community.
What kind of intelligence is most important?	Crime intelligence (study of particular crimes or series of crimes).	Criminal intelligence (information about the activities of individuals or groups).
What is the essential nature of police accountability?	Highly centralized; governed by rules, regulations, and policy directives; accountable to the law.	Emphasis on local accountability to community needs.
What is the role of headquarters?	To provide the necessary rules and policy directives.	To preach organizational values.
What is the role of the press liaison department?	To keep the "heat" off operational officers so they can get on with the job.	To coordinate an essential channel of communication with the community.
How do the police regard prosecutions?	As an important goal.	As one tool among many.

Source: Malcolm K. Sparrow, "Implementing Community Policing" (Washington, D.C.: U.S. Department of Justice, National Institute of Justice: U.S. Government Printing Office, November 1988), pp. 8–9.

A MAJOR STEP FORWARD:
PROBLEM-ORIENTED POLICING

EARLY BEGINNINGS

Problem solving is not new; police officers have always tried to solve problems. The difference is that in the past, officers received little guidance, support, or technology from police administrators for dealing with problems. But the routine application of problem-solving techniques *is* new. It is premised on two facts: that problem solving can be applied by officers throughout the agency as part of their daily work and that routine problem-solving efforts can be effective in reducing or resolving problems.

Problem-oriented policing (POP) was grounded on different principles than COP, but they are complementary. POP is a strategy that puts the COP philosophy into practice. It advocates that police examine the underlying causes of recurring incidents of crime and disorder. The problem-solving process helps officers identify problems, analyze them completely, develop response strategies, and assess the results.

Herman Goldstein is considered by many to be the principal architect of POP. Goldstein first coined the term problem-oriented policing in 1979 out of frustration with the dominant model for improving police operations: "More attention [was] being focused on how quickly officers responded to a call than on what they did when they got to their destination."[4]

As a result, Goldstein argued for a radical change in the direction of efforts to improve policing. The first step in POP is to move beyond just handling incidents; to recognize that incidents are often overt symptoms of problems. It requires that officers take a more in-depth interest in incidents by acquainting themselves with some of the conditions and factors that cause them. The expanded role of police officers under problem-oriented policing is discussed later.

THE PROBLEM-SOLVING PROCESS: S.A.R.A.

POP has at its nucleus a four-stage problem-solving process known as **S.A.R.A.**, for *scanning, analysis, response*, and *assessment*.[5]

Scanning: Problem Identification

Scanning involves problem identification. As a first step, officers should identify problems on their beats and look for a pattern, or persistent repeat incidents. At this juncture, the question might well be asked, "What is a 'problem'?" A problem has been defined as

> a group of two or more incidents that are similar in one or more respects, causing harm and, therefore, being of concern to the police and the public.

Once disorder begins to descend on a location, crime soon follows, and the police will become involved. (*Courtesy Reno, Nevada, Police Department*)

Incidents may be *similar* in various ways, including

- *Behaviors*—this is the most frequent indicator and includes activities such as drug sales, robberies, thefts, and graffiti.
- *Location*—problems occur in area hot spots, such as downtown, and in housing complexes plagued by burglaries, and parks in which gangs commit crimes.
- *Persons*—can include repeat offenders or victims; both account for a high proportion of crime.
- *Time*—seasonal, day of week, hour of day; examples include rush hours, bar closings, tourist activity.
- *Events*—crimes may peak during events such as university spring break, rallies, and gatherings.

There does not appear to be any inherent limit on the types of problems patrol officers can face, but several types of problems are particularly appropriate for problem solving.

Numerous resources are available to the police to help them identify problems, including calls for service (CFS) data, especially repeat calls from the same location or a repeated series of similar incidents. Other ways include citizen complaints, census data, data from other government agencies, newspaper and media coverage of community issues, officer observations, and community surveys.

The primary purpose of scanning is to conduct a preliminary inquiry to determine if a problem really exists and whether further analysis is needed. During this stage, priorities should be established if multiple problems exist, and a specific officer or team of officers should be assigned to handle the problem. Scanning initiates the problem-solving process.

Analysis: The Heart of Problem Solving

The second stage, *analysis*, is the heart of the problem-solving process. Crime analysis has been defined as "a set of systematic, analytical processes providing timely and pertinent information to assist operational and administrative personnel."[6]

Effective, tailor-made responses cannot be developed unless people know what is causing the problem. Thus, the purpose of analysis is to learn as much as possible about problems in order to identify their causes. Complete analysis includes identifying the seriousness of the problem, identifying all the persons/groups involved and affected, identifying all of the causes of the problem, and assessing current responses and their effectiveness.

Over time, several methods have been developed for analyzing crime and disorder. We examine some of them now, including the problem-analysis triangle; analyses of crime maps, offense reports, and calls for service; and the use of surveys.

Problem-Analysis Triangle. One tool that may be used for analyzing problems is the problem-analysis triangle, which helps officers to visualize the problem and understand the relationship between the three elements of the triangle. Additionally, it helps officers analyze problems, suggests where more information is needed, and helps with crime control and prevention. Generally, three elements must be present before a crime or harmful behaviors—problems—can occur: an *offender* (someone who is motivated to commit harmful behavior), a *victim* (a desirable and vulnerable target must be present), and a *location* (the victim and offender must both be in the same place at the same time; we discuss locations later) (see Figure 6–1). If these three elements show up over and over again in patterns and recurring problems, removing one of these elements can stop the pattern and prevent future harm.[7]

Mapping and Offense Reports. Computerized crime mapping (discussed in greater detail in Chapter 14) also assists with crime analysis. Mapping combines geographic information from global positioning satellites with crime statistics gathered by the department's computer-aided dispatching (CAD) system and demographic data provided by private companies or the U.S. Census Bureau.

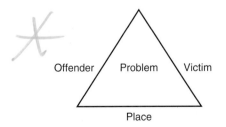

FIGURE 6–1 Problem-Analysis Triangle. (*Source:* U.S. Department of Justice, Bureau of Justice Assistance, *Comprehensive Gang Initiative: Operations Manual for Implementing Local Gang Prevention and Control Programs* [Draft, October 1993], p. 3.)

Police offense reports can also be useful, analyzed for suspect characteristics, MOs, victim characteristics, and many other factors. Offense reports are also a potential source of information about high-crime areas and addresses, since they capture exact descriptions of locations. In a typical department, however, patrol officers may write official reports on only about 25 to 30 percent of all calls to which they respond. Another limitation is that there may be a considerable lag between when the officer files a report and when the analysis is complete.[8]

Computer software now exists for COPPS, to assist with beat profiling and demographics, finding patterns of problems, helping plan daily officer activities, balancing beat and officer workloads, and identifying current levels of performance. Such software can scan through hundreds of millions of pieces of data for patterns, trends, or clusters in beats and neighborhoods while ranking and reranking problems. In the field, the officer simply highlights the neighborhood, beat, or grid under consideration, then selects the problem(s) to be worked from a menu on the computer.

Call for Service Analysis. With the advent of CAD systems, a more reliable source of data on CFS has become available. CAD systems, containing information on all types of calls for service, add to information provided by offense reports, yielding a more extensive account of what the public reports to the police.[9] The data captured by CAD systems can be sorted to reveal hot spots of crime and disturbances—specific locations from which an unusual number of calls to the police are made.

A study in Minneapolis on hot spots analyzed nearly 324,000 calls for service for a one-year period over all 115,000 addresses and intersections. The results showed relatively few hot spots accounting for the majority of calls to the police:

> Fifty percent of all calls came from 3 percent of locations.
> All robbery calls came from 2.2 percent of locations.
> All rape calls came from 1.2 percent of locations.
> All auto thefts came from 2.7 percent of locations.[10]

Many police agencies have the capability to use CAD data for repeat call analysis (which is related to repeat victimization, discussed later). The repeat call locations identified in this way can become targets of directed patrol efforts, including problem solving. For example, a precinct may receive printouts of the top 25 calls-for-service areas to review for problem-solving assignments. In Houston, the police and Hispanic citizens were concerned about violence at cantinas (bars). Through repeat call analysis, police learned that only 3 percent of the cantinas in the city were responsible for 40 percent of the violence. The data narrowed the scope of the problem and enabled a special liquor control unit to better target its efforts.[11]

Repeat-alarm calls are another example of how CAD data can be used to support patrol officer problem solving. In fact, when an experiment began in Baltimore County, Maryland, some commanders preferred that officers start with alarm projects. Data documenting repeat alarm calls by address were readily available, and commanders anticipated that solving alarm problems would be relatively simple and would bring considerable benefits compared to the investment of time.[12]

Surveys. Not to be overlooked in crime analysis is the use of community surveys to analyze problems. For example, an officer may canvass all the business proprietors in shopping centers on his or her beat. In Baltimore, an officer telephoned business owners to update the police department's after-hours business contact files. Although the officer did not conduct a formal survey, he used this task to also inquire about problems the owners might want to bring to police attention.

On a larger scale, a team of officers may survey residents of a housing complex or neighborhood known to have particular crime problems. The survey could assist in determining residents' priority concerns, acquiring information about hot spots, and learning more about residents' expectations of police.[13] Residents will also be more likely to keep the police abreast of future problems when officers leave their cards and encourage residents to contact them directly.

Many police agencies now provide citizens with crime analysis information via the Internet. As an example, Exhibit 6–1 shows the Tempe, Arizona, Police Department Crime Analysis Unit's Website homepage.

Response: Formulating Tailor-Made Strategies

After a problem has been clearly defined and analyzed, the officer confronts the ultimate challenge in problem-oriented policing: the search for the most effective way of dealing with it. The response may be quite simple (such as reprogramming a public telephone at a convenience store where drug dealers conduct their "business" in front of patrons to make only outgoing calls) or quite involved (e.g., screening and evicting some tenants from a housing complex; cleaning up a

Citizen surveys provide the police with valuable information about citizen concerns and police performance. (*Courtesy Reno, Nevada, Police Department*)

EXHIBIT 6–1

Tempe, Arizona, Police Department Crime Analysis Unit's Web Site Homepage

Tempe Police Department's Crime Analysis Unit

ABOUT CRIME ANALYSIS

Historically, the causes and origins of crime have been the subject of investigation by varied disciplines. Some factors known to affect the volume and type of crime occurring from place to place are:

Population density and degree of urbanization with size locality and its surrounding area.

Variations in composition of the population, particularly youth concentration.

Stability of population with respect to residents' mobility, commuting patterns, and transient factors.

Modes of transportation and highway system.

Economic conditions, including median income, poverty level, and job availability.

Cultural factors and educational, recreational, and religious characteristics.

Family conditions with respect to divorce and family cohesiveness.

Climate.

Effective strength of law enforcement agencies.

Administrative and investigative emphases of law enforcement.

Policies of other components of the criminal justice system (e.g., prosecutorial, judicial, correctional, and probational).

Citizens' attitudes toward crime.

Crime reporting practices of the citizenry.

Crime Analysis is defined as . . .

A set of systematic, analytical processes directed at providing timely and pertinent information relative to crime patterns and trend correlations to assist the operational and administrative personnel in planning the deployment of resources for the prevention and suppression of criminal activities, aiding the investigative process, and increasing apprehensions and the clearance of cases. Within this context, Crime Analysis supports a number of department functions including patrol deployment, special operations, and tactical units, investigations, planning and research, crime prevention, and administrative services (budgeting and program planning).—Steven Gottlieb et al., 1994, "Crime Analysis: From First Report to Final Arrest."

Types of Crime Analysis

> **Tactical crime analysis**: An analytical process that provides information used to assist operations personnel (patrol and investigative officers) in identifying specific and immediate crime trends, patterns, series, sprees, and hotspots, providing investigative leads, and clearing cases. Analysis includes associating criminal activity by method of the crime, time, date, location, suspect, vehicle, and other types of information.

> **Strategic**: Concerned with long-range problems and projections of long-term increases or decreases in crime (crime trends). Strategic analysis also includes the preparation of crime statistical summaries, resource acquisition, and allocation studies.

> **Administrative**: Focuses on provision of economic, geographic, or social information to administration.

Crime Analysis Personnel

The Tempe Police Department's Crime Analysis Unit is comprised of three full-time crime analysts and a full-time crime analysis clerk.   and   are the current Crime Analysts. The Tempe Police Department's Crime Analysis Unit performs all three types of Crime Analysis: Tactical, Strategic, and Administrative.

INTERESTING STATISTICS

The following reflect 2000 figures unless otherwise noted:

> The population of Tempe is 163,000.
>
> There were a total of 122,830 citizen generated calls for service 2000, up 3% from 1999.
>
> The citizen generated calls for service rate per person is 754 per 1,000 persons. It was 734 per 1,000 persons in 1999.
>
> The most common type of citizen generated call for service is the "burglary alarm" call. These calls are 11.6% of the total calls for service. Approximately 84.5% of alarm calls are false, 1.7% result in a report, and the remaining 15.6% of alarm calls are unknown in outcome (no evidence of a false alarm or of a crime committed). Officers spend an average of 18 minutes on an alarm call.
>
> The average amount of time it took an officer to respond to an emergency call for service is 6 minutes and 28 seconds.
>
> Approximately one report is generated for every five calls for service (not necessarily a criminal report).
>
> Tempe's crime rate is 9,353 Part I crimes per 100,000 persons.

REPORTS AND BULLETINS

Thematic map which shades neighborhoods according to number of Part I crimes reported that month. Part I crimes include Murder, Rape, Robbery, Aggravated Assault Burglary, Motor Vehicle Theft, Larceny, and Arson. Available monthly after the 20th for the previous month. Recently, the crime data have been somewhat behind due to processing which is beyond the control of the Crime Analysis Unit.

This page begins with a map of all the beats in the city. From there, you can click on the beat you are interested in and a "Beat Information" page about that beat will come up. These individual beat pages include a detailed map of reporting districts (RDs), links to annual information, and the Monthly Part I Crime report for the most recent month (available).

Thematic map which shades neighborhoods according to number of calls for service that month. A call for service is any request for police service. Available monthly after the 10th for the previous month.

Monthly ranking of apartment communities including location and number of units rated according to the ratio of calls for service per unit. Available monthly after the 10th for the previous month.

Note that the rankings included in this report do not imply that any one apartment community is "worse" than the others. These rankings are only a relative measure of calls for service reported to the police and are not a measure of the safety or quality of the apartment community.

2000 Reports!!! 1) Ranking of apartment communities according to calls for service per unit, also includes location and number of units. 2) Ranking of apartment communities according to Part I Property Crime per unit 3) Ranking of apartment communities according to Part I Violent Crime per unit.

Note that the rankings included in this report do not imply that any one apartment community is "worse" than the others. These rankings are only a relative measure of crime and other activity reported to the police and are not a measure of the safety or quality of the apartment community.

This report is a list of the top five types of calls for service and the frequencies for each of the 173 apartment communities (included in the monthly and annual ranking reports). This is a more detailed report than the Apartment Community Rankings that do not distinguish between types of calls for service at a community.

This report is a list of ALL of the Part I crime and the frequencies for each of the 173 apartment communities (included in the monthly and annual ranking reports). This is a more detailed report than the Apartment Community Rankings that do not distinguish between types of Part I crime at a community.

List of mobile home communities including location and number of units rated according to the ratio of calls for service per unit.

Note that the rankings included in this report do not imply that any one apartment community is "worse" than the others. These rankings are only a relative measure of calls for service reported to the police and are not a measure of the safety or quality of the apartment community.

List of the number of calls for service at each elementary, middle, and high school in Tempe.

List of the number of calls for service at each Park in Tempe.

1999 REPORT ON POLICING

The 1999 Report on Policing in Tempe is here!!! This report includes information about 1999 crime and calls for service in Tempe. It also contains the results of the **1999 Citizen Survey**! It contains historical information as well as includes analyses of specific types of crimes, e.g. theft from vehicle, motor vehicle theft, domestic violence, and calls for service, e.g. burglary alarms, traffic accidents. Additionally, it contains information on specific types of locations like parks, schools, apartment communities, mobile home communities, the Arizona Mills Mall and others. Finally, individual beat analyses are included in the report that contain the number of crimes, number of calls for service, top five types of each and a month chart for each beat. The file is in PDF Format and requires Acrobat Reader.

The 1998 Report on Policing in Tempe is here!!! This report includes information about 1998 crime and calls for service in Tempe. It contains historical information as well as includes analyses of specific types of crimes, e.g. robbery, theft from vehicle, motor vehicle theft, domestic violence, and calls for service, e.g. burglary alarms, traffic accidents. Additionally, it contains information on specific types of locations like parks, schools, apartment communities, mobile home communities, the Arizona Mills Mall and others. Finally, individual beat analyses are included in the report that contain the number of crimes, number of calls for service, top five types of each and a month chart for each beat.

neighborhood that is overcome with graffiti, debris, and junk cars; taking legal action to create a curfew; or condemning and razing a drug house). A number of examples of responses are provided in the two COPPS case studies provided later.

This stage of the S.A.R.A. process focuses on developing and implementing responses to the problem. Before entering this stage, an agency must overcome the temptation to implement a response prematurely and be certain that it has thoroughly analyzed the problem; attempts to fix problems quickly are rarely effective in the long term.

To develop tailored responses, problem solvers should review their findings about the three sides of the crime triangle—victims, offenders, and location—and develop creative solutions that will address at least two sides of the triangle.[14] It is also important to remember that the key to developing tailored responses is making sure the responses are very focused and *directly linked* to the findings from the analysis phase of the project.

Responses may be wide ranging and often require arrests (however, apprehension may not be the most effective solution), referral to social service agencies, or changes in ordinances.

Assessment: Evaluating Overall Effectiveness

Finally, in the *assessment* stage, officers evaluate the effectiveness of their responses. A number of measures have traditionally been used by police agencies and community members to assess effectiveness. These include numbers of arrests; levels of reported crime; response times; clearance rates; citizen complaints; and various workload indicators, such as CFS and the number of field interviews conducted.[15] Several of these measures may be helpful in assessing the impact of a problem-solving effort.

A number of nontraditional measures will also shed light on whether a problem has been reduced or eliminated, such as the following:

- Reduced instances of repeat victimization
- Decreases in related crimes or incidents
- Neighborhood indicators, which can include increased profits for businesses in the target area, increased usage of the area, increased property values, less loitering and truancy, and fewer abandoned cars
- Increased citizen satisfaction regarding the handling of the problem, determined through surveys, interviews, focus groups, electronic bulletin boards, and so on
- Reduced citizen fear related to the problem[16]

Assessment is obviously key in the S.A.R.A. process; knowing that we must assess the effectiveness of our efforts emphasizes the importance of documentation and baseline measurement. Supervisors can help officers assess the effectiveness of their efforts.

A COLLABORATIVE APPROACH: "COPPS"

BASIC PRINCIPLES

The two concepts of community oriented policing and problem-oriented policing are separate but complementary notions that can work together.

Both COP and POP share some important characteristics: (1) decentralization (to encourage officer initiative and the effective use of local knowledge); (2) geographically defined rather than functionally defined subordinate units (to encourage the development of local knowledge); and (3) close interactions with local communities (to facilitate responsiveness to, and cooperation with, the community).[17] Following is a definition that accurately captures the essence of this concept:

> **Community Oriented Policing and Problem Solving (COPPS)** is a proactive philosophy that promotes solving problems that are criminal, affect our quality of life, or increase our fear of crime, as well as other community issues. COPPS involves identifying, analyzing, and addressing community problems at their source.[18]

In order for COPPS to succeed, the following measures are required:

- Conducting accurate community needs assessments
- Mobilizing all appropriate players to collect data and brainstorm strategies
- Determining appropriate resource allocations and creating new resources where necessary
- Developing and implementing innovative, collaborative, comprehensive programs to address underlying causes and causal factors
- Evaluating programs and modifying approaches as needed[19]

IMPLEMENTING COPPS

DEPARTMENTWIDE VERSUS EXPERIMENTAL DISTRICTS

Since COPPS came into being, most police executives have implemented the strategy throughout the entire agency; some executives, however, have attempted to introduce the concept in a small unit or an experimental district,[20] often in a specific geographic area of the jurisdiction.

It is strongly argued that the departmentwide implementation of COPPS be used. When COPPS has been established as a distinct unit within patrol rather than departmentwide, the introduction of this "special unit" seems to exacerbate the conflict between community policing's reform agenda and the more traditional outlook and hierarchical structure of the agency. A perception of elitism is created—a perception that is ironic, because COPPS is meant to close the gap between patrol and special units and to empower and value the rank-and-file patrol officer as the most important functionary of police work.

The key lesson from research in implementation, however, is that there is no golden rule or any universal method to ensure the successful adoption of COPPS. Two general propositions are important, however, for implementing the concept: the role of the rank-and-file officer and the role of the environment (or "social ecology") where COPPS is to be implemented.[21] The social ecology of COPPS includes both the internal/organizational and external/societal environments. Both of these factors are discussed next.

PRINCIPAL COMPONENTS OF SUCCESSFUL IMPLEMENTATION

Moving an agency from the reactive, incident-driven mode to COPPS is a complex endeavor, often requiring a complete change in the culture of the police organization. Four principal components of **implementation** profoundly affect the way agencies do business: leadership and administration, human resources, field operations, and external relations.[22]

Leadership and Administration

It is essential that chief executives communicate the idea that COPPS is departmentwide in scope. To get the whole agency involved, the chief executive must adopt four practices as part of the implementation plan:

1. Communicate to all department members the vital role of COPPS in serving the public. They must describe why handling problems is more effective than just handling incidents.
2. Provide incentives to all department members to engage in COPPS. This includes a new and different personnel evaluation and reward system as well as positive encouragement.
3. Reduce the barriers to COPPS that can occur. Procedures, time allocation, and policies all need to be closely examined.
4. Show officers how to address problems. Training is a key element of COPPS implementation. The executive must also set guidelines for innovation. Officers must know they have the latitude to innovate.[23]

Middle managers (captains and lieutenants) and first-line supervisors (sergeants) also play a crucial role in planning and implementing COPPS and in encouraging their officers to be innovative, take risks, and be creative.[24] First-line supervisors and senior patrol officers seem to generate the greatest resistance to community policing, largely because of long-standing working styles cultivated from years of traditional police work and because these officers can feel disenfranchised by a management system that takes the best and brightest out of patrol and that (they often believe) has left them behind.

Furthermore, the mechanisms that motivate, challenge, reward, and correct employees' behaviors must be compatible with the principles of COPPS. These include *recruiting, selection, training, performance evaluations, promotions, honors and awards*, and *discipline*, all of which should be reviewed to ensure that they promote and support the tenets of COPPS. First, recruiting literature should reflect the principles of COPPS. A job task analysis identifying the new knowledge, skills, and abilities should be conducted and become a part of the testing process for entry-level employees. COPPS should also be integrated into academy training, field-training programs, and in-service training. Performance evaluations and reward systems should reflect new job descriptions and officers' application of their COPPS training. Finally, promotion systems should be expanded from their usual focus on tactical decision making and include knowledge of the research on community policing and test an officer's ability to apply problem solving to various crime and neighborhood problems.

External Relations

Collaborative responses to neighborhood crime and disorder are essential to the success of COPPS. This also requires new relationships and the sharing of information and resources between the police and *the community, local government agencies, service providers*, and *businesses*. The *media* also provide an excellent opportunity for police to educate the community about COPPS and crime and disorder.

Political support is another essential consideration when implementing COPPS. The political environment varies considerably, say, with the strong-mayor and council-manager forms of government. These and other rapidly changing political environments make the implementation of COPPS more difficult—especially when we add to the mix the at-will employment of most police executives.

A BROADER ROLE FOR THE STREET OFFICER

A major departure of problem-oriented policing from the conventional style lies with POP's view of the line officer, who is given much more discretion and decision-making ability and is trusted with a much broader array of responsibilities.

Problem-oriented policing values "thinking" officers, urging that they take the initiative in trying to deal more effectively with problems in the areas they serve. This concept effectively uses the potential of college-educated officers, "who have been smothered in the atmosphere of traditional policing."[25] It also gives officers a new sense of identity and self-respect; they are more challenged and have opportunities to follow through on individual cases—to analyze and solve problems, which will give them greater job satisfaction. Using patrol officers in this manner also allows the agency to provide sufficient challenge for both better-educated officers and those who remain unpromoted patrol officers throughout their entire careers.[26]

Under problem-oriented policing, officers continue to handle calls, but they also do much more. They combine the information gathered in their responses to incidents with information obtained from other sources to get a clearer picture of the problem. They then address the underlying conditions. If they are successful in ameliorating these conditions, fewer incidents may occur; those that do occur may be less serious. The incidents may even cease. At the very least, information about the problem can help police to design more effective ways of responding to each incident.

POP also challenges those personnel who remain street-level patrol officers throughout their entire careers. Therefore, police administrators ought to be recruiting people as police officers who can "serve as mediators, as dispensers of information, and as community organizers."[27]

DID IT SUCCEED? EVALUATING COPPS

Rigorous **evaluation** is also an essential component of the COPPS initiative. Until rigorous evaluations are completed, there will be no clear verdict about whether the COPPS approach made a difference in controlling crime and disorder. An evaluation will also help to determine whether a crime-prevention initiative has achieved such goals as reducing crime and the fear of crime, has raised the community's quality of life, and is worthy of continued funding.

Evaluation of outcomes of problem-solving projects is not the same as the assessment stage of the S.A.R.A. problem-solving process. Evaluation is the more overarching concept, and it involves large projects—surveys, performance evaluations, and so on.

Broken down into simplified steps, evaluation asks some or all of the following questions:

- What is the problem? (Define it by such measures as community indicators, police data, and public surveys.)
- How does the project intend to address the problem? (Look at the project's goal statements.)
- What does the project do to resolve the problem? (Look at the project's objectives.)
- How does the project carry out its objectives? (Look, for example, at collaborative efforts among the police, other governmental agencies, and private businesses.)
- What impact (over time) does the prevention project have on the problem? (For example, over a specified time period, reported crimes increased or decreased by what percent?)[28]

The evaluative criteria employed in the professional policing model, such as crime rates, clearance rates, and response times, have been problematic when applied to the professional model itself and are even less appropriate for the COPPS model. These measures do not gauge the effect of crime-prevention efforts. A decrease in the reliance on these quantitative measures of police success is also important because communities will differ in the services they desire, depending on the particular characteristics of individual communities.[29]

Several criteria can assist in assessing the success of a COPPS effort. An effective COPPS strategy has a positive impact on reducing neighborhood crime, allays citizen fear of crime, and enhances the quality of life in the community.[30] Assessing the effectiveness of COPPS efforts includes determining whether problems have indeed been solved and how well the managers and patrol officers have used community partnerships.

COPPS evaluations can include *outcome* measures and *impact* measures. Outcome measures can include the following:

- Control of crime—compare, for example, the present rates of serious and violent crime with those of an earlier time period. Included in this category are behavioral changes, such as increased use of the once drug-infested park or increases in students attending drug-education programs.
- Citizen satisfaction with police services and a decrease in levels of fear of crime, which are measured with basic survey techniques (discussed later). Content analysis of media portrayals of police work, analysis of letters from citizens, and analysis of citizen complaints can also be used to evaluate citizen satisfaction with police services.[31] Outcome measures can also reflect environmental changes (such as number of street lights installed, traffic patterns altered, graffiti removed).

Impact measures describe and monitor changes in community indicators (such as community efforts to prevent or deal with crime and violence, community perception of quality of life, level of business, and other activity in the area).[32]

AN IMPORTANT COROLLARY OF COPPS: CRIME PREVENTION

An important corollary of COPPS—a very important and rapidly developing concept that all police officers should understand—is crime prevention. It stands to reason that it is preferable as well as much less expensive (in terms of both financial and human resources) to prevent a crime from occurring in the first place rather than having to try to solve the offense and arrest, prosecute, and possibly incarcerate the offender. Crime prevention shifts a police organization's purpose. Once the question becomes "How can we prevent the next crisis?" all kinds of approaches become possible.[33]

Crime prevention once consisted primarily of exhorting people to "lock it or lose it" and giving advice to citizens about door locks and window bars for their homes and businesses. It typically was (and often still is) an add-on program or appendage to the police agency, which normally included a few officers who were trained to go to citizens' homes and perform security surveys or engage in public speaking on the subject of target hardening. But, again, times are rapidly changing in this regard.

Crime prevention and COPPS are close companions, attempting to define the problem, identify the contributing causes, seek out the proper people or agencies to assist in identifying potential solutions, and work as a group to implement the solution. The problem drives the solution.[34] At its heart, COPPS is about preventing crime.

Next we briefly discuss two important aspects of crime prevention: crime prevention through environmental design and repeat victimization.

CRIME PREVENTION THROUGH ENVIRONMENTAL DESIGN

Crime prevention through environmental design (CPTED) is defined as the "proper design and effective use of the environment that can lead to a reduction in the fear and incidence of crime, and an improvement in the quality of life."[35] At its core are three principles that support problem-solving approaches to crime:

- *Natural Access Control*—employing elements such as doors, shrubs, fences, and gates to deny admission to a crime target and to create a perception among offenders that there is a risk in selecting the target.
- *Natural Surveillance*—placing windows, lighting, and landscaping properly to increase the ability to observe intruders as well as regular users, allowing observers to challenge inappropriate behavior or report it to the police or the property owner.
- *Territorial Reinforcement*—using elements such as sidewalks, landscaping, and porches to distinguish between public and private areas and to help users exhibit signs of ownership that send hands-off messages to would-be offenders.[36]

Five types of information are needed for CPTED planning:

1. *Crime analysis information:* This can include crime mapping, police crime data, incident reports, and victim and offender statistics.
2. *Demographics:* This should include statistics about residents such as age, race, gender, income, and income sources.
3. *Land use information:* This includes zoning information (such as residential, commercial, industrial, school, and park zones) as well as occupancy data for each zone.
4. *Observations:* These should include observations of parking procedures, maintenance, and residents' reactions to crime.
5. *Resident information:* This includes resident crime surveys and interviews with police and security officers.[37]

STUDYING PREY: REPEAT VICTIMIZATION

Our society—including the police—gives far greater attention to criminal offenders than to crime victims. Just as at the zoo, where there always seem to be more spectators around the lions and tigers than around wildebeests and antelope, more attention is focused on the predators than their prey. There is, however, an evolving body of research in Great Britain indicating that crime victims should be placed on the national agenda. In the United States—where POP has spread across

The glass stairwell in this parking garage demonstrates how natural surveillance can be designed into a facility.

the country—patterns of **repeat victimization (RV)** have not been examined or assimilated into problem solving. Therefore, police officers in the United States would benefit from this developing body of knowledge, which can play a major role in crime prevention and analysis.

The premise underlying RV is that if the police want to know where a crime will occur next, they should look where it happened last. RV is not new; police officers have always been aware that the same people and places are victimized again and again. What is new, however, are attempts abroad to incorporate repeat victimization knowledge into formal crime-prevention efforts.

One in three burglaries reported in the United States is a repeated burglary of a household. Furthermore, a 48 percent RV was found for sexual incidents (including grabbing, touching, and assault), 43 percent for assaults and threats, and 23 percent for vehicle vandalism.[38] A study of white-collar crime[39] indicated that the same people are victims of fraud and embezzlement time and time again and that banks that have been robbed also have high rates of repeat victimization.[40]

These data show that providing crime-prevention assistance to potential victims is not only morally justifiable in most instances, it is also an efficient and practical way of allocating limited police resources.

Why would, say, a burglar return to burgle the same household again? One could argue that, for several reasons, the burglar might be stupid *not* to return: Temporary repairs to a burgled home will make a subsequent burglary easier, the burglar is familiar with the physical layout and surroundings of the property, the burglar knows what items of value were left behind at the prior burglary, and the burglar also knows that items that were taken at an earlier burglary are likely to have been replaced through insurance policies.

Repeat victimization is arguably the best single predictor routinely available to the police in the absence of specific intelligence information. A small percentage of victims account for a disproportionate number of victimizations.[41]

A RELATED ATTEMPT AT CRIME PREVENTION: DARE

A related attempt to prevent crime should be discussed briefly: the well-known Drug Abuse Resistance and Education program, or DARE. The program, launched in 1983 and now administered by the police in 80 percent of schools, has not fared well among researchers. Over the past decade, studies have repeatedly shown that the $226 million program has little effect on keeping kids from abusing drugs. Indeed, the mayor of Salt Lake City recently halted the city's budget for DARE, declaring it "a complete waste of money, a fraud on the American people."[42] One of its key flaws, researchers allege, has been that students are taught that all drugs are equally dangerous; when students find that not to be true, the DARE message is undercut.[43]

Now DARE is admitting it needs a new direction. Officials are revamping the program, reducing the lecturing role of local officers, and involving kids in a more active way. A new curriculum is being developed for use in some schools that will show brain scans after drug use to demonstrate the harm, shift officers into more of a coaching role, and have kids engage in role playing about peer pressure.[44]

COPPS CASE STUDIES

Following are two excellent case studies of COPPS efforts using the S.A.R.A. model, the first in Tulsa, Oklahoma,[45] and the second in San Diego, California.[46] Note in each case the variety of police responses that were used to combat crime and disorder instead of the police merely reactively showing up at crime scenes, taking offense reports, and leaving the scene.

AMELIORATING JUVENILE PROBLEMS IN TULSA

Scanning: North Tulsa experienced consistently higher crime rates than the rest of the city. Nearly half of the violent crimes that were reported occurred in this section of the city—a depressed, low-income area lacking in adequate services. The Tulsa Housing Authority was established to support the city's low-income public housing. In an attempt to determine the nature of the crime problem in North

Tulsa, a special management team of Tulsa police officials conducted a study and decided to concentrate on five public-housing complexes where high crime rates and blatant street dealing existed.

Analysis: A residential survey conducted by patrol officers revealed that 86 percent of the occupants lived in households headed by single females. Officers in the target area noticed large groups of school-age youth in the housing complexes who appeared to be selling drugs during school hours. A comparison of the dropout and suspension rates in North Tulsa schools with those in other areas of the city determined that the city's northernmost high school, serving most of the high school–age youth in the five complexes, had the highest suspension (4.4 percent) and dropout (10 percent) rates of any school in the city. It also reported the highest number of pregnant teenagers in the school system. Few of the juveniles observed in the complexes had legitimate jobs, and most of them appeared to be attracted to drug dealing by the easy money.

Supervisors at Uniform Division North placed volunteers into two-officer foot teams, assigned to the complexes on eight-hour tours. The teams established a rapport with residents and assured them that police were present to ensure their safety. Within a month, officers verified juvenile involvement in drug trafficking. A strategy was needed to provide programs to deter youth from selling or using drugs.

Response: Officers S. and N., assigned to the foot patrol at one of the complexes, believed that the youth needed programs that would improve their self-esteem, teach them values, and impart decision-making skills. Eighty-six percent of the boys came from homes without fathers. To provide positive role models for young men, the officers started a Boy Scout troop in the complex for boys between 11 and 17 years of age. Officer N., a qualified Boy Scout leader, and Officer S. began meeting with the boys on Saturdays in a vacant apartment.

Officers J. and E. also organized a Boy Scout troop. In addition, they started a group that worked to raise money for needy residents and police-sponsored youth activities. Officer J. spoke at civic group meetings and local churches throughout the city to solicit donations and increase awareness of the needs of young people on the city's north side. Volunteers came from the churches and the civic groups where Officer J. spoke.

Officers B. and F., foot patrol officers at another complex, developed plans for unemployed young people. Officer B. organized a group called the Young Ladies Awareness Group that hosted weekly guest speakers invited to come and teach different job-related skills. Programs instructed young women how to dress for job interviews and employment; role-playing officers demonstrated proper conduct during interviews. The women were also instructed in resume writing and makeup, hair care, and personal hygiene. Officer F. worked with a government program that sponsored sessions on goal setting and self-esteem building to prepare young people to enter job training programs. Officer F. also helped area youth apply for birth certificates and arranged for volunteers from the Oklahoma Highway Patrol and a local school to help teach driver's education. Officers even provided funds for some young people who were unable to pay the fees to obtain birth certificates or driver's licenses. Officers F. and B. also tried to explain the value of an education and persuade youth in their complex to return to school. Unfortunately, too often parents appeared unconcerned when their children were missing classes.

The foot patrol officers became involved in a day camp project conducted at the "Ranch," a 20-acre northside property confiscated by police from a convicted drug dealer. The project used the property for a day camp for disadvantaged youth recruited from the target projects. Tulsa's mayor and chief of police came to the Ranch to meet with the youth, as did psychologists, teachers, ministers, and celebrities. Guests tried to convey the value of productive and drug-free lives, among other ethical values.

To combat dropout and suspension problems, a program called Adopt a School involved police officers patrolling the schools during classes, not to make arrests but rather to establish rapport with the students. The program was intended to reduce the likelihood of student involvement in illegal activities.

Assessment: The police noted a decline in street sales of illegal drugs in the five target complexes. Youth reacted positively to the officers' efforts to help them, and the programs seemed to deter them from drug involvement. The police department continued to address the problems of poor youth in North Tulsa. Foot patrol officers met with the Task Force for Drug Free Public Housing to inform the different city, county, and statewide officials of the needs of youth in public housing. Other social service agencies began working with the police department, establishing satellite offices on the north side of the city, scheduling programs, and requesting police support in their efforts.

ADDRESSING DRUGS AND GUNS IN SAN DIEGO

Scanning: Located in the southeast section of San Diego in a residential neighborhood of mostly apartment buildings, a privately owned three-story complex contained more than 75 units. The complex was divided into four buildings with a small grassy commons area between the buildings. An on-site manager, employed by the building management company, was responsible for maintenance. Shortly after the complex opened, serious problems with drug using and drug dealing became evident.

Initially, officers focused on making arrests, but suspects often evaded police by running into the apartments or by running through a nearby canyon, open field, or row of apartments. Police determined that drug users and dealers congregated in and around two complex laundry rooms because officers had found small plastic bags, glass pipes, and used matchbooks—all signs of crack cocaine use—in these rooms. Police considered each of the locations an easy place to make an arrest, as there was usually someone around who was either in possession of drugs or under their influence.

Soon, gunfire and gang-related violence had become common occurrences at the complex during evening hours. During one three-week period, police records indicated nightly shootings. Police learned through an informant and confirmed through undercover surveillance that the complex had become a major supply source of crack cocaine for several area gangs. Arrests provided only temporary relief, so the officers decided to implement problem-oriented policing.

Analysis: Officers A. and W. spoke with the apartment manager and suggested that he place locks on the laundry room doors and install additional lighting in the center of the complex. Next, the officers evaluated their arrest and field interview data, which helped them to identify many dealers, gangs, and tenants, who were collaborating with dealers. Evidence indicated that the manager had been dissuading residents from contacting police, saying he would deal with the problem. The officers then contacted the complex's management company and requested a key to a vacant apartment so they could observe drug dealers. The officers later uncovered a gun-running operation and in a subsequent investigation seized a large weapons cache of mostly handguns, an Uzi machine gun, a MAC-10 automatic pistol, and several sawed-off shotguns. Officers A. and W. sought search warrants for the apartments where they had observed drug dealing. When they informed the management company of their findings, its representatives offered to cooperate, agreeing to evict problem tenants and install security doors on the laundry room.

Response: With the assistance of the Special Weapons and Tactics (SWAT) unit, five search warrants were executed simultaneously in the complex. Numerous guns and large quantities of drugs and drug paraphernalia were seized and numerous eviction notices were immediately served on the problem tenants. During the legal eviction process, Officers A. and W. continued to work with the private management company and maintain their surveillance, but drug users and sellers continued to congregate on the apartment grounds and in the general vicinity. The patrol officers went directly to the owner of the complex and learned that he was unaware of the situation. Once informed, he fired the apartment manager and subsequently replaced the management company. The new company hired security guards, improved the apartment grounds, and initiated a new tenant screening process.

Assessment: At the time of this report, the complex was virtually drug free. Residents were mostly families with children, and the security guards reported no drug or gang activity.

SUMMARY

This chapter has set out the basic principles and strategies of the community polic-ing and problem-oriented policing concepts. It also offered a combined approach, which is the current era of policing and represents the best strategies for the future: community oriented policing and problem solving, or COPPS. Blending these two concepts results in a better, more comprehensive approach to providing quality police service, combining the emphasis on forming a police-community partnership to fight crime with the use of the problem-solving S.A.R.A. process. It was shown that two very important components of this philosophy are the expanded role of the street officer and the focus on crime analysis.

ITEMS FOR REVIEW

1. Define community policing and the major areas in which this concept differs from traditional policing.
2. List and explain the four parts of the problem-solving process for police.
3. Briefly describe the importance of implementation and evaluation of COPPS.
4. Explain what is meant by crime prevention and how it relates to community policing and problem solving.
5. Provide an overview of successful case studies of COPPS at work with juvenile and drug problems.

NOTES

1. W. L. Melville Lee, *A History of Police in England* (London: Methuen, 1901), Chapter 12.
2. Robert Trojanowicz and Bonnie Bucqueroux, *Community Policing: A Contemporary Perspective* (Cincinnati: Anderson, 1990), p. 154.
3. Mark H. Moore and Robert C. Trojanowicz, *Corporate Strategies for Policing*. U.S. Department of Justice, National Institute of Justice (Washington, D.C.: U.S. Government Printing Office, 1988), pp. 8–9.
4. Herman Goldstein, "Problem-Oriented Policing" (paper presented at the Conference on Polic-ing: State of the Art III, National Institute of Justice, Phoenix, Ariz., June 12, 1987).
5. *Ibid.*, pp. 43–52.
6. Noah Fritz, "Crime Analysis" (Tempe, AZ: Tempe Police Department, no date), p. 9.
7. John Eck, "A Dissertation Prospectus for the Study of Characteristics of Drug Dealing Places" (College Park, University of Maryland, November 1992).
8. Barbara Webster and Edward F. Connors, "Community Policing: Identifying Problems" (Alexandria, VA: Institute for Law and Justice, March 1991), p. 9.
9. See Lawrence W. Sherman, Patrick R. Gartin, and Michael E. Buerger, "Hot Spots of Predatory Crime: Routine Activities and the Criminology of Place," *Criminology* 27 (1989): 27.
10. *Ibid.*, p. 36.
11. William Spelman, *Beyond Bean Counting: New Approaches for Managing Crime Data* (Washington, D.C.: Police Executive Research Forum, January 1988).
12. Webster and Connors, "Community Policing," p. 11.
13. For an example of this type of survey process, see William H. Lindsey and Bruce Quint, *The Oasis Technique* (Fort Lauderdale, FL: Florida Atlantic University/Florida International Univer-sity Joint Center for Environmental and Urban Problems, 1986).

14. Rana Sampson, "Problem Solving," in *Neighborhood-Oriented Policing in Rural Communities: A Program Planning Guide* (Washington, D.C.: U.S. Department of Justice, Office of Justice Programs, Bureau of Justice Assistance, 1994), p. 4.

15. Darrel Stephens, "Community Problem-Oriented Policing: Measuring Impacts," in Larry T. Hoover, *Quantifying Quality in Policing* (Washington, D.C.: Police Executive Research Forum, 1995).

16. U.S. Department of Justice, Office of Community Oriented Policing Services, *Problem Solving Tips: A Guide to Reducing Crime and Disorder Through Problem-Solving Partnerships*, p. 20.

17. Moore and Trojanowicz, *Corporate Strategies for Policing*, p. 11.

18. Kenneth J. Peak and Ronald W. Glensor, *Community Policing and Problem Solving: Strategies and Practices*, 3d ed. (Upper Saddle River, NJ: Prentice Hall, 2002), p. 99.

19. *Ibid.*

20. Goldstein, *Problem-Oriented Policing* (New York: McGraw-Hill, 1990), p. 172.

21. Gregory Saville and D. Kim Rossmo, "Striking a Balance: Lessons from Community-Oriented Policing in British Columbia, Canada," unpublished manuscript, June 1993, pp. 29–30.

22. Ronald W. Glensor and Kenneth J. Peak, "Implementing Change: Community-Oriented Policing and Problem Solving," *FBI Law Enforcement Bulletin 7* (July 1996): 14–20.

23. John E. Eck and William Spelman, *Problem-Solving: Problem-Oriented Policing in Newport News* (Washington, D.C.: Police Executive Research Forum, 1987), pp. 100–101.

24. *Ibid.*, p. 9.

25. Herman Goldstein, "Toward Community-Oriented Policing," *Crime and Delinquency* 33 (1987): 6–30.

26. *Ibid.*, p. 21.

27. *Ibid.*

28. Adapted from *ibid.*, p. 4.

29. Community Policing Advisory Committee, *Community Policing Advisory Committee Report* (Victoria, B.C. Canada, Author, 1993), p. 61.

30. U.S. Department of Justice, Bureau of Justice Assistance, The Community Policing Consortium, *Understanding Community Policing: A Framework for Action* (Washington, D.C.: Author, 1993), p. 86.

31. California Department of Justice, Attorney General's Office, Crime Prevention Center, *COPPS: Community Oriented Policing and Problem Solving* (Sacramento, CA: Author, November 1992), pp. 90–91.

32. *Ibid.*, pp. 4–5.

33. Jim Jordan, "Shifting the Mission: Seeing Prevention as the Strategic Goal, Not a Set of Programs," *Subject to Debate* (Washington, D.C.: Police Executive Research Forum, December 1999), pp. 1–2.

34. *Ibid.*, p. 8.

35. C. R. Jeffrey, *Crime Prevention Through Environmental Design* (Beverly Hills, CA: Sage, 1971).

36. National Crime Prevention Council, *Designing Safer Communities: A Crime Prevention Through Environmental Design Handbook* (Washington, D.C.: Author, 1997), pp. 7–8.

37. *Ibid.*, p. 3.

38. G. Farrell and W. Sousa, "Repeat Victimization in the United States and Ten Other Industrialized Countries" (paper presented at the National Conference on Preventing Crime, Washington, D.C., October 13, 1997).

39. *Ibid.*

40. *Ibid.*

41. G. Farrell, "Preventing Repeat Victimization." In *Building a Safer Society*, ed. M. Tonry and D. P. Farrington (Chicago: University of Chicago Press, 1995), pp. 469–534.

42. Rocky Anderson, quoted in Claudia Kalb, "DARE Checks into Rehab," *Newsweek*, February 26, 2001, p. 56.

43. *Ibid.*

44. *Ibid.*

45. U.S. Department of Justice, Bureau of Justice Assistance, "Problem-Oriented Drug Enforcement: A Community-Based Approach for Effective Policing." Washington, D.C.: Police Executive Research Forum, October 1993, pp. 45–47.

46. *Ibid.*, pp. 27–28.

CHAPTER 7

Criminal Investigation

The Science of Detection

Key Terms and Concepts

Bertillon system
Corpus delicti
Criminalistics
Cybercrooks
Dactylography
Detective

DNA testing
Firearms identification
Forensic entomology
Forensic science
Frye Standard

Investigative stages
Modus operandi
Polygraph examination
Profiling
Stalking

Murder though it hath no tongue will speak . . .—*Shakespeare,* Hamlet, *Act II, Scene 2*

And the Lord said unto Cain, "Where is thy brother Abel"? [the first known recorded instance of a criminal interrogation] And he said, I know not; am I my brother's keeper? [the first known recorded instance of perjury] And He said, what hast thou done? The voice of thy brother's blood crieth unto me from the ground. [the first known recorded instance of criminal evidence]—*Genesis 4:9–10*

The truth is rarely pure and never simple.—*Oscar Wilde*

INTRODUCTION

To these chapter opening quotes, one more by Ludwig Wittgenstein could have been added: "How hard I find it to see what is *right in front of my eyes!*"[1] Investigating crimes has indeed become a complicated art as well as a science, as will be seen in this chapter.

The art of sleuthing has long fascinated the American public. Certainly the O. J. Simpson trial of 1995, which received more media attention than any other case before or after, heightened America's opportunity and fervor to become educated about detective work and forensic science. For decades, Americans have feasted on the exploits of dozens of well-known television- and novel-based detectives. Some masterminds, such as Sherlock Holmes and Agatha Christie's Hercule Poirot, epitomized the intellectual detective, using a minimum of clues and a lot of psychology and logic toward effecting an arrest. Others, such as Clint Eastwood's Dirty Harry character, relied more on weaponry and machismo, capturing bad guys—often using very violent means—while paying little attention to the rule of law (as will be covered in Chapter 9).

In reality, investigative work is largely misunderstood, often boring, and often overrated; it results in arrests only a fraction of the time; and it relies strongly on the assistance of witnesses and even some luck. Nonetheless, the related fields of forensic science and criminalistics are the most rapidly developing areas of policing—and probably in all of criminal justice. This is an exciting time to be in the investigative or forensic disciplines.

This chapter begins by defining forensic science and criminalistics and looking at their origins. Then we discuss the evolution of criminal investigation, emphasizing the identification of persons and firearms. Next we analyze the application of forensic science within the larger context of the criminal justice system, followed by a view of the necessary qualities possessed by detectives.

We then consider undercover police work, the use of polygraph testing, and DNA analysis. We examine behavioral science applications in investigations next, including criminal profiling, psychics, and hypnosis. Then we examine some recent investigative developments in the field, including forensic entomology, using criminals' own videos against them, and investigating stalkers. The chapter concludes with a review of methods and problems involving the Internet.

THE SCOPE OF FORENSIC SCIENCE AND CRIMINALISTICS

The terms "forensic science" and "criminalistics" are often used interchangeably. **Forensic science** is the broader term; it is that part of science used to answer legal questions. It is the examination, evaluation, and explanation of physical evidence in law. Forensic science encompasses pathology, toxicology, physical anthropology, odontology (development of dental structure and dental diseases), psychiatry, questioned documents, ballistics, tool work comparison, and serology (the reactions and properties of serums), among other fields.[2] **Criminalistics** is one branch of forensic science; it deals with the study of physical evidence related to crime. From such a study, a crime may be reconstructed.

Criminalistics is interdisciplinary, drawing upon mathematics, physics, chemistry, biology, anthropology, and many other scientific fields.[3] Dr. Paul Kirk, a leader in the criminalistics movement in this country, once remarked that "criminalistics is an occupation that has all of the responsibilities of medicine, the intricacy of the law, and the universality of science."[4] Criminalistics has occasionally reached plateaus, but on the whole it is a dynamic and progressive discipline.

Basically, the analysis of physical evidence is concerned with identifying traces of evidence, reconstructing criminal acts, and establishing a common origin of samples of evidence. DeForest and colleagues[5] described the types of information that physical evidence can provide:

1. Information on the **corpus delicti** (or the "body of the crime"): physical evidence showing that a crime was committed—tool marks, broken doors and windows, ransacked homes, and missing valuables in a burglary or a victim's blood, a weapon, and torn clothing in an assault.

2. Information on the **modus operandi** (or method of operation): physical evidence showing means used by the criminal to gain entry, tools that were used, types of items taken, and other signs, such as urine left at the scene, an accelerant used at arson scenes, the way crimes are committed, and so forth. Many well-known criminals left their "calling card" at their crimes, either in terms of what they did to their victims or the physical condition of the crime scene.

3. Linking a suspect with a victim: one of the most important linkages, particularly with violent crimes. It includes hair, blood, clothing fibers, and cosmetics that may be transferred from victim to perpetrator. Items found in a suspect's possession, such as bullets or blood on a knife, can also be linked to a victim.

4. Linking a person to a crime scene: also a common and significant linkage. It includes fingerprints, gloveprints, blood, semen, hairs, fibers, soil, bullets, cartridge cases, toolmarks, footprints or shoeprints, tire tracks, and objects that belonged to the criminal. Stolen property is the most obvious example.

5. Disproving or supporting a witness's testimony: evidence can indicate whether or not a person's version of events is true. An example is a driver whose car matched the description of a hit-and-run vehicle. If blood is found on the underside of the car and the driver claims that he hit a dog, tests on the blood can determine whether the blood is from an animal or from a human.

6. Identification of a suspect: one of the best forms of evidence for identifying a suspect is fingerprints, which prove "individualization." Without a doubt, that person was at the crime scene.

ORIGINS OF CRIMINALISTICS

The study of criminalistics began in Europe. The first major book describing the application of scientific disciplines to criminal investigations was written in 1893 by Hans Gross, a public prosecutor and later a judge from Graz, Austria.[6] Translated into English in 1906, it remains a highly respected work in the field. Two prominent aspects of criminalistics, personal identification and firearms analysis, are discussed next.

PERSONAL IDENTIFICATION: ANTHROPOMETRY AND DACTYLOGRAPHY

Historically there have been two major systems for personal identification of criminals: anthropometry and dactylography. The former did not survive long; the latter, better known as fingerprint identification, is widely used throughout the world today.

Anthropometry was developed by Alphonse Bertillon (1853–1914) in 1882. This method, the first attempt at criminal identification that was thought to be reliable and accurate, was based on the facts that every human being differs from every other one in the exact measurements of their bodies and that the sum of these measurements yields a characteristic formula for each individual.[7] Bertillon, largely a failure at everything he tried, performed menial tasks in 1879 for the Paris police department, filing cards that described criminals so vaguely as to have little meaning—"stature: average . . . face: ordinary."[8]

Bertillon became increasingly frustrated with the senselessness of his work. He began comparing photographs of criminals and taking measurements of those who had been arrested, eventually concluding that if 11 physical measurements of a person were taken, the chances of finding another person with the same 11 measurements were 4,191,304 to 1.[9]

Bertillon's report to his superiors outlining his criminal identification system was not met with enthusiasm. His chief said, "Bertillon? If I am not mistaken, you are a clerk of the twentieth grade and have been with us for only eight months, right? And already you are getting ideas? Your report sounds like a joke."[10] In 1883, however, Bertillon's "joke" was given worldwide attention when it was implemented on an experimental basis; Bertillon correctly made his first criminal identification.[11]

Around the turn of the century, many countries abandoned anthropometry, or the **Bertillon system**, adopting the simpler and more reliable system of fingerprinting identification. Bertillon realized the potential of fingerprinting when, in 1902, he solved a murder case when he discovered the defendant's prints on a glass cupboard. Yet he persisted in his view that anthropometry was superior to dactylography. He explored other ideas as well, developing the prototype of the "mug shot" (to which he reluctantly added fingerprints) and pioneering police photography. But fingerprinting was eventually found superior to the Bertillon system (discussed later). After Bertillon's death in 1914, France became the last major country to replace anthropometry with dactylography as its system of criminal identification.[12] However, Bertillon's pioneering work in personal identification earned him a place in history; today Bertillon is considered the "father of criminal investigation."[13]

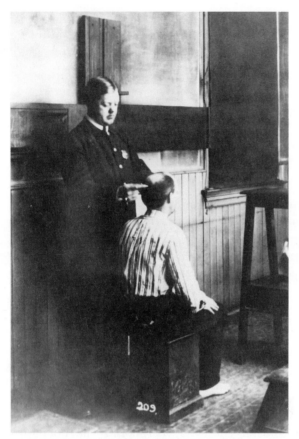

A police officer taking Bertillon measurements.
(*Courtesy St. Louis, Missouri, Police Department*)

FINGERPRINTS PREVAIL OVER BERTILLONAGE

Dactylography was first used in 1900 in England as a system of criminal identification; however, fingerprints have a long legal and scientific history. In the first century, a Roman lawyer named Quintilianus introduced a bloody fingerprint at trial, successfully defending a child accused of murdering his father.[14] Fingerprints were also used on contracts during the Chinese T'ang Dynasty in the eighth century. In 1684, in England, Dr. Nehemiah Grew first called attention to the system of pores and ridges in the hands and feet. And in 1823, John Perkinje, a professor at the University of Breslau, named nine standard types of fingerprint patterns and outlined a broad method of classification.[15] In spite of these developments, fingerprinting would not emerge as a prominent means of criminal investigation for another 75 years.

Beginning in 1858, William Herschel, a British official in India, requested the palmprints and fingerprints of those with whom he did business in an attempt to get people to keep their agreements. Over a 20-year period, Herschel noticed that the patterns of the lines on the fingerprints for an individual never changed, despite growth and other physical changes. Excited by the prospects of using fingerprints to identify criminals, he wrote in 1877 to the Inspector General of the Prisons of Bengal. The reply implied in kind terms that Herschel's idea was insane. Discouraged by this, Herschel never pursued his discovery.[16] Meanwhile, Henry Faulds, a Scottish physician teaching in Tokyo, had been interested in fingerprints for several years. After clearing a theft suspect by comparing his fingerprints with those found in a home, Faulds reported his findings in the journal *Nature* in 1880. Herschel read Faulds's article and published a reply, claiming credit for the discovery over 20 years before. A controversy arose that was never resolved; and still no official interest in using fingerprints was shown.

In 1888, Sir Francis Galton (1822–1911), a cousin of Charles Darwin, became interested in criminal investigation and contacted Herschel, who unselfishly turned over all of his files in the hope that Galton could apply fingerprinting to practical police uses.[17] In 1892, Galton published the first definitive book on dactylography, *Finger Prints*, wherein he presented statistical proof of the uniqueness of fingerprints and its applications.[18] In Argentina, in 1894, Juan Vucetich outlined his method of fingerprint classification. A disciple of Vucetich's, one Inspector Alvarez, obtained the first American criminal conviction based on fingerprints in 1892. They helped convict a woman who beat her two sons to death.[19]

The major breakthrough for fingerprints was made by Edward Henry (1850–1931), who went to India at the age of 23 and became Inspector General of Police in Nepal in 1891. Henry developed an interest in fingerprints. He instituted Bertillon's system, but he added fingerprints to the cards. In 1893, Henry began working on a simple, reliable method of classification, and in 1897 he recommended that anthropometry be dropped in favor of his own fingerprint classification system. Six months later, his recommendations were followed throughout British India. In 1901, Henry published his *Classification and Use of Finger Prints* and was appointed assistant police commissioner of London, rising to the post of commissioner two years later.[20]

THE JONES AND WEST CASES

In 1904, New York City Detective Sergeant Joseph Faurot was sent to England to study fingerprints, becoming the first foreigner trained in the Henry classification system. Upon his return to New York when he reported his findings, Faurot was told by his superiors to forget such scientific nonsense; he was transferred to a walking beat. In 1906, Faurot arrested a man dressed in formal evening wear but

FEDERAL BUREAU OF INVESTIGATION
UNITED STATES DEPARTMENT OF JUSTICE
J. Edgar Hoover, Director

History of the
"West Brothers" Identification..

Bertillon Measurements are not always a Reliable Means of Identification

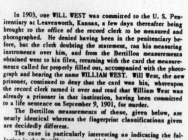

In 1903, one WILL WEST was committed to the U. S. Penitentiary at Leavenworth, Kansas, a few days thereafter being brought to the office of the record clerk to be measured and photographed. He denied having been in the penitentiary before, but the clerk doubting the statement, ran his measuring instruments over him, and from the Bertillon measurements obtained went to his files, returning with the card the measurements called for properly filled out, accompanied with the photograph and bearing the name WILLIAM WEST. Will West, the new prisoner, continued to deny that the card was his, whereupon the record clerk turned it over and read that William West was already a prisoner in that institution, having been committed to a life sentence on September 9, 1901, for murder.

The Bertillon measurements of these, given below, are nearly identical whereas the fingerprint classifications given are decidedly different.

The case is particularly interesting as indicating the fallacies in the Bertillon system, which necessitated the adoption of the fingerprint system as a medium of identification. It is not even definitely known that these two Wests were related despite their remarkable resemblance.

Their Bertillon measurements and fingerprint classifications are set out separately below:

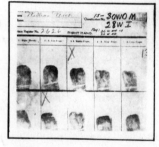
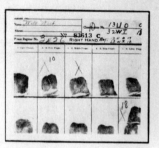

177.5; 188.0; 91.3; 19.8; 15.9; 14.8; 6.5; 27.5; 12.2; 9.6; 50.3
15- 30 W OM 13 ‹Ref: 30 W OM 13
28 W I   26 U OO

178.5; 187.0; 91.2; 19.7; 15.8; 14.8; 6.6; 28.2; 12.3; 9.7; 50.2
10- 13 U O O Ref: 13 U O 17
32 W I 18   28 W I 18

The West Brothers Case. (*Courtesy FBI*)

no shoes who was creeping out of a suite at the Waldorf-Astoria Hotel; the man claimed to be a respected citizen named James Jones. Faurot sent the man's fingerprints to Scotland Yard and learned that "James Jones" was actually Daniel Nolan, who had 12 prior convictions for hotel thefts and was wanted for burglary in England. Nolan, confronted with this information, confessed to several thefts in the Waldorf-Astoria and was sent to prison for seven years. Publicity surrounding this case greatly advanced the credibility of fingerprinting in America.[21]

But an even more important and amazing incident furthering the use of fingerprints in America was the West case. In 1903, Will West arrived at the federal penitentiary in Leavenworth, Kansas. While West was being processed into the institution, a staff member said a photograph was already on file for him, along with Bertillon measurements. West denied ever having been in Leavenworth. A comparison of fingerprints showed that despite nearly identical physical appearance and Bertillon measurements, the identification card on file belonged to a

William West who had been in Leavenworth since 1901. The incident served to establish the superiority of fingerprints over anthropometry as a system of personal identification.

FIREARMS IDENTIFICATION

Firearms are perennially the leading instrument of death for Americans, causing about 70 percent of the nation's 22,000 homicide deaths each year.[22] Furthermore, the Centers for Disease Control recently found in an eight-year-long study that 93 percent of all firearms-related deaths involve the intentional use of guns.[23] It is plain that the frequency of shootings in this country has made **firearms identification** very important.

In 1835, Henry Goddard, one of the last of the Bow Street Runners, made the first successful attempt to identify a murderer from a bullet recovered from the body of a victim. Goddard noticed that the bullet had a distinctive slight gouge on it and seized a bullet mold with a corresponding gouge at the home of a suspect; the defendant confessed.[24] In 1889, Professor Lacassagne removed a bullet from a corpse in France; upon examining it, he found seven grooves made as the bullet passed through the barrel of a gun. Shown the guns of a number of suspects, Lacassagne identified the one that could have left seven grooves. With this evidence, a man was convicted of the murder.[25]

In 1898, a German chemist named Paul Jeserich was given a bullet taken from the body of a man murdered near Berlin. After firing a test bullet from the defendant's revolver, Jeserich took microphotographs of the fatal and test bullets, later testifying that the defendant's revolver fired the fatal bullet. Although he was at the threshold of eminence, Jeserich did not pursue his technique further.[26] Attention gradually shifted to other aspects of firearms. In 1913, Professor Balthazard published a major article on firearms identification, noting that the firing pin, breechblock, extractor, and ejector all leave marks on cartridges and that these markings vary among different types of weapons. With World War I approaching, Balthazard's article did not receive wide attention.

Chicago witnessed the St. Valentine's Day Massacre in 1929. A special grand jury inquiring into the matter noted that there were no facilities for analyzing the numerous bullets and cartridge cases that were strewn about. As a result, several influential jury members raised funds to establish a permanent crime laboratory. Colonel Calvin Goddard (1858–1946), a retired army physician, was appointed director of the lab, which was established at the Northwestern University Law School.[27] Goddard is the person most responsible for raising the status of firearms identification to a science and for perfecting the bullet comparison microscope.[28]

Firearms identification goes beyond comparing a bullet found in the victim and a test bullet fired from the defendant's weapon. It also includes identification of types of ammunition, designing firearms, restoring obliterated serial numbers on weapons, and estimating the distance between a gun's muzzle and a victim when the weapon was fired.[29]

CONTRIBUTIONS OF AUGUST VOLLMER AND OTHERS

August Vollmer's contributions to the development of criminalistics and investigative techniques were considerable. In 1907, as police chief of Berkeley, California, he enlisted the services of a University of California chemistry professor named Loeb to identify a suspected poison during a murder investigation. Vollmer instituted a formal training program to ensure that his officers properly collected and preserved criminal evidence. Vollmer called upon scientists on campus on several other occasions. His support helped John Larson produce the first workable polygraph in 1921.

Vollmer also established the first full forensic laboratory in Los Angeles in 1923; the concept soon spread to other cities, including Sacramento (a state laboratory), San Francisco, and San Diego. Although they were small, the influence of these laboratories is still seen today. Furthermore, Vollmer's efforts in establishing a relationship between his police department and the university led to other scientists getting involved in forensic science. For example, Paul Kirk became so interested in criminalistics that he soon began offering courses in it as part of the biochemistry curriculum at the University of California at Berkeley.[30] Many of the graduates of that program went on to become criminalists and crime laboratory directors.

Other contributors included Albert Osborn, who in 1910 wrote *Questioned Documents*, a definitive work; Edmond Locard, who maintained a central interest in locating microscopic evidence; and Leone Lattes, who in 1915 developed a blood-typing procedure from dried blood—a key event in serology.[31]

THE EVOLUTION OF CRIMINAL INVESTIGATION

INVESTIGATIVE BEGINNINGS: THE ENGLISH CONTRIBUTION

To fully understand the development of criminal investigation in America, it is important to first understand the social, economic, political, and legal contexts in which it evolved. To do so, we will briefly discuss the impact of (1) the agricultural and industrial revolutions, (2) the Fielding brothers and their Bow Street Runners, and (3) London's Metropolitan Police.

It is probably difficult to see any relationship between agriculture and fingerprinting or surveillance, but as Swanson et al. observed,

> During the eighteenth century two events, an agricultural and industrial revolution, began a process of change that profoundly affected how police services were delivered and investigations conducted. Improved agricultural methods gave England increased agricultural productivity in the first half of the eighteenth century. Improvements in agriculture [led to] the Industrial Revolution, in the second half of the eighteenth century, because they freed people from farm work for city jobs. As the population of England's cities grew, slums also grew, crime increased and disorders became more frequent. Consequently, public demands for government to control crime grew louder.[32]

In 1748, Henry Fielding became chief magistrate of London's Bow Street, and in 1750 he established a small band of volunteer "thief takers," known as the Bow Street Runners. These homeowners hurried to the scenes of reported crimes and began investigations. In 1752, Fielding began publishing *The Covent Garden Journal*, which included descriptions of wanted persons. When Henry Fielding died in 1754, his blind brother, John, carried on his work for the next 25 years (although less successfully; the effectiveness of the amateur force declined during John's tenure). By 1785, the government was so impressed by the work of the Runners that each of them was paid a salary, and an attempt was made to expand the detective concept to other areas of London.

Following the creation of London's Metropolitan Police in 1829, the British public became suspicious and even hostile toward its new force. The French, they knew, had experienced oppression under their centralized police force. Thus, high standards were set for the London bobbies (discussed in Chapter 1); there were 5,000 dismissals and 6,000 forced resignations during the first three years of operation. High standards of conduct were expected and maintained. Despite the eventual reputation for fairness and professionalism the London force earned, the fear remained that the use of police "spies"—detectives in plain clothes—would reduce civil liberties.[33] Occasionally, bobbies were temporarily removed from patrol duties to investigate crimes on their beats, but the public was uneasy. As an example, in 1833, a sergeant was dismissed from duty after Parliament learned that he had infiltrated a radical group, acquired a leadership position, and argued for the use of violence.

Until 1842, Metropolitan constables competed with Bow Street Runners to investigate crimes; in that year, a regular detective branch was opened at Scotland Yard, superseding the Runners.[34] The detective force was to have no more than 16 investigators, and its operations were restricted because of concern about "clandestine methods."[35] Following a scandal in which three of four chief inspectors of detectives were convicted of taking bribes, in 1878 a separate, centralized Criminal Investigation Division (CID) was established at Scotland Yard. It was headed by an attorney; uniformed constables who had shown an aptitude for investigation were recruited to become CID detectives.[36]

INVESTIGATIVE TECHNIQUES COME TO AMERICA

Meanwhile, the success of the Metropolitan Police in London was not going unnoticed in America. Beginning with a paid, 24-hour professional police force in New York City in 1844, by 1880 virtually all major cities had a police force that used Peel's principles. Ironically, while the bobbies were striving for public acceptance that would allow them to work out of uniform, American police administrators were having difficulty getting their officers to accept uniforms (see Chapter 1). Only after the Civil War did the wearing of a uniform—which was invariably blue—become widely accepted by American police officers.[37]

Following the advent of private police in America in the mid-1840s (most notably Pinkerton's National Detective Agency), the concept of plainclothes officers spread in America. As early as 1845, New York City had 800 such officers; however, not until 1857 was the department authorized to appoint 20 patrol officers as "detectives."[38] In November 1857, the New York Police Department established a rogues' gallery—photographs of known offenders arranged by criminal specialty and offender height. By June 1858, it had over 700 photographs for detectives to study for future recognition.[39] Photographs in the rogues' gallery showed some offenders grimacing, puffing their cheeks, rolling their eyes, and otherwise trying to distort their appearance to thwart detectives and lessen the chance of later recognition.[40]

To assist its detectives, in 1884 Chicago established the nation's first municipal Criminal Investigation Bureau.[41] The Atlanta Police Department's Detective Bureau was organized in 1885 with a staff of one captain, one sergeant, and eight detectives.[42] In 1886, Thomas Byrnes, chief detective of New York City, published *Professional Criminals in America*, which included pictures, descriptions, and methods of all criminals known to him.[43] To supplement the rogues' gallery, Byrnes instituted the Mulberry Street Morning Parade. At 9:00 every morning, all criminals arrested in the past 24 hours were marched before his detectives, who were expected to take notes and to recognize the criminals later.[44] Byrnes had personal shortcomings, however; in 1894, he was forced to leave the department when he admitted that he had grown wealthy by tolerating gambling dens and brothels.[45]

STATE AND FEDERAL DEVELOPMENTS

From its earliest days, the federal government employed investigators to detect revenue violations, but their responsibilities were narrow and their numbers few.[46] In 1865, Congress created the U.S. Secret Service to combat counterfeiting. In 1903—two years after the assassination of President McKinley—the guarding of the president was made a permanent Secret Service responsibility.[47] In 1905, the California Bureau of Criminal Identification was established to share information about criminal activity; that same year, the state of Pennsylvania created a state police force. Serving as a prototype for modern state police organizations, one of the functions of the Pennsylvania State Police was to provide local police with assistance in investigations; that tradition continues among today's state police agencies.[48]

The forerunner of what was to become the Federal Bureau of Investigation (FBI) (discussed in Chapter 2) was created in 1908. In 1924, J. Edgar Hoover assumed leadership of the Bureau of Investigation; eleven years later Congress enacted legislation giving the FBI its present designation. Under Hoover, who understood the importance and uses of information, records, and publicity, the FBI became known for investigative efficiency. In 1932, the FBI established a crime laboratory and made its services free—they remain free of charge today to state and local police. In 1935, it opened its National Academy, providing training courses for state and local police as well. And in 1967, the National Crime Information Center (NCIC) was operationalized by the FBI, providing data on wanted persons and property stolen in all 50 states. These developments gave the FBI considerable influence over policing in America; Hoover and the FBI vastly improved policing practices in this country, keeping crime statistics and assisting investigations.[49]

FORENSIC SCIENCE AND THE CRIMINAL JUSTICE SYSTEM

INVESTIGATIVE STAGES AND ACTIVITIES

The police, more specifically investigators and criminalists, operate on the age-old theory that there is no such thing as a perfect crime; criminals either leave a bit of themselves (such as a hair or clothing fiber) at the crime scene or take a piece of the crime scene away with them. Thus it is the job of the police and the crime lab to unify their efforts and find that incriminating piece of evidence, which they can

A homicide investigation is the most important and challenging work performed by detectives. (*Courtesy NYPD Photo Unit*)

use in conjunction with other pieces of evidence to determine "whodunnit" and bring the guilty party to justice.

In the apprehension process, when a crime is reported or discovered, police officers respond, conduct a search for the offender (it may be a "hot" crime scene search, where the offender is likely present, a "warm" search in the general vicinity, or a "cold" investigative search), and check out suspects. If the search is successful, evidence for charging the suspect is assembled and the suspect is apprehended.[50] Cases not solved in the initial phase of the apprehension process are assigned either to an investigative specialist or, in smaller police agencies, to an experienced uniformed officer who functions as a part-time investigator. According to Weston and Wells,[51] what follows are the basic **investigative stages**:

> *The Preliminary Investigation:* This work is crucial, involving the first police officer at the scene. Duties to be completed include establishing whether a crime has been committed; securing from any witnesses a description of the perpetrator and his or her vehicle; locating and interviewing the victim and all witnesses; protecting the crime scene (and searching for and collecting all items of possible physical evidence); determining how the crime was committed and what the resulting injuries were, as well as the nature of property taken; recording in field notes and sketches all data about the crime; and arranging for photographs of the crime scene.
>
> *The Continuing Investigation:* This stage, which begins when preliminary work is done, includes follow-up interviews; developing a theory of the crime; analyzing the significance of information and evidence; continuing the search for witnesses; beginning contact with crime lab technicians and assessing their analyses of the evidence; conducting surveillances, interrogations, and polygraph tests as appropriate; and preparing the case for the prosecutor.
>
> *Reconstructing the Crime:* The investigator seeks a rational theory of the crime. Most often, inductive reasoning is used: The collected information and evidence are carefully analyzed to develop a theory. Often, a rational theory of a crime is developed with some assistance from the careless criminal. *Verbrecherpech*, or "criminal's bad luck," is an unconscious act of self-betrayal. One of the major traits of criminals is vanity; their belief in their own cleverness, not chance, is the key factor in their leaving a vital clue. Investigators look for mistakes.
>
> *Focusing the Investigation:* When this stage is reached, all investigative efforts are directed toward proving that one suspect, and any accomplices, are guilty of the crime. This decision is based upon the investigator's analysis of the relationship between the crime, the investigation, and the habits and attitudes of the suspect.

Arrest and Case Preparation

A lawful arrest brings the investigation into even greater focus and provides the police with several investigative opportunities. The person arrested can be searched and booked at the police station and fingerprinted for positive identification and future use. Evidence may be found at these stages. The prisoner may wish to talk to the police. Here, the officer(s) must obviously know and understand the laws of arrest and search and seizure as well as the laws of evidence (especially the "chain of custody"). Any evidence found at the arrest stage must be collected, marked, transported, and preserved as carefully as that found at a crime scene.

"Case preparation is organization."[52] For an investigation to succeed at trial, all reports, documents, and exhibits must be arrayed in an orderly manner. This package must then be forwarded to the prosecutor. At this point, the investigator neither intrudes nor injects personal opinions and conclusions into the case. The identification of the accused leads to an array of witnesses and physical evidence. The corpus delicti of the crime has been established and the combination of "what happened" and "who did it" has occurred, at least in the mind of the investigator. The investigator must also prepare for the almost inevitable negative evidence that must be countered at trial, where the accused may contend that he or she did not commit the crime. (He or she may try to attack the investigative work, use an alibi, or get the evidence suppressed.) The defendant may offer an affirmative defense, admitting that he or she committed the acts charged, but claiming that he or she was coerced, acted in self-defense, was legally insane, and so forth. Or the defendant may attack the corpus delicti, contending that no crime was committed or that there was no intent present.[53]

In the prosecutor's office, the case is reviewed, assigned for further investigation, and, if warranted, prepared for trial. Conferences with the investigator and witnesses are usually held. The prosecutor may waive prosecution if the case appears too weak to result in a conviction; if the accused will inform on other (usually more serious) offenders; if a plea bargain is more attractive than a trial; or if there are mitigating circumstances in the case (such as emotional disturbance).

An investigation is successful when the crime being investigated is solved and the case closed. Often a case is considered cleared even if no arrest has been made, as when the offender dies, the case is found to be a murder-suicide, the victim refuses to cooperate with police/prosecutor, or the offender has left the jurisdiction and the cost of extradition is not justified.[54]

QUALITIES OF DETECTIVES

The detective function is now well established within the police community. A survey by the RAND Corporation revealed that every city with a population of more than 250,000, along with 90 percent of the smaller cities, have officers specially assigned to investigative duties.[55]

Several myths surround police **detectives**, who are often portrayed in movies as rugged, confident (sometimes overbearing), independent, streetwise individualists who bask in glory and are adorned with big arrests and beautiful women. Detective work carries a strong appeal for many patrol officers, young and veteran alike. In reality, detective work is seldom glamorous or exciting. Investigators, like their bureaucratic cousins, often wade in paper work and spend many hours on the telephone. Furthermore, studies have not been kind to detectives, showing that their vaunted productivity is overrated. Not all cases have a good or even a 50-50 chance of being cleared by an arrest. Indeed, a RAND research team learned in a study of over 150 large police departments that only about 20 percent of their crimes could have been solved by detective work.[56] Another study, involving the Kansas City, Missouri, Police Department, found that fewer than 50 percent of all reported crimes received more than a minimal half-hour's investigation by detectives. In many of these cases, detectives merely reported the facts discovered by the patrol officers during the preliminary investigations.[57]

Yet the importance and role of detectives should not be understated. They know that a criminal is more than a criminal. As Paul Weston and Kenneth Wells said,

> John, Jane and Richard are not just burglar, prostitute and killer. John is a hostile burglar and is willing to enter a premises that might be occupied. Jane is a prostitute who wants a little more than pay for services rendered, is suspected of working with a robbery gang and enticing her customers to secluded areas. Richard is an accidental, a person who, in a fit of rage, killed the girl who rejected him.[58]

To be successful, the investigator must possess four personal attributes to enhance the detection of crime: an unusual capability for observation and recall; extensive knowledge of the law, rules of evidence, scientific aids, and laboratory services; power of imagination; and a working knowledge of social psychology.[59] Successful detectives (and even patrol officers) also appear to empathize with the suspect; if a detective can appear to understand why a criminal did what he did ("You robbed that store because your kids were hungry, right?"), a rapport is often established that results in the suspect's telling the officer his or her life history—including how and why he or she committed the crime in question.

Perhaps first and foremost, however, detectives need *logical skills*, the ability to exercise deductive reasoning, to assist in their investigative work. (An interesting example of the logical skills needed for police work was provided by Al Seedman, former chief of detectives for the New York Police Department, in Chapter 3.)

OFFICERS WHO "DISAPPEAR": WORKING UNDERCOVER

A HIGHLY SOUGHT-AFTER ASSIGNMENT

A highly sought and valued type of police work of an investigative nature is undercover work. Undercover work may be defined as those assignments of police officers to investigative roles in which they adopt fictitious civilian identities for a sustained period of time in order to discover criminal activities that are not usually reported to police.[60] (For more about undercover work, see the "Practitioner's Perspective" by former FBI special agent George Togliatti in Chapter 8.)

Undercover police operations have increased greatly since the 1970s, owing largely to expanded drug investigations. The selection process typically is intense and very competitive. Since only a few officers are actually selected for undercover assignments, these officers enjoy a professional mystique, in large measure because of wide discretionary and procedural latitude in their roles, minimal departmental supervision, ability to exercise greater personal initiative, and a higher degree of professional autonomy than regular patrol officers.[61]

PROBLEMS WITH THE ROLE

However, such conditions may lessen officer accountability and lower adherence to procedural due process and confidence in the rule of law.[62]

One of the most critical requirements is the ability to cultivate informants for information on illegal activities and for contacts with active criminals. Deals and bargains must be struck and honored. Therefore, close association with criminals—both the informants and targeted offenders—heightens the challenges of the undercover role considerably. Undercover officers must sustain a deceptive front over extended periods of time, thereby facing increased risk of stress-induced illness, physical harm, or corruption. One study determined that the more undercover assignments undertaken, the more drug, alcohol, and disciplinary problems federal officers had during their careers.[63]

Undercover agents can experience profound changes in their value systems, often resulting in overidentification with criminals and a questioning of certain criminal statutes they are sworn to enforce.[64] These isolated assignments may also involve a separation of self, disrupting or interfering with officers' family relationships and activities and perhaps even leading to a loss of identity and the adoption of a criminal persona as they distance themselves from a conventional lifestyle:[65]

> A good example of this is the case of a Northern California police officer who participated in a "deep cover" operation for eighteen months, riding with the Hell's Angels. He was responsible for a very large number of arrests, including previously almost untouchable higher-level drug dealers. But this was at the cost of heavy drug use, alcoholism, brawling, the break-up of his family, an inability to fit back into routine police work, resignation from the force, several bank robberies, and a prison term.[66]

The stressful nature of this work is discussed further in Chapter 12.

RETURNING TO PATROL DUTIES

Ending an undercover assignment and returning to patrol duty is an awkward experience for many officers, who experience difficulty in adjusting to the everyday routine of traditional police work. These officers may suffer from such emotional problems as anxiety, loneliness, and suspiciousness, and they may experience marital problems. Officers will quickly have less autonomy and diminished initiative in job performance; they are no longer working with a tight-knit unit with expanded freedom and control. They no longer feel as though they are behind enemy lines in the battle against crime, where their work experiences are intense and inherently dangerous. The return to routine patrol may be analogous to coming down from an emotional high, and officers in this position may feel depressed and lethargic.[67]

USES OF THE POLYGRAPH

The polygraph has been used by the police in the investigation of serious crimes since at least the early 1900s. The modern polygraph is a briefcase-sized device or computerized model that records changes in skin resistance (perspiration), blood pressure, pulse rate, and breathing activity. Activity in each of these physiological measurements is monitored and recorded on a paper chart.[68]

The polygraph, commonly referred to as a lie detector, cannot actually detect when a lie is told; therefore, the **polygraph examination** is an inferential process in which "lying" is inferred from comparisons of the aforementioned physiological responses to questions that are asked during polygraph testing. In police work, the two major uses of polygraph testing are specific issue testing and in preemployment screening. In specific issue testing, the polygraph is used to investigate whether a particular person is responsible for or involved in the commission of a specific offense. The use of the polygraph for preemployment screening is very controversial and is, in fact, illegal in some jurisdictions. When used, however, polygraph testing can help to verify information collected during traditional background investigations and to uncover information not otherwise available. Studies show that polygraph procedures may yield an accuracy of about 90 percent.[69]

The commonly held belief that polygraph examination results are not admitted into court is untrue. Some courts admit polygraph evidence even over the objection of counsel. In other jurisdictions, polygraph results are admitted by stipulation. At the federal level, there is no single standard governing admissibility. It is also common for prosecutors to use polygraph results to decide whether and which charges to file, and defense attorneys rely on polygraph testing to plan their defense and to negotiate pleas. Some judges also use polygraph results in sentencing decisions.[70]

DNA ANALYSIS

METHODS

As mentioned earlier, O. J. Simpson's criminal trial in 1995 brought **DNA testing** into the minds of most Americans and heightened their interest and possibly their concern about its use. More recent events concerning the administration of the death penalty across the country, however, have caused DNA to be cast in a more positive light.

"DNA fingerprinting" has been used in legal cases in the United States since 1987; it is viewed as the greatest advancement since fingerprinting and has been called the "magic bullet" of criminal investigations.[71] Proponents hope that DNA will end the search for the perfect means of personal identification.

DNA (deoxyribonucleic acid) is the chemical dispatcher of genetic information found in almost all human cells. Except for identical twins, no two people share the same DNA sequences. Developed in England in 1984 by Alec Jeffreys, this process uses purified DNA obtained from whole blood, sperm, hair roots, skin, or other tissues (blood and semen are the richest sources).[72] Actually, two DNA testing methods are used for creating the DNA profile. The most widely used method, known as RFLP (restriction fragment length polymorphism) uses highly dangerous radioactive material to produce a DNA image on an X-ray film. The second method, PCR (polymerase chain reaction), employs no radioactive material. PCR is the principal method for forensic science DNA testing.[73] (See Figure 7–1.)

DNA profiling has rapidly spread across the country. The FBI began DNA analysis in December 1988, and in 1990 it began developing a national DNA identification index.[74] DNA testing by the FBI is performed free of charge for all police agencies; further, agencies do not incur any travel expenses for DNA examiners who must testify in court. In addition, more than 50 police laboratories now perform DNA analysis.

The chances that two people (except for identical twins) could have the same DNA profile range anywhere from 1 in 50,000 to 1 in 5 million, according to scientists' most conservative estimates.[75] For some people, the lesser odds and other factors (discussed later) combine to make the test vulnerable to legal attack; for others, these higher odds make any question about the value of DNA testing a nonissue.

THE DEBATE ABOUT FORCED TESTING

Although offenders have been convicted and even executed largely on DNA evidence (the first such execution occurring in Virginia in April 1994),[76] DNA profiling has been under judicial scrutiny since 1988.[77] Questions about population statistics, matching, contamination, and interpretation are commonly raised. Defense attorneys argue that the statistics are inaccurate or overstated and that they do not weigh enough variations between races and ethnicities.

Even with this reluctance on the part of some juries, courts have recognized that genetic profiles developed from DNA are reliable and objective.[78] At issue is the ability of crime laboratories to match similar DNA profiles reliably and to assess the frequency that the matched profile is expected to occur in the U.S. population.

What Is DNA?

Macro Level

1. DNA is the chemical substance that makes up our chromosomes and controls all inheritable information (i.e., eye, hair, and skin color).

2. DNA is different for every individual except identical twins.

Micro Level
Body Cells

Nucleus Nucleus Nucleus

Chromosomes

Body Cells

DNA

3. DNA is found in all cellular material (white blood cells, tissue cells, bone cells, hair root cells, and spermatozoa).

4. Half of an individual's DNA/chromosomes comes from the father, the other half from the mother.

5. DNA is a double-stranded macromolecule.

6. The DNA strands are chemically made of four different building blocks:

| (A) Adenine | (T) Thymine |
| (G) Guanine | (C) Cytosine |

7. The four building blocks and their sequence in DNA make up the letters of the genetic code.

8. In specific regions on a DNA strand, each person has a unique sequence of building blocks or genetic code.

9. It is a person's unique genetic code that allows forensic scientists to identify an individual to the exclusion of others.

Molecular Level

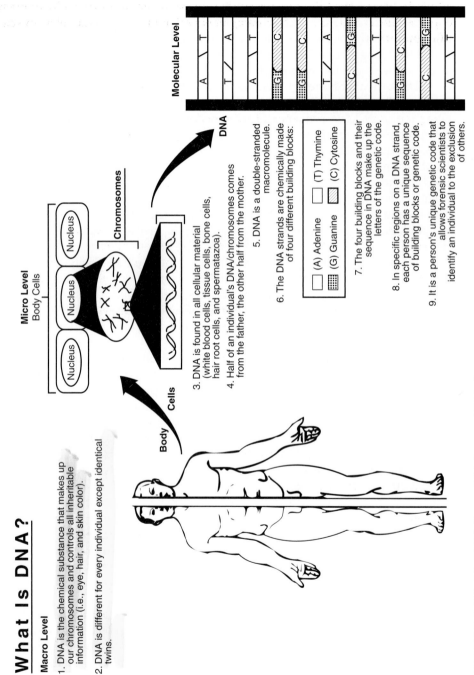

FIGURE 7–1 What Is DNA? (*Source:* FBI Laboratory Division)

Traditionally, the *Frye* **standard**, established in 1923, was the primary test of admissibility of scientific evidence; under this standard, a court would admit scientific evidence only after it had gained general acceptance in the relevant scientific community.[79] In 1993, the U.S. Supreme Court set a new standard, significantly easing the rules for the admissibility of scientific evidence in *Daubert v. Merrell Dow Pharmaceuticals, Inc.*[80] *Daubert*, which superseded the much stricter *Frye* standard, allows the following factors to determine whether any form of scientific evidence is reliable: whether it has been subject to testing and peer review, whether there are known or potential rates of error, and whether there are standards controlling application of the techniques involved. Using this standard, the Court held that DNA testing at the FBI's crime laboratory easily met the necessary criteria for scientific acceptability.

A majority of courts have admitted DNA profiling results from the three major laboratories involved in DNA analysis: the FBI, Cellmark, and Lifecodes. Courts in at least 49 states have admitted DNA evidence; more than 30 appellate courts, including 11 state supreme courts, have reported favorable decisions after reviewing DNA profiling.[81] When defense challenges have arisen against DNA evidence or the courts have balked at admitting it, such reluctance is usually on the grounds of how common or rare the reported DNA profile is in the U.S. population. In other words, the prosecution must demonstrate the significance of the DNA match based on population genetics. The FBI estimates that the likelihood of two DNA samples matching each other is as low as one in a trillion. And in *United States v. Jakobetz*,[82] 1990, population statistics produced by the FBI indicated that the DNA profile of the defendant was extremely rare—it was expected to occur only once in every 300 million persons.[83] DNA critics, however, charge that the FBI has incorrectly applied the theories of population biology underlying such calculations and, as a result, the chances of a mismatch are artificially minimized in the courtroom.

In August 1993, the Washington Supreme Court became the first state high court to endorse forced genetic testing of all persons convicted of sexual offenses or violent crimes. Such testing is to generate a DNA database to identify and prosecute future sex offenders or violent criminals. Washington is one of more than 30 states that have enacted similar statutes. Although opponents of such forced testing believe it is a waste of money and an unnecessary invasion of privacy with the sole purpose of investigating future crimes, the Washington Supreme Court found a "rational connection between the DNA testing statute and law enforcement."[84]

ESTABLISHING INNOCENCE

In addition to establishing guilt in criminal acts, DNA profiling can serve the cause of justice by establishing innocence as well. It has been said that "one of the surprises of DNA testing is how often the police get the wrong man."[85] This "surprise" led Illinois Governor George Ryan to impose a moratorium on executions in January 2000, after 13 inmates—one of whom came within two days of being executed—were proved innocent.[86] With increasing evidence from DNA that the

American justice system is capable of making some horrendous mistakes, the debate about the morality of capital punishment was revived with a vengeance in early 2000. And with good reason: For example, when local police arrest someone for rape or rape-murder and submit their DNA samples to the FBI laboratory for analysis, the DNA test excludes the principal suspect in 26 percent of the cases.[87]

Each passing year brings new advances that expand the universe of cases where DNA analysis can help to free inmates from prison and even death row. Examples such as that involving Jeffrey Pierce are getting more and more common; new DNA evidence released Pierce from an Oklahoma prison in May 2001 after he served 15 years of a 65-year sentence for a rape he did not commit. During the 10-year span of 1992 to 2001, DNA testing helped free 88 innocent convicts, some of whom had served nearly 20 years in prison.[88]

But controversy remains, even though DNA crime-fighting techniques are still gaining favor over fingerprinting kits and magnifying glasses. New twists on DNA evidence are turning up every day. For example, a Wisconsin prosecutor filed rape charges recently against an unknown assailant based solely on his DNA profile in hope of beating the rapidly approaching six-year statute of limitations for filing charges. Other DNA-related investigative nuances are arising, such as that of the Chicago Police Department. A killer who murdered three women on the city's South Side, identified only by his DNA and nicknamed "Pattern D," was placed on the department's most-wanted list.[89]

BEHAVIORAL SCIENCE IN CRIMINAL INVESTIGATION

CRIMINAL PROFILING

Criminal **profiling** of serial killers has captured the public's fancy more than any other investigative technique used by the police. Profiling depends on the profiler's ability to draw upon investigative experience, training in forensic and behavioral science, and empirically developed information about the characteristics of known offenders. It is more art than science. The focus of the analysis is the behavior of the perpetrator within the crime scene.[90]

There are various types of investigative profiles and profiling. Drug courier profiles have been developed from collections of observable characteristics that experienced investigators believe indicate a person who is carrying drugs. Other types of profiles include store loss-control specialists' profiles of shoplifters and threat assessments, such as the Secret Service's profiles of potential presidential assassins. Criminal profiling of violent offenders, however, is the area for which the most descriptive information has been collected and analyzed and the most extensive training programs developed.[91] Unfortunately, most people associate criminal profiling with the psychic profiler seen on television's *The Profiler* or with Agent Starling in the film *The Silence of the Lambs*—both of which are inaccurate portrayals.[92]

Profiling is not a new discovery; indeed, Sir Arthur Conan Doyle's fictional character, Sherlock Holmes, often engaged in profiling. For example, in "A Study of Scarlet," published in 1887, Holmes congratulated himself on the accuracy of his psychological profile: "It is seldom that any man, unless he is very full-blooded, breaks out in this way through emotion, so I hazarded the opinion that the criminal was probably a robust and ruddy-faced man. Events proved that I had judged correctly."[93]

Profiling was used by a psychiatrist to study Adolph Hitler during World War II and predict how he might react to defeat.[94] Psychological profiling, while not an exact science, is obviously of assistance to investigators; however, it does not replace sound investigative procedures. Profiling works in harmony with the search for other physical evidence. Victims play an important role in the development of a profile, as they can provide the investigator with the offender's exact conversation. Other items needed for a complete profile include photographs of the crime scene and any victim(s), autopsy information, and complete reports of the incident, including the weapon used. From this body of information the profiler looks for motive.[95]

Serial murderers—killers who are driven by a compulsion to murder again and again—are also profiled. Many psychologists believe that serial murderers fulfill violent sexual fantasies they have had since childhood. They satisfy their sexual needs by thinking about their killings, but when the satisfaction wears off, they kill again. Most serial murderers, the FBI has learned, are solitary males; an alarming number are doctors, dentists, or other health care professionals. Almost one-third are ex-convicts and former mental patients. Many are attracted to policing (as was Kenneth Bianchi, the Los Angeles Hillside Strangler, a security guard when he was finally caught in Washington state who often wore a police uniform during his crimes). They seem normal, and they principally attack lone women, children, older people, the homeless, hitchhikers, and prostitutes.[96]

Another psychology-related investigative tool is psycholinguistics, which provides an understanding of those who use criminal coercion and provides strategies for dealing with threats. The 1932 kidnapping case of Charles Lindbergh's infant son (perpetrated by a German-born illegal alien, whose notes revealed his background and ethnicity) marked the beginning of this investigative method. Concentrating on evidence obtained from a message, spoken or written, the psycholinguistic technique microscopically analyzes the threats or messages for clues to the origins, background, and psychology of the maker. Every sentence, syllable, phrase, word, and comma are computer-scanned. A "threat dictionary" containing more than 350 categories and 15 million words is consulted; these "signature" words and phrases are then used to identify possible suspects.[97]

Clearly, profiling can be useful in criminal inquiries in several ways: focusing the investigation on more likely types of offenders, suggesting proactive strategies, suggesting trial strategies, and preventing violent crimes. From 1984 to 1991, the FBI trained 33 state and local police investigators in the profiling process. The program required 12 months of intensive training and hands-on profiling experience and consisted of an academic phase and an application phase.[98]

PSYCHICS AND HYPNOSIS

Psychics are also brought in from time to time, usually as a last resort when an investigation has stalled and no solution is imminent. Psychics are persons believed to possess extrasensory perception (ESP) or paranormal powers. National interest was focused upon parapsychology—the discipline dealing with ESP, clairvoyance, and so forth—with publicity accompanying the aid provided to the police in solving three sex murders of youths in South Gate, California, in 1978. After two fruitless years of investigation, police resorted to the services of a local psychic, who helped an artist sketch a drawing of a face that had appeared to the psychic in a vision. With this information, police apprehended a suspect. However, Martin Reiser, a prominent LAPD psychologist, studied the use of psychics and found no support for the "contention that psychics can provide significant information leading to the solution of major crimes."[99]

In the early 1970s, policing turned to hypnosis as an investigative tool. A number of police practitioners were trained in how to place witnesses, victims, arresting officers, and even investigators in a hypnotic state.[100] Hypnosis has been used to help people recall license plate numbers, names, places, or details of an incident. Although a trend toward acceptance of hypnosis by appellate courts began to develop in the early and mid-1970s, a 1976 California ruling reflected current skepticism with the technique. Hypnosis was used in a murder case to secure details of the suspect's identity and obtain his conviction.[101] Following this case, the courts, increasingly concerned with the suggestibility of potential witnesses while under hypnosis, began rendering adverse rulings against such testimony. It then became clear to police and prosecutors that witnesses whose recall had been enhanced through hypnosis could not be put on the witness stand. Thus, today hypnosis is used sparingly, generally reserved for people who would otherwise be useless as material witnesses.

Today, only two states—Nevada and Texas—expressly legalize such evidence; officials in those states believe that forensic hypnotists can help otherwise frustrated investigators solve crimes and that investigators with information obtained through hypnotism are free to either confirm, corroborate, or refute that information through the use of other evidence.[102]

RECENT DEVELOPMENTS IN FORENSIC SCIENCE

ADVANCES ON SEVERAL FRONTS

Technological opportunities are rapidly developing for federal, state, and local police practitioners. Police and prosecutors can use advances in forensic science, sophisticated information systems, and newly developed investigative techniques to arm themselves more effectively than ever before. Cutting-edge technological research sponsored by the NIJ is now under way in the following areas: computer profiling to identify and track serial killers, ongoing efforts to set national standards for DNA profiling, analysis of blood splattering to help reconstruct crimes, computerized age progression, voice analysis with a spectrograph to generate "pictures" of recorded voices, and other technologies to aid in the search for and recovery of missing children.

New laboratory technologies make the analysis of bodies and biological evidence, weapons, trace metals, fibers, documents, and other physical evidence more precise than ever before. NIJ is funding a study that is developing a genetic algorithm (mathematical model) system to construct a computerized composite image of a suspect's face. Crime victims and witnesses sometimes take chance videotapes or photographs of bank robbers, firearms, and license plates; however, because of poor equipment or haste, the resulting pictures may be unclear and useless as evidence. A digital computer process has been developed that can convert unclear photographic images on a computer screen and, through mathematical modeling, enhance the images so the objects and people can be identified.

It should be noted that along with new technologies in the crime lab come new problems as well. Laboratory technicians and police officers of all ranks deal with a variety of body fluids, coming in contact with blood, saliva, and urine. Three diseases deserving special precautions by police and forensic science personnel are AIDS, hepatitis B, and tuberculosis. The AIDS virus has been isolated from blood, bone marrow, saliva, lymph nodes, brain tissue, semen, cell-free plasma, vaginal secretions, tears, and human milk. Accidental inoculation by contact with AIDS-infected needles and blood have occurred.[103] Hepatitis B is a viral infection that can result in jaundice, cirrhosis, and sometimes cancer. The virus may be found in human blood, urine, semen, vaginal secretions, and saliva. Injection into the bloodstream and contact with broken skin are the primary hazards. Tuberculosis can be transmitted through saliva, blood, urine, and other body fluids. It can enter the body through droplets that are inhaled, and it primarily causes lung infections. Lab personnel should wear disposable gloves, surgical masks, and protective eyewear, and all sharp objects should be handled carefully.

FORENSIC ENTOMOLOGY: USING "INSECT DETECTIVES"

An example of how the field of forensic science is vastly developing is the use of new "investigators": insects. Anyone involved in death investigations is aware of the connection between dead bodies and insects, especially maggots. As one forensic entomologist stated,

> Insects are major players in nature's recycling effort, and in nature a corpse is simply organic matter to be recycled. Left to its own devices, nature quickly populates a corpse with a diverse community of organisms, all dedicated to reducing the body to its basic components.[104]

Until recently, most death investigators regarded insects as merely a sign of decay to be washed away rather than as potentially significant evidence. Yet the

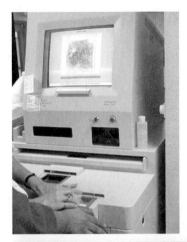
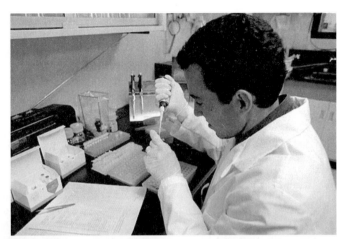

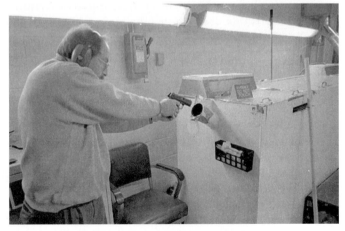

Technology is rapidly advancing in forensic laboratories. Here, in the upper left photo, a technician uses an automated fingerprinting system. At the upper right, a technician performs a DNA analysis. In the middle left photo, a technician is looking into a compound comparison microscope used to compare items such as fibers and hairs. A drug analysis is shown in the middle right photo, as are serological (at lower left), and ballistics (lower right) analyses. (*Courtesy Washoe County, Nevada, Sheriff's Office*)

application of insect evidence to criminal investigations is not a new idea. In 1255, a Chinese "death investigator" named Sung Ts'u wrote a book entitled *The Washing Away of Wrongs*. Sung tells of a murder in a Chinese village in which the victim was repeatedly slashed. The local magistrate thought the wounds might have been inflicted by a sickle and ordered all the village men to assemble, each with his own sickle. In the hot summer sun, flies were attracted to one sickle because of the residue of blood and small tissue fragments still clinging to the blade and handle. The owner of the sickle confessed.[105]

Not until several centuries later, in 1668, was the link between fly eggs and maggots discovered in the West. Before then, people did not realize that maggots hatched from the eggs flies laid on exposed meat or decomposing bodies. Unfortunately, it would not be until much later, in 1855, that this discovery was used in a forensic investigation. During a remodeling of a house outside Paris, the mummified body of an infant was discovered behind a mantelpiece. Suspicion centered on a young couple then occupying the house. An autopsy determined that a flesh fly had exploited the body during the first year and that mites had laid their eggs on the dried corpse the following year. The logical suspects were the occupants of the house in 1848—the young couple—who were subsequently arrested and convicted.[106]

In the mid-1930s, entomological evidence came to the fore in a brutal murder case. A woman saw a severed human arm while looking over a bridge spanning a small stream in Scotland. Ultimately, over 70 pieces of two badly decomposed corpses—the wife of a local physician and her personal maid—were recovered from the area. A group of maggots feeding on the decomposing body parts was sent to the laboratory; they were identified as belonging to a type of blow fly and between 12 and 14 days old. For a number of reasons, suspicion fell on the physician: the entomological estimate of the time of death, the skillful manner of the dismemberment, and the fact that the doctor had been seen with a cut finger. He was convicted of both murders.[107]

Obviously, today **forensic entomology** can be of tremendous assistance to investigators. When a person dies, hundreds of species of insects descend upon the corpse and crime scene. Attracted by what scientists think may be a "universal death scent," common green flies, ground beetles, and other insects that thrive on flesh migrate to a corpse to lay eggs or to feast. The death of a human triggers very predictable patterns of insect activity that can be traced backward through time. There are several waves of insects, and their presence is informative. The life cycles of these creatures are so fixed and precise that they act as a natural clock to the trained eye.[108]

There is also a phenomenon that occurs known as succession; as each organism feeds on a body, it changes the body. This change in turn makes the body attractive to another group of organisms, which changes the body for the next group, and so on until the body has been reduced to a skeleton. This is a predictable process.[109] This process is now being studied scientifically at the so-called Body Farm at the University of Tennessee in Knoxville—the only place in the world where corpses rot in the open air to advance human knowledge (see also

---

EXHIBIT 7-1

A New Crime Scene Academy

A number of assessments have found that the quality of evidence collection and identification could be improved, particularly at the local level. Now help is on the way. The Knoxville, Tennessee, Police Department is spearheading a project to create a National Forensic Academy for in-service criminalists. The 400-hour curriculum includes 180 hours of class work and 220 hours of field practicum. The academy session ends with 10 hours of testing. It covers arson, autopsy, trace evidence (including DNA and serology), mass fatalities, death investigations, cold case studies, emerging trends, fingerprinting, blood splatter, child fatality investigations, and computer forensics, among other topics. Applicants for 14 available slots must be first-line responders, computer literate, willing to participate in group activities, and be prepared to work with human cadavers. The first academy class was held in September 2001.

Source: Adapted from "Is Your Crime-Scene Work a Crime? Help Will Soon Be on the Way," *Law Enforcement News*, March 15, 2001, p. 1. Used with permission from *Law Enforcement News*, John Jay College of Criminal Justice. (CUNY) 555 West 57th St., New York, NY 10019.

Exhibit 7–1). William Bass III, the Body Farm's founder, trains anthropologists and FBI agents in human composition. More than 20 corpses are decomposing there as maggots do their work efficiently; over the years more than 300 people have decayed at this location, some in car trunks, others under water or under the earth or hung from scaffolds. The overarching goal of the study is to produce an atlas for the police that will provide a "gold standard" for decomposition—a page-by-page, color-by-color, insect-by-insect depiction of the process of human decay on a time and temperature line. The study also includes burying multiple bodies under four pads of concrete to assist the FBI's tests of its latest ground-penetrating radar.[110]

Often the strongest circumstantial evidence in a criminal prosecution is the fact that the victim and a suspect were seen together. For example, the body of a nine-year-old girl who had been missing for three weeks was found in Kenosha, Wisconsin; insect evidence put her death at midnight 21 days before—the exact day she had gone to a carnival with the suspect, who was convicted.[111]

There are other uses of insects as well. A man claiming to have found his girlfriend dead was convicted of her murder when none of the usual creatures were present on the body; the man had thrown open the windows before police arrived. A Florida murderer who dropped his victim in a swamp was convicted when mayflies, only active in a swamp region a few days of the year, were found in his car radiator. In a rape case, the offender was convicted when his ski mask was found to contain larvae from caterpillars that emerged only in late summer and were known to dwell in the woods next to the rape scene. Furthermore, insects feeding on infected tissues where children's diapers have not been changed or under bandages and bedsheets of the elderly have often been the only evidence that neglect has occurred.[112]

A forensic entomology laboratory at Simon Fraser University in British Columbia is the first university laboratory in North America founded for the sole purpose of refining the ways in which insect biology can help solve crimes. Scientists at the laboratory are attempting to greatly expand the horizons of forensic entomology: They train police officers to collect specimens properly and correctly, provide greater advances in DNA technology (for example, blood contained in an insect that migrates from a suspect to a dead body can be matched), and even submerge pig carcasses in water to develop guidelines for how saltwater animals such as crabs and shrimp interact with corpses. The basic premise held by such scientists is that the dead have rights, too.[113]

CATCHING CRIMINALS WITH THEIR OWN VIDEOS

In Sparta, Michigan, a 16-year-old high school dropout with a criminal record bludgeons a man to death, then cuts off his head. At home, the youth repeatedly slashes the severed head with a butcher knife and removes the brain. In Fort Lauderdale, Florida, five teenagers vandalize and burglarize eight homes and a school. During their escapades, they blow up a live sea trout in a microwave and get a dog high on marijuana. In both of these incidents—and in many others across the nation as well—the offenders were convicted by evidence they created themselves: They had videotaped their exploits, either to immortalize their actions or to achieve "stardom." They also provided incontrovertible evidence of their crimes.[114]

Such cases indicate that criminals seem more interested in publicizing their crimes than in remaining discreet. An increasing number of cases involve videotaping drug parties. Other perpetrators have videotaped their involvement in hate-related crimes. Even murderers have memorialized some of their activities on video.

Such videotaping has important implications for the police. For example, investigators in sex offense units have long recognized the need to include in the search warrant any photographic or recording devices, photograph albums, videotapes, and audiotapes; however, when investigators apply for warrants to search suspects' homes where burglaries, robberies, or assaults have been committed, they tend not to include requests to seize such materials. Without sufficient justification, magistrates probably would not sign an original warrant to include the seizure of audio- or videotapes. Therefore, investigators must establish probable cause of the existence of video evidence related to a particular crime. To establish probable cause, at the outset investigators should routinely ask victims, complainants, and witnesses whether the offender used some electronic device during the crime's commission.[115]

STALKING INVESTIGATIONS

Stalking was recently described as "the crime of the 1990s."[116] California enacted the first stalking law in 1990, following the stalking and subsequent death of Rebecca Shaeffer, the costar of the television series *My Sister Sam*. Today all but two states have their own stalking laws (Arizona and Maine utilize their harassment and terrorizing statutes to combat stalking).[117]

There are four types of stalking situations:

1. *Erotomanic:* The stalker has a delusional disorder wherein the victim, usually a person of higher status and opposite gender (often a celebrity or a public figure), is believed to be in love with the subject. The victim does not know the stalker, and the stalking is usually short lived, lasting one to four months.

2. *Love obsessional:* Similar to the erotomanic, here the subject does not know the victim except through the media and has a psychiatric diagnosis; subjects usually write letters and make telephone calls in a campaign to make their existence known to the victim. This behavior is much longer term, often up to or longer than 10 years.

3. *Simple obsessional:* The stalker had a prior relationship with the intended victim, as a former spouse or employer, for example, and the stalking began after the relationship soured. The stalking usually lasts about 5 months.

4. *The false victimization syndrome:* This is the rarest form, in which a person claims that someone is stalking him or her in order to become a victim, apparently for attention. There is no stalker, and this phenomenon is similar to Munchausen Syndrome by proxy (people intentionally produce physical symptoms of illness in their children in order to gain attention and sympathy).[118]

Evidence collection for crimes of stalking begins with the victims. Victims should record each time they see the stalker or when any contact is made. Further, victims should document specific details, such as time, place, location, and any witnesses. Messages on answering machines, faxes, letters, and computer e-mail messages provide useful resources to build a case against the offender. Police agencies should also consider providing the victim with a small tape recorder to facilitate the collection of this information and should encourage victims to report in a journal how the stalking has affected them and their lifestyle, to later help convince a jury of the victim's fear and trauma. Another investigative strategy is surveilling suspects at times when they would likely stalk the victim. Executing a search warrant for the suspect's personal and work computers, residence, and vehicle can prove useful in many circumstances. Officers should look for spying equipment (such as binoculars and cameras), photos, and any property belonging to the victim.[119]

INVESTIGATING "CYBERCROOKS"

More than 150 million Americans are plugged into cyberspace, and thousands more enter the online world every day.[120] But because the Internet is so vast and uncharted, the full scope of its dark side has never been explored. And the amount of crimes that occur online is staggering. It is no wonder that the "crime scene" of this millennium might well be evidence that is found within computers of the activities of **cybercrooks**.

Indeed, during 1999, the Federal Trade Commission received more than 18,000 Internet-related complaints. That figure was more than double the volume for 1998, and by mid-2000 the number had already reached 11,000. The FBI opened 1,500 online child sex cases during 1999, up from 700 during 1998. And according to a recent survey, 70 percent of companies experienced "cyberattacks"—including such crimes as credit card thefts and identity thefts—during 1999; 300 companies reported losses of more than \$265 million.[121]

Child pornography, counterfeiting (of driver's licenses, school transcripts, passports, social security cards, and price stickers), car thefts, drug offenses, stock market fraud, and myriad other crimes can be committed more easily with a computer than with a gun. Even "the cruelest con"—adoption fraud—has been made much easier by the Internet. Electronic chatrooms are trolled by scammers promising babies if only the adopters will send money. Less-than-honest adoption agencies also post pictures of cute babies who are, in fact, not available for adoption.[122]

Cybercops are investigators who have the task of navigating computer hard drives in their search for evidence. As computers have become more user friendly, criminals have gotten used to storing more and more data. In the absence of court rulings, cybercops—who possess software programs that enable them to get through password-protect features—seize those data and retrieve deleted files.[123]

A common ploy by police investigators is to pose on the Internet as young girls to catch predators. Since going online in February 1999 until September 2000, a Colorado Springs, Colorado, task force caught 22 offenders nationwide—many of these arrests occurred after children had been molested. Investigators spend many hours scouring the Web, looking for certain profiles based on leads and complaints.[124] Offenders often take on a "wolf in sheep's clothing" mien, such as the 26-year-old Stanford University student who often spent his free time volunteering at elementary schools in San Jose, California, cautioning young children to avoid sexual predators on the Internet. In May 2001, the student was arrested in a parking lot where he was supposed to meet a 10-year-old girl for a sexual rendezvous, after befriending her in one of his classes and seducing her online. Detectives had used carefully crafted e-mails for a month, decorated with references to The Backstreet Boys and finger painting.[125]

Some researchers have indicated serious concern with the current ability of the police to investigate computer-related crime; in fact, Mark Correia and Craig Bowling flatly stated that "law enforcement, specifically on a local level, is not prepared to effectively handle computer-related crime, nor, for the most part, is it preparing for such crime."[126] Clearly there is much more work to be done in the way of recruiting and training personnel before the police are to be deemed capable of handling "digital disorder."

In a related vein, we discuss liability issues concerning the handling of computer evidence in Chapter 10.

SUMMARY

The evolution of forensic evidence, criminalistics, and criminal investigation is the product of a successful symbiosis of science and policing. This chapter has shown the truly interdisciplinary nature of police work; in it we discussed not only the influence of the so-called hard sciences—computer science, chemistry, biology, and physics—but also the assistance of psychology.

There is little doubt that the future holds greater advances in the realm of forensic science. This discipline has traveled a great distance, especially since the advent of the computer. The potential of computer hardware and software is limited only by our funds and imagination. Thus, policing should continue to reap the benefits of continually progressing scientific aids for the solution of crime.

ITEMS FOR REVIEW

1. Distinguish between the terms "forensic science" and "criminalistics."
2. Explain the origin of criminalistics, including a comparison of anthropometry and dactylography.
3. How did August Vollmer and other pioneers in law enforcement contribute to the development of criminal investigation techniques?
4. What qualities do detectives need?
5. Describe the contributions of behavioral science to criminal investigation, providing actual case studies.
6. Explain the fundamental advantages of forensic entomology for criminal investigators.
7. Explain (in lay terms) how DNA analysis operates. What is its future outlook, both legally and scientifically?
8. Review the basic functions of the polygraph and it legal status in the courts.
9. Describe how the Internet has provided new opportunities for criminals; include examples.
10. Review the four types of stalking situations and how detectives must go about addressing them.

NOTES

1. Quoted in *Random House Webster's Quotationary*, ed. Leonard Roy Frank (New York: Random House, 1999), p. 761.
2. Marc H. Caplan and Joe Holt Anderson, *Forensic: When Science Bears Witness* (Washington, D.C.: GPO, 1984), p. 2.
3. Charles R. Swanson Jr., Neil C. Chamelin, and Leonard Territo, *Criminal Investigation*, 4th ed. (New York: Random House, 1988), pp. 223–224.
4. Paul L. Kirk, "The Ontogeny of Criminalistics," *Journal of Criminology and Police Science* 54 (1963): 238.
5. Peter R. DeForest, R. E. Gaensslen, and Henry C. Lee, *Forensic Science: An Introduction to Criminalistics* (New York: McGraw-Hill, 1983), p. 29.
6. Richard Saferstein, *Criminalistics* (Englewood Cliffs, NJ: Prentice Hall, Inc., 1977), p. 5.
7. Jurgen Thorwald, *Crime and Science* (New York: Harcourt, Brace & World, 1967), p. 4.

8. Jurgen Thorwald, *The Century of the Detective* (New York: Harcourt, Brace & World, 1965), p. 7.
9. *Ibid.*, pp. 9–10.
10. *Ibid.*, p. 12.
11. Swanson, Chamelin, and Territo, *Criminal Investigation*, p. 13.
12. *Ibid.*, pp. 14–15.
13. *Ibid.*, p. 15.
14. Anthony L. Califana and Jerome S. Levkov, *Criminalistics for the Law Enforcement Officer* (New York: McGraw-Hill, 1978), p. 20.
15. Frederick R. Cherrill, *The Finger Print System of Scotland Yard* (London: Her Majesty's Stationery Office, 1954), p. 3.
16. Thorwald, *The Century of the Detective*, p. 18.
17. *Ibid.*, p. 33.
18. Saferstein, *Criminalistics*, p. 4.
19. Jurgen Thorwald, *The Marks of Cain* (London: Thames and Hudson, 1965) p. 81.
20. Thorwald, *Century of the Detective*, p. 62.
21. Thorwald, *The Marks of Cain*, pp. 78–79.
22. U.S. Department of Justice, Federal Bureau of Investigation, *Crime in the United States–2000* (Washington, D.C.: Government Printing Office, 2001), p. 17.
23. Reported in *Newsweek*, September 30, 1991, p. 10.
24. Thorwald, *The Marks of Cain*, pp. 87–88.
25. Thorwald, *The Century of the Detective*, pp. 418–419.
26. Thorwald, *The Marks of Cain*, p. 164.
27. DeForest, Gaensslen, and Lee, *Forensic Science: An Introduction to Criminalistics*, p. 14.
28. Swanson, Chamelin, and Territo, *Criminal Investigation*, p. 19.
29. Saferstein, *Criminalistics*, p. 30.
30. DeForest, Gaensslen, and Lee, *Forensic Science: An Introduction to Criminalistics*, pp. 13–14.
31. *Ibid.*, p. 19.
32. Swanson, Chamelin, and Territo, *Criminal Investigation*, p. 1.
33. *Ibid.*, p. 3.
34. Thomas A. Reppetto, *The Blue Parade* (New York: The Free Press, 1978), pp. 26–28.
35. *Ibid.*, p. 29.
36. John Coatman, *Police* (New York: Oxford, 1959), pp. 98–99.
37. Swanson, Chamelin, and Territo, *Criminal Investigation*, p. 4.
38. Augustine E. Costello, *Our Police Protectors* (Montclair, NJ: A. Patterson Smith, 1972 reprint of an 1885 edition), p. 402.
39. James F. Richardson, *The New York Police* (New York: Oxford, 1970), p. 37.
40. Swanson, Chamelin, and Territo, *Criminal Investigation*, p. 8.
41. William J. Bopp and Donald Shultz, *Principles of American Law Enforcement and Criminal Justice* (Springfield, IL: Charles C. Thomas, 1972), pp. 70–71.
42. William J. Mathias and Stuart Anderson, *Horse to Helicopter* (Atlanta: Community Life Publications, Georgia State University, 1973), p. 22.
43. Thorwald, *The Marks of Cain*, p. 136.
44. *Ibid.*
45. *Ibid.*, p. 137.
46. Reppetto, *The Blue Parade*, p. 26.
47. *Ibid.*, p. 267.
48. Swanson, Chamelin, and Territo, *Criminal Investigation*, p. 10.
49. *Ibid.*, p. 11.
50. President's Commission on Law Enforcement and the Administration of Justice, *Task Force Report: Science and Technology* (Washington, D.C.: Government Printing Office, 1967), pp. 7–18.
51. Paul B. Weston and Kenneth M. Wells, *Criminal Investigation: Basic Perspectives,* 4th ed. (Englewood Cliffs, NJ: Prentice Hall, 1986), pp. 5–10.
52. *Ibid.*, p. 207.
53. *Ibid.*, pp. 207–209.
54. *Ibid.*, p. 214.
55. Peter W. Greenwood and Joan Petersilia, *The Criminal Investigation Process*, vol. 1, *Summary and Policy Implications* (Santa Monica, CA: Rand Corporation, 1975). The entire report is found in Peter W. Greenwood, Jan M. Chaiken, and Joan Petersilia, *The Criminal Investigation Process* (Lexington, MA: D. C. Heath, 1977).

56. *Ibid.*

57. *Ibid.*, p. 19

58. *Ibid.*, p. 10.

59. *Ibid.*, p. 11.

60. Mark R. Pogrebin and Eric D. Poole, "Vice Isn't Nice: A Look at the Effects of Working Undercover," *Journal of Criminal Justice* 21 (1993): 383–394.

61. *Ibid.*, pp. 383–384.

62. Peter K. Manning, *The Narc's Game: Organizational and Informational Limits on Drug Enforcement* (Cambridge, MA: MIT Press, 1980).

63. M. Girodo, "Drug Corruption in Undercover Agents: Measuring the Risk," *Behavioral Sciences and the Law* 3 (1991): 299–308; also see David L. Carter, "An Overview of Drug-Related Conduct of Police Officers: Drug Abuse and Narcotics Corruption," in *Drugs, Crime, and the Criminal Justice System*, ed. Ralph Weisheit (Cincinnati, OH: Anderson, 1990).

64. U.S. Department of Justice, Federal Bureau of Investigation, *The Special Agent in Undercover Investigations* (Washington, D.C.: U.S. Department of Justice, 1978).

65. A. L. Strauss, "Turning Points in Identity," in *Social Interaction*, ed. C. Clark and H. Robboy (New York: St. Martin's, 1988).

66. Gary T. Marx, "Who Really Gets Stung? Some Issues Raised by the New Police Undercover Work," in *Moral Issues in Police Work*, ed. F. Ellison and M. Feldberg (Totowa, NJ: Bowman and Allanheld, 1988), pp. 99–128.

67. G. Farkas, "Stress in Undercover Policing," in *Psychological Services for Law Enforcement*, ed. J. T. Reese and H. A. Goldstein (Washington, D.C.: U.S. Government Printing Office, 1986).

68. Frank Horvath, "Polygraph," in *The Encyclopedia of Police Science*, 2d ed., ed. William G. Bailey (New York: Garland, 1995), pp. 640–642.

69. *Ibid.*, p. 643.

70. *Ibid.*, p. 642.

71. Sharon Begley, Ginny Carroll, and Karen Springen, "Blood, Hair, and Heredity," *Newsweek*, July 11, 1994, p. 24.

72. Jay V. Miller, "The FBI's Forensic DNA Analysis Program," *FBI Law Enforcement Bulletin* 60 (July 1991): 11–15.

73. Richard M. Rau, "Forensic Science and Criminal Justice Technology: High-Tech Tools for the 90's," *National Institute of Justice Reports* (Washington, D.C.: GPO, 1991), pp. 6–10.

74. John R. Brown, "DNA Analysis: A Significant Tool for Law Enforcement," *The Police Chief*, March 1994, p. 51.

75. LynNell Hancock and Mark Miller, "Testing the Gene Fit," *Newsweek*, January 9, 1995, p. 64.

76. Begley, Carroll, and Springen, "Blood, Hair, and Heredity," p. 24.

77. See *People v. Wesley*, 533 N.Y.S.2d 643 (Sup. Ct., 1988) (the first reported decision passing on the admissibility of forensic DNA profiling).

78. For a complete listing of court decisions, see John T. Sylvester and John H. Stafford, "Judicial Acceptance of DNA Profiling," *FBI Law Enforcement Bulletin* 60 (July 1991): 26–31.

79. *Frye v. United States*, 293 F. 1013, 1014 (D.C. Cir. 1923).

80. *Daubert v. Merrell Dow Pharmaceuticals, Inc.*, 113 S.Ct. 2786, decided June 28, 1993.

81. William J. Cook, "Courtroom Genetics," *U.S. News and World Report*, January 27, 1992, pp. 60–61.

82. *United States v. Jakobetz*, 747 F. Supp. 250, 254–255 (D.Vt. 1990).

83. Sylvester and Stafford, "Judicial Acceptance of DNA Profiling," pp. 28–29.

84. Victoria Slind-Flor, "Court OKs Forced DNA Test," *The National Law Journal*, September 6, 1993, p. 10.

85. Jonathan Alter and Mark Miller, "A Life or Death Gamble," *U.S. News and World Report*, May 29, 2000, p. 25.

86. *Ibid.*

87. *Ibid.*

88. "Free at Last, Thanks to DNA," *U.S. News and World Report*, May 21, 2001, p. 13.

89. Warren Cohen, "The Uncharted Future of DNA Detective Work," *U.S. News and World Report*, October 25, 1999, p. 33.

90. Patrick E. Cook and Dayle L. Hinman, "Criminal Profiling: Science and Art," *Journal of Contemporary Criminal Justice* 15 (August 1999): 230.

91. *Ibid.*, p. 232.

92. Steven A. Egger, "Psychological Profiling," *Journal of Contemporary Criminal Justice* 15 (August 1999): 243.

93. *Ibid.*, p. 242.

94. Walter C. Langer, *The Mind of Adolph Hitler* (New York: World Publishing Company, 1978).
95. Swanson, Chamelin, and Territo, *Criminal Investigation*, pp. 601–602.
96. Brad Darrach and Joel Norris, "An American Tragedy," *Life*, August 1984, p. 58.
97. Swanson, Chamelin, and Torrito, *Criminal Investigation*, pp. 606–607.
98. Cook and Hinman, "Criminal Profiling," p. 234.
99. Martin Reiser, Louise Ludwig, Susan Saxe, and Clare Wagner, "An Evaluation of the Use of Psychics in the Investigation of Major Crimes," *Journal of Police Science and Administration* 7 (1979): 18–25.
100. Joseph Deladurantey and Daniel Sullivan, *Criminal Investigation Standards* (New York: Harper and Row, 1980), p. 72.
101. *People v. Quaglino*, Docket No. 109525, Superior Court, Santa Barbara, California, 1976.
102. Bill O'Driscoll, "Hypnotist Keeping Busy Since Law Let His Evidence into State Courts," *Reno Gazette Journal*, June 21, 1999, p. 1c.
103. Paul D. Bigbee, "Collecting and Handling Evidence Infected with Human Disease-Causing Organisms," *FBI Law Enforcement Bulletin* 56 (July 1987): 1–5.
104. M. Lee Goff, *A Fly for the Prosecution: How Insect Evidence Helps Solve Crimes* (Cambridge, MA: Harvard University Press, 2000), p. 9. An excellent source of information concerning forensic entomology, this book contains not only the effects (and contributions) of insects to forensic analyses but also the consequences of predator, air, fire, and water activity and provides a number of case studies in these regards.
105. *Ibid.*, p. 10.
106. *Ibid.*, p. 11.
107. *Ibid.*, pp. 12–13.
108. Joannie M. Schrof, "Murder, They Chirped," *U.S. News and World Report*, October 14, 1991, pp. 67–68.
109. Goff, *A Fly for the Prosecution*, p. 14.
110. Daniel Pedersen, "Down on the Body Farm," *Newsweek*, October 23, 2000, pp. 50–52.
111. *Ibid.*, p. 67.
112. *Ibid.*, p. 68.
113. "Dead Men Tell No Tales—But Bugs Do," *Time*, March 19, 2001, p. 58.
114. Edward F. Davis and Anthony Pinizzotto, "Caught on Tape: Using Criminals' Videos Against Them," *FBI Law Enforcement Bulletin* (November 1998): 13.
115. *Ibid.*, pp. 14–15.
116. Harvey Wallace, *Victimology: Legal, Psychological, and Social Perspectives* (Boston: Allyn and Bacon, 1998), p. 333.
117. *Ibid.*
118. *Ibid.*, pp. 333–334.
119. George E. Wattendorf, "Stalking: Investigation Strategies," *FBI Law Enforcement Bulletin* (March 2000): 11.
120. Margaret Mannix, "The Web's Dark Side," *U.S. News and World Report*, August 28, 2000, p. 36.
121. *Ibid.*, p. 38.
122. *Ibid.*, p. 39.
123. *Ibid.*
124. Becca Blond, "Officers Scouring Web to Thwart Sexual Predators," *Reno Gazette Journal*, September 20, 2000, p. 4A.
125. *Reno Gazette Journal*, May 18, 2001, p. 2A.
126. Mark E. Correia and Craig Bowling, "Veering Toward Digital Disorder: Computer-Related Crime and Law Enforcement Preparedness," *Police Quarterly* 2 (June 1999): 225–241.

CHAPTER 8

Extraordinary Problems and Methods

Key Terms and Concepts

Border Patrol
Deinstitutionalization
Gangs
Graffiti
Hate crimes

Homeless
Mafia
Militias
Rural policing
Terrorism

In utrumque paratus / Seu versare dolos seu certae occumbere morti.
[Prepared for either event / to set his traps or to meet with certain death.]—*Virgil, 70–19 B.C.*

We can no longer overwhelm our problems; we must master them with imagination, understaNding, and patience.—*Henry Kissinger*

INTRODUCTION

Certain types of people in the United States, through their actions or status, pose unique challenges for the police. Some problems (for example, terrorism, hate groups, the militia movement, the Mafia, illegal immigration, and gangs) often compel the police to engage in clandestine operations in order to arrest those persons wishing to avoid detection and apprehension. Other circumstances—such as policing in small towns or working with the homeless—require a less structured and formal approach. This chapter examines each of these unique challenges to police and their methods.

The common thread in all of these circumstances is that the police must constantly attempt to develop new methods and practices to deal with them. As contemporary American society becomes more complex and rife with problems and violence, the police will increasingly be compelled to approach such matters differently.

TERRORISTS, HATE GROUPS, MILITIAS

Terrorism has obviously become a major point of vulnerability and concern for the police since the mid-1990s. Hate crimes have also become more commonplace. A hate group may commit a violent act of terrorism; such an act may also constitute a hate crime. People who belong to militias can also compose a hate group and commit terrorist acts. Therefore, we treat those topics together in this section.

TERRORISM ON OUR SOIL

The threat of **terrorism** has changed and become more deadly. Over the last several years, a new trend has developed in terrorism within the United States: a transition from more numerous low-level incidents to less frequent but more destructive attacks, with the goal of producing mass casualties and attracting

World Trade Center after September 11, 2001. (*Courtesy Brian Hayes*)

intense media coverage. While the number of terrorist attacks in the United States declined during the 1990s, the number of persons killed and injured increased.[1]

Such attacks in the United States are caused by both foreign and domestic terrorists. Examples of the former are the attacks in September 2001 with hijacked jetliners on the World Trade Center building in New York and the Pentagon in Virginia, with more than 3,000 people killed or missing, as well as the bombing of the World Trade Center in New York City in February 1993; an example of the latter is the April 1995 bombing of the Murrah Building in Oklahoma City by Timothy McVeigh, killing 168 people.[2]

While chasing terrorists is expensive, the resources available for the task have greatly increased. From 1993 to 2001, the FBI's counterterrorism budget increased from $77 million to $376 million,[3] or 388 percent (the most significant increase following the Oklahoma City bombing); surely the costs of trying to secure the nation's people and places will skyrocket even more in the aftermath of the World Trade Center/Pentagon attacks in September 2001.

Terrorism can take many forms, and it does not always involve bombs and guns. For example, the Earth Liberation Front (ELF) and a sister organization, the

Destruction of the Murrah Building in Oklahoma City, left, and the World Trade Center (below) in New York City in 1993 demonstrated for all Americans—especially the police—that domestic terrorism is a constant threat.

Animal Liberation Front, have been responsible for the majority of terrorist acts committed in the United States for several years. These "ecoterrorists" have burned greenhouses, tree farms, logging sites, ski resorts, and new housing developments. The groups' members have eluded capture, have no centralized organization or leadership, and are so secretive as to be described as "ghosts" who are more difficult to infiltrate than organized crime.[4]

U.S. embassies abroad have also been singled out for terrorist attacks; in 1998, bombings of embassies in Kenya and Tanzania left 224 dead and resulted in the placement of Saudi millionaire Osama bin Laden at the top of the FBI's most wanted list as well as the conviction in June 2001 of four of bin Laden's foot soldiers.[5]

The FBI divides the current international terrorist threat into three categories:

1. *Foreign sponsors of international terrorism.* Seven countries—Iran, Iraq, Syria, Sudan, Libya, Cuba, and North Korea—are designated as such sponsors and view terrorism as a tool of foreign policy. They fund, organize, network, and provide other support to formal terrorist groups and extremists.

2. *Formalized terrorist groups.* Autonomous organizations (such as bin Laden's al Qaeda, Afghanistan's Taliban, the Iranian-backed Hezbollah, Egypt's Al-Gama'a al-Islamiyya, and Palestinian HAMAS) have their own infrastructures, personnel, finances, and training facilities.

3. *Loosely affiliated international radical extremists.* As noted earlier, examples of this type are the persons who attacked the World Trade Center towers and the Pentagon in 2001 and bombed the World Trade Center in 1993. They may or may not represent a particular nation, but they pose a most urgent threat to the United States because they remain relatively unknown to law enforcement agencies.[6]

These international groups now pose another source of concern to law enforcement agencies: They are cooperating. For example, a terrorist manual is circulating among some of them, and al Qaeda is known to maintain close ties with Hezbollah.[7] Another concern is their choice of weapons; although terrorists continue to rely on such conventional weapons as bombs and small arms, there are indications that terrorists and other criminals may consider using unconventional chemical or biological weapons in an attack in the United States at some point in the future.[8]

With regard to domestic terrorism, many people living in the United States are determined to use violence to advance their agendas. Domestic terrorism seems to have hit a high-water mark from 1982 to 1992, when a majority of the 165 terrorist incidents were conducted by domestic terrorist groups. One current troubling branch of right-wing extremism is the militia or patriot movement (discussed later). Their members are generally law-abiding citizens who have become intolerant of what they perceive as violations of their constitutional rights.[9]

WHAT CAN THE POLICE DO?

What can law enforcement agencies do about domestic terrorism? Many avenues are available, although not all of them will be pursued or supported by the American people. Furthermore, many, if not most, of the possible solutions to terrorism involve political, as opposed to strictly law enforcement, support and approval.

Some suggested means for addressing domestic terrorism, particularly following the 2001 attacks in New York City, include lifting a 1976 executive order banning U.S. involvement in foreign assassinations; loosening restrictions on FBI surveillance; more power for law enforcement agencies to conduct wiretaps, detain foreigners, and track money-laundering cases; recruiting and paying overseas agents linked to terrorist groups, for the purpose of infiltration; having a counterterrorism czar within the White House; and expanding the intelligence community's ability to intercept and translate messages in Arabic, Farsi, and other languages.[10] Other measures involve the aviation industry, such as the fortification of cockpits to prevent access by hijackers, placing federal air marshals on commercial flights, eliminating curbside check-in, and more intensive screening of luggage.[11]

Other joint efforts are being taken to prevent acts of terrorism and to respond to such acts when they do occur. For example, 16 formalized Joint Terrorism Task Forces (JTTFs), composed of federal, state, and local law enforcement and police agencies, are located around the country. Today the JTTF includes more than 140 members in its 16 task forces from numerous federal and local agencies. Investigations by the JTTF have resulted in a number of successes, including arrests of 15 persons following the 1993 World Trade Center attack.[12]

Several fundamental lessons have been learned, and "articles of faith" have been developed concerning critical incident management following terrorist acts. A federal report stated the following: "Emergencies can only be managed by people at the site. They can't be managed back in Washington. Expect the unexpected and be prepared to adjust accordingly."[13]

POLICING HATE

In the past, the police often tended to dismiss the context in which crimes occurred; an assault was simply an assault, and so on. In 1990, however, Congress passed the Hate Crimes Statistics Act, which forced the police to collect statistics on hate crimes. Several states have enacted statutes that place higher penalties on crimes that have a "hate motive." **Hate crimes** are those that "manifest evidence of prejudice based on certain group characteristics." Table 8–1 shows the distribution of hate crimes by victim group.

Numerous hate groups exist in the United States. They include the American Nazi Party, the Arizona Patriots, the Aryan Brotherhood and the Aryan Nations,

TABLE 8–1 Incidents, Offenses, Victims, and Known Offenders by Bias Motivation, 2000[a]

Bias Motivation	Incidents	Offenses	Victims[b]	Known Offenders[c]
Total	**8,063**	**9,430**	**9,924**	**7,530**
Single-Bias Incidents	**8,055**	**9,413**	**9,906**	**7,520**
Race:	**4,337**	**5,171**	**5,397**	**4,452**
Anti-White	875	1,050	1,080	1,169
Anti-Black	2,884	3,409	3,535	2,799
Anti-American Indian/Alaskan Native	57	62	64	58
Anti-Asian/Pacific Islander	281	317	339	273
Anti-Multiracial Group	240	333	379	153
Religion:	**1,472**	**1,556**	**1,699**	**577**
Anti-Jewish	1,109	1,161	1,269	405
Anti-Catholic	56	61	63	33
Anti-Protestant	59	62	62	23
Anti-Islamic	28	33	36	20
Anti-Other Religious Group	172	187	210	77
Anti-Multireligious Group	44	46	52	18
Anti-Atheism/Agnosticism/etc.	4	6	7	1
Sexual Orientation:	**1,299**	**1,486**	**1,558**	**1,443**
Anti-Male Homosexual	896	1,023	1,060	1,088
Anti-Female Homosexual	179	211	228	169
Anti-Homosexual	182	210	226	153
Anti-Heterosexual	22	22	24	18
Anti-Bisexual	20	20	20	15
Ethnicity/National Origin:	**911**	**1,164**	**1,216**	**1,012**
Anti-Hispanic	557	735	763	694
Anti-Other Ethnicity/National Origin	354	429	453	318
Disability:	**36**	**36**	**36**	**36**
Anti-Physical	20	20	20	22
Anti-Mental	16	16	16	14
Multiple-Bias Incidents[d]	**8**	**17**	**18**	**10**

[a]Because hate crime submissions have been updated, data in this table may differ from those published in *Crime in the United States, 2000*.

[b]The term *victim* may refer to a person, business, institution, or society as a whole.

[c]The term *known offender* does not imply that the identity of the suspect is known, but only that the race of the suspect is identified which distinguishes him/her from an unknown offender.

[d]A *multiple-bias incident* is a hate crime in which two or more offense types were committed as a result of two or more bias motivations.

Source: U.S. Department of Justice, Federal Bureau of Investigation. Hate Crimes [Online] Available http://www.fbi.gov/ucr/cius_00/hate00.pdf, February 1, 2002.

the Christian Defense League, the Identity Church Movement, the Ku Klux Klan, the Posse Comitatus, and the Skinheads, to name a few. In addition, there are more than 200 militias operating in 39 states.[14] (Hate groups and militias are discussed later.)

The rationale for police involvement in hate crimes is that one of their primary roles is the enforcement or reinforcement of community values.[15] Even though a particular hate crime may be relatively minor and unorganized (e.g., graffiti, simple assault, or disorderly conduct), it may attack the very fiber of a community. In reality, it may be a much graver offense because it may result in special or unusual effects on the victim or the community as a whole. Hate crimes have a compounding effect that touches a group of people, a neighborhood, or a whole community. As such, the police must react decisively to hate crimes to ensure that public confidence is maintained and that deterioration of the community does not occur. A primary task of the police in this regard is the reporting and documentation of such crimes. Furthermore, the police must exhibit a commitment to eradicating hate crimes by thoroughly investigating them. The police must protect every citizen's rights regardless of culture, race, gender, or sexual orientation.[16]

In this age of globalization, hate groups—especially white supremacists— are becoming international in scope. Until the 1980s, America's postwar white supremacists were a ragtag collection of local Ku Klux Klansmen and neo-Nazis, who had little exposure to people or events overseas. Fueled by the Internet and inexpensive jet travel, neo-Nazi leaders are exchanging speakers and literature and forming chapters of their groups abroad. They have better funding and more hiding places and are capable of greater violence. And their numbers are rising; between 100,000 and 200,000 Americans are thought to be tied to the extreme right political fringe.[17] Among the most active neo-Nazi groups is Hammerskin Nation, a federation of so-called skinheads whose members sport swastikas along with their shaved heads and steel-toed boots; known for their violence, members run chapters in Australia, New Zealand, and across Europe and North America (often visiting the United States). The Ku Klux Klan has also ventured abroad, setting up chapters in Britain and Australia.[18]

The unofficial ambassador for the American radical right is William Pierce, leader of the National Alliance and author of the notorious *Turner Diaries*, a crude novel depicting an American race war in which the U.S. government is overthrown and Jews and minorities are systematically slain. Timothy McVeigh was one of the novel's fans, and his bombing of the federal building in Oklahoma City closely resembled a scene from the book. Under Pierce's leadership, the National Alliance has established chapters in 11 countries and wholeheartedly embraced the Internet (the *Turner Diaries*, once banned in Germany, can now be read on the alliance's Web site in German as well as French).[19]

Despite its expansion, however, the neo-Nazi movement remains widely fractured, and groups often hate each other. The danger, though, is that individuals—such as McVeigh—can cause enormous damage.[20]

The police must be prepared for all types of incidents, from both natural and human causes. Disasters such as floods bring several different agencies together (left), while incidents such as hostage situations require a special weapons unit. (*Courtesy Washoe County, Nevada, Sheriff's Department*)

THE MILITIA MOVEMENT

Today there exists in America a secretive, paranoid, and alienated political subculture that increasingly constitutes a threat to law and order. This subculture, whose political genealogy can be traced in part to notorious white-supremacist groups like those named earlier, has spawned a nationwide movement of heavily armed "patriot," "militia," or "citizen guerrilla" groups that are loosely connected except that they share a common distrust and even a deep hatred of The Enemy—the federal government and its law enforcement agencies. To some true believers, Washington is "the beast," while others consider it ZOG (Zionist Occupation Government). Their numbers may reach 100,000 in the 30 states where they exist, and they are potentially violent,[21] spending at least part of their time dressing up in camouflage, undergoing automatic weapons training, and preparing for final battle with the federal government.[22]

But at its root, the **militia** movement is about guns, and it staunchly opposes any form of gun control. To militias, the Brady bill and the 1994 federal ban on assault weapons are harbingers of totalitarianism. So was the bloody debacle of the federal assault on the Branch Davidians in Waco, Texas, on April 19, 1993, a siege in which 86 men, women, and children perished. For many people (including Timothy McVeigh), Waco was a wake-up call for Americans to realize that their government is working against its people. These individuals also point to the April 19, 1992, aborted ATF raid on Randy Weaver, a white supremacist whose wife was killed by a sniper in Ruby Ridge, Idaho. April 19, 1775, also holds symbolic significance for true believers in the movement—the date of the Battle of Lexington. It was no accident that McVeigh's bombing in Oklahoma City occurred on April 19, 1995; he had an obsession with April 19, and even his forged South Dakota driver's license showed the date of issue as April 19, 1993.[23]

The Militia of Montana, or MOM, is said to be the "Mother of all Militias" and the most prolific disseminator of support materials in the country. Among the dozens of these "unorganized" armed civilian groups that have mushroomed across the country, MOM is considered to be one of the most radical and also serves as an organizational model nationwide. Many, if not most, of the country's militias owe their existence in part to MOM's aggressive proselytizing and organizing campaign since the spring of 1994.[24]

Militia devotees often use the media, short-wave radios, and other means to spread their views. Some devotees advocate violence, such as shooting federal agents. A newsletter from the Militia of Montana exhorts its members to "GET YOUR THREE B'S [boots, beans, and bullets] PUT AWAY."[25]

POLICING THE MAFIA

ORIGIN AND ORGANIZATION

Imagine spending an entire police career and making only a few or even no arrests, devoting your career to the collection and analysis of intelligence information. Visualize yourself investigating crime by watching newspaper obituaries and appearing at funerals to log license plates and take photographs from a safe distance away from the mourners, then tracing names to the fifth cousins. Consider spending years to investigate a single case, combing through records and files that go back 20 years. Picture yourself going undercover for several years to investigate a crime organization, never knowing when your cover will be blown and you will be targeted for death (see the Practitioner's Perspective).

Enter the world of organized crime. These are only some of the unusual methods and schemes that are employed by police agents who are assigned to the underworld. Organized crime is defined by the FBI as "any group having some manner of formalized structure and whose primary objective is to obtain money through illegal activities." Under this definition, several organized-crime syndicates or organizations exist in America, such as street youth gangs, prison gangs, those from several Asian countries (such as the Chinese Triads, the Japanese Boryokudan, and Vietnamese Tongs), bikers, and Gypsies. Probably the oldest, most profitable, and dangerous form of organized crime in America is the **Mafia**, also known as "La Cosa Nostra," or "this thing of ours."

Mafia origins can be traced to thirteenth-century Sicily. The history of the Mafia is a confusing hodgepodge of lore and legend, but the one having the most appeal (probably due to its romantic nature) involves the French occupation of Sicily and Italy in 1282. The local people used a form of guerrilla warfare to fight the French military, engaging in hit-and-run tactics and using their knowledge of the rugged terrain to fight the superior French army. A legend holds that a young French soldier and an Italian maiden fell in love during this military occupation. Unfortunately, they were discovered and assassinated by the French army. Legend holds that following the deaths of these two young people, a vindictive cry arose

PRACTITIONER'S PERSPECTIVE

Going Undercover: An FBI Agent's Two-Year Experience

George Togliatti

George Togliatti joined the FBI in 1973 and was subsequently assigned to offices in Boise, Detroit, and Las Vegas. In Las Vegas, he supervised the White Collar Crime, the Property Crime, and the Organized Crime Investigations and Drug Programs. He holds a bachelor's degree in economics from Iona College and a master's in criminal justice administration from the University of Detroit. He retired from the FBI in 1996. The following is a case study of his role as a supervisory special agent.

George Togliatti is soft spoken, part Italian and part Irish. He obtained degrees in economics and criminal justice administration and flew search-and-rescue helicopter missions for five years in North Vietnam. He then worked in the computer industry and took graduate accounting courses. Not happy with his career path, he joined the FBI at age 28. After short assignments in Butte, Montana, and Boise, Idaho, he acquired his first undercover experience in Detroit. But none of that could have prepared Togliatti for his undercover role as "George Dario"—the undercover Mafioso who loved partying and the criminal lifestyle in the Las Vegas underworld.

Information was obtained that mobsters from Detroit and New York were attempting to gain a major foothold in Las Vegas. Wiretaps also revealed serious attempts to skim casino profits before they were counted for tax purposes, some Mafiosi were killing one another, and ties between organized labor and the mob were deepening. The FBI conceived Operation Desert Fox, where Togliatti would operate on the fringe of mob activities and gather information for later prosecutions. Although it was originally intended to be a two-week assignment, Togliatti would be undercover for two years.

"George Dario" sported a Rolex watch, drove a Cadillac Seville, and had access to a $500,000 home in a country club addition. He had little trouble assuming his new identity. He was readily accepted by the mob. He quickly made new friends, his glib, streetwise New York background serving him well. He assumed a phony background as a sales manager for a legitimate New York music company. Cruising casinos, bars, and restaurants, he noted criminal activities, such as extortion cases and drug deals. He was once approached by a man with an offer of $10,000 to beat up a girlfriend's ex-husband.

To fit into the underworld, where crime, sex, and drugs were easily available, Dario's values were often compromised, as when he would have to fake snorting cocaine. But he realized that one improper move would blow his cover and possibly cost him his life. He avoided committing several illegal and unsafe acts by turning ugly, telling associates he preferred alcohol to drugs and saying he wanted a clear head during his crimes. He saw rampant drug addiction among his associates, though, and today many of Dario's former associates who cheated on their wives wonder if that information is still tucked away in some FBI file.

Two close friends—fellow FBI agents—kept Dario informed of developments with the information he had uncovered. The trio often met in hideaway restaurants to discuss goings-on in the bureau office and the other world in general. The friends also performed support work, such as checking license plate numbers or hotel registrations to identify people.

Eventually two years had passed and Dario realized that it was time to become Togliatti again; the mob had learned he was a fake and the telephone lines were abuzz. Operation Desert Fox was a huge success. Would Togliatti do it again? Today he says no; he can never recover the two years lost from his family and feels it was too high a price to pay. Yet he says it made him a better special agent; indeed, several promotions came his way in the next five years. He plays down the danger surrounding what he did, saying a lot of people in the FBI could have done the job. And, he says, "Vietnam was tougher, hands down."

Portions excerpted from, and used with permission of, the *Reno Gazette-Journal*, November 1, 1987, pp. 1A, 14A.

among the people: "*Morte alle Francia Italia anela,*" meaning "death to the French is Italy's cry." The acronym of this cry is M.A.F.I.A.

Many Sicilian and Italian mafiosi immigrated to the United States in the late 1800s and early 1900s, maintaining their criminal lifestyle of intimidation and force. It has been argued that organized crime existed even in the Wild West, focusing on the crimes of gambling, prostitution, robbery, and cattle rustling.[26]

Mafia families have a formally organized nature, consisting of the El Capo (or "godfather"), Sotto Capo ("underboss"), the Consigliere ("counselor"), caporegimas ("lieutenants," or enforcers), and the soldatos (or "soldiers") (see Figure 8–1). The major obstacle to investigating and apprehending these criminals is that they shun the public spotlight. In fact, their primary, centuries-old code of honor is that of *omerta*—silence.

Mafiosi loathe unwelcomed attention for the most part; it has been said that it is almost as if the mafiosi are products of parthenogenesis—as if they just arose out of the dust, like the mythical phoenix. When confronted by a police officer and

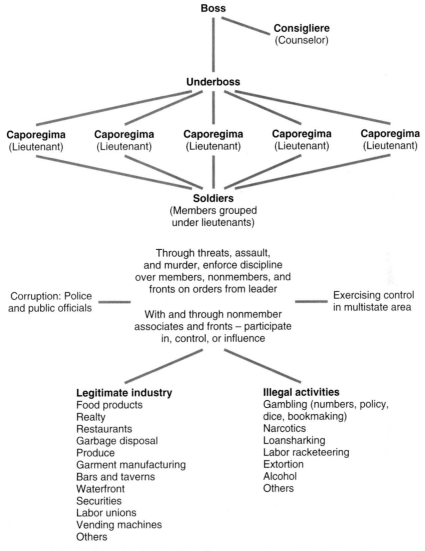

Boss

Consigliere
(Counselor)

Underboss

Caporegima
(Lieutenant)

Caporegima
(Lieutenant)

Caporegima
(Lieutenant)

Caporegima
(Lieutenant)

Caporegima
(Lieutenant)

Soldiers
(Members grouped
under lieutenants)

Through threats, assault,
and murder, enforce discipline
over members, nonmembers, and
fronts on orders from leader

Corruption: Police
and public officials

Exercising control
in multistate area

With and through nonmember
associates and fronts – participate
in, control, or influence

Legitimate industry
Food products
Realty
Restaurants
Garbage disposal
Produce
Garment manufacturing
Bars and taverns
Waterfront
Securities
Labor unions
Vending machines
Others

Illegal activities
Gambling (numbers, policy,
dice, bookmaking)
Narcotics
Loansharking
Labor racketeering
Extortion
Alcohol
Others

FIGURE 8–1 An Organized Crime Family. (*Source:* The President's Commission on Law Enforcement and Administration of Justice, *Task Force Report: Organized Crime* [Washington, D.C.: Author, 1967], p. 9.)

asked questions about himself (the Mafia is an all-male organization), he seems to have no roots, parentage, or any other background that the police can maintain as intelligence information.

SUCCESSFUL POLICE OFFENSIVES

For 40 years, FBI Director J. Edgar Hoover denied that the Mafia even existed. Until very recently the "mob" had a veritable chokehold on America; it was involved in killing for hire, extortion, loan sharking, money laundering, narcotics, prostitution, smuggling, bookmaking, bribery, business infiltration, embezzlement, hijacking, horse-race fixing, racketeering, pornography, bank fraud, and fencing. By the 1960s, the Mafia's influence extended from America's largest labor unions into trucking, construction, longshoring, waste disposal, gambling, and garment making; it had grown into a multibillion-dollar syndicate of criminal enterprises run by 26 "families" nationwide.[27]

Beginning in the mid-1980s, however, the FBI spearheaded an unprecedented assault on the Mafia, putting away two generations of godfathers on racketeering charges. That initiative continues today and has met with tremendous success toward dealing with this very elusive group.

How has this success been obtained? Through tenacity, hard work, and some luck—and for three fundamental, practical reasons:

- *The expanded use of electronic eavesdropping (wiretapping).* When federal agents were listening in on nearly a million conversations a year, mobsters were forced to operate in a climate of constant suspicion. Cooperation between federal and state law enforcement authorities also speeded up convictions. Furthermore, the Internal Revenue Service can legally share information, enabling it to assist in complicated tax cases in almost two-thirds of all organized crime prosecutions.

- *The use of informants.* Even more significant than bugs in the toppling of mob kingpins has been the use of former gangsters who have proved willing to violate the code of *omerta*, turning state's evidence against their old cronies. This younger generation of turncoats, or "stool pigeons," is largely the result of a lesser degree of dedication to the old Sicilian-bred mores. More youthful gangsters do not possess the level of loyalty to the family their ancestors did. Many feel that a long prison sentence is too stiff a price to pay for the family. Government prosecutors have managed to exploit this new, less-than-loyal breed of mafiosi. With lieutenants and captains in crime families becoming informants, organized crime is finding that it cannot rely on its membership. The willingness of some hoodlums to defy the Mafia is partly due to the existence of the federal Witness Protection Program within the U.S. Marshals Service; since 1970, the program has helped several thousand people move to different locations and acquire new identities and jobs.

- *The Racketeer Influenced and Corrupt Organizations Act (RICO).* The single most important piece of legislation ever enacted against organized crime, RICO defines racketeering in a very broad manner. It includes many offenses that do not ordinarily violate a federal statute and attempts to prove a pattern of crimes conducted through an organization. Under RICO, it is a separate crime to *belong* to an enterprise that is involved in a pattern of racketeering.

The number of Mafia families has fallen to just nine today. And while the estimated number of full-fledged (sworn, or "made") members has remained at about 1,100 for more than three decades, half of those are now in jail or inactive. Experts believe the Mafia's Commission—a ruling body of godfathers who mediate disputes—has not met since 1996. Once-thriving Mafia families in Cleveland, Detroit, Kansas City, and Milwaukee are "down to two or three guys doing low-level scams."[28]

The FBI can take well-earned credit for the Mafia's change in fortune. During the tenure of Director Louis Freeh (1993–2001), the agency compiled 274 convictions of Mafia figures, including 20 top bosses. During the year 2000 alone, the FBI and New York City police arrested more than 70 people in the New Jersey crime family, decimating its leadership.[29]

But the Mob has proved to be remarkably resilient and is moving into sophisticated white-collar crimes such as stock swindles. And while the Mafia may be in relative latency, other organized ethnic groups have risen: Latin-American drug traffickers, Nigerian con artists, Chinese gambling bosses, and Russian mobsters. The FBI needs linguists and ethnic agents who can penetrate the new wave of mobsters.[30]

POLICING STREET GANGS

DEFINITION AND EXTENT OF THE PROBLEM

A youth **gang** is an association of individuals who have a name and recognizable symbols, a geographic territory, a regular meeting pattern, and an organized, continuous course of criminal activity.[31]

The formation of youth gangs in the United States is not a recent phenomenon. Hispanic youth gang activity was first recognized in the early 1900s in the southern California area. Indeed, a major study of gangs in 1927 found that most gangs were small (6 to 20 members) and formed spontaneously in poor and socially disorganized neighborhoods, the result of disintegration of family life, inefficiency of schools, formalism and externality of religion, corruption and indifference in local politics, low wages and monotony in occupational activities, unemployment, and lack of opportunity for wholesome recreation.[32] This explanation of causes of gang affiliation still applies in large measure today.

According to U.S. Justice Department estimates,[33] there are more than 16,000 gangs and over half a million gang members in the United States.[34] Nearly half of all gang members (47.8 percent) are African American youth, while Hispanic youngsters accounted for 42.7 percent and Asians totaled 5.2 percent. (Specific types of ethnic and racial gangs are discussed later.)

Today gangs are a substantial problem in America, and their members are becoming younger. Most research on youth gangs in the United States has concluded that the most typical age range of gang members has been approximately 14 to 24; researchers are aware of gang members as young as 10 years of age, however, and in some areas (such as southern California, where some Latino gangs originated more than 100 years ago), one can find several generations in the same family who are gang members with active members in their 30s. Youngsters generally begin hanging out with gangs at 12 or 13 years of age, join the gang at 13 or 14, and are first arrested at 14.[35]

ORGANIZATION, REVENUES, MORTALITY

Studies of individual gangs have found a leadership class (typically a leader and three officers), a foot soldier class (ranging from 25 to 100 members age 16 to 25), and rank-and-file members (usually 200 persons younger than high school age). More than 70 percent of a gang's total annual revenue of approximately $280,000 can be generated from the sale of crack cocaine. Dues can provide some additional gang revenue (about 25 percent), and extortion directed at local businesses and entrepreneurs represents about 5 percent. The gang generally operates in a neighborhood of roughly four city blocks.[36]

When a gang member is killed, the gang pays funeral expenses (about $5,000 per funeral) and typically provides compensation to the slain member's family. Leaders in the studies were found to retain about $4,200 per month from the gang's revenues, for an annual wage of about $50,000. The longer a member was in a leadership position, however, the greater his "salary," which could go as high as $130,000. Officers (runners, enforcers, or treasurers) earned about $1,000 per month, and foot soldiers were paid about $200 per month for working about 20 hours per week (but foot soldiers were also allowed to sell drugs outside the gang structure).[37]

The annual death rate among gangs' foot soldiers in the studies was 4.2 percent—more than 40 times the national average for African American males age 16 to 25. Most alarmingly, gang members who were active for a four-year period had roughly a 25 percent chance of dying; members also averaged more than two nonfatal injuries (mostly from gunshots) and nearly six arrests over a four-year period.[38]

ETHNIC AND RACIAL TYPES OF GANGS

Gang members usually join the gang either by committing a crime or undergoing an initiation procedure. Members use automatic weapons and sawed-off shotguns in violent drive-by shootings while becoming more sophisticated in their criminal activities and more wealthy.

Gangs are comprised of three types of members: *hard core* (those who commit violent acts and defend the reputation of the gang); *associates* (members who frequently affiliate with known gang members for status and recognition but who move in and out on the basis of interest in gang functions); and *peripherals* (those who are not gang members, but identify with gang members for protection—usually the dominant gang in their neighborhood). Most females fall into this latter category.

There are several levels of gun-toting "gang-bangers" wanting to earn their stripes. The "wannabe" begins by having target practice and handling the guns; he may shoot but does not actually aim. The next level is the gang-involved youth who wants a tough-guy reputation; he will eventually kill somebody, but he's not seen as a hard-core crazy person. When a teenager reaches that level—the crazed killer—he doesn't care about himself or his victims; his violence is random and cold blooded.

The most prominent African American gangs are the Crips and the Bloods. The Crips began in Los Angeles in 1969, reportedly on the campus of Washington High School, supposedly as Community Resources for an Independent People; one of the school's colors was blue, which is now the color of gang identification. Another popular belief is that their name derived from "crypt," from the then-popular *Tales from the Crypt* horror movie. Crips address each other with the nickname "Cuzz." Crip graffiti can be identified by the symbol "B/K," which stands for "Blood Killers." The Bloods are reported to have formed as a means of protection against the Crips in and near Compton, California. Bloods use the color red and address each other as "Blood." Gang graffiti frequently uses the terms "BS" for "Bloodstone" or "C/K" for "Crip Killers." Both Crips and Bloods refer to fellow gang members as "homeboys" or "homeys."[39]

Hispanic gangs invariably name their gangs after a geographical area, or turf, that they believe is worth defending. Hispanic gang activity often becomes a family affair; young boys (10–13) are the "pee wees," the 14- to 22-year-olds are the hard core, and those living beyond age 22 become a *vetrano*, or veteran. Headgear (knit cap or monickered bandanna), shirts (Pendleton or T-shirt), pants (highly starched khaki or blue jeans), tattoos (of gang identification), and vehicles (preferably older model Chevrolets, lowered and with extra chrome and fur) are standard fare.

White gangs are also expanding their membership, as growing numbers of young neo-Nazi Skinheads are linking up with old-line hate groups in the United States. This unity has bolstered the morale and criminal activity of the Ku Klux Klan and other white supremacist organizations.

GRAFFITI AND HAND SIGNALS

Nonverbal forms of communication not only allow gang members to communicate with each other and with rival gangs; they also bring recognition that the members need and seek out. The style and quality of **graffiti** can create and enhance the gang's image; thus, those who draw it are given additional status. Graffiti also serves to mark the gang's turf; if a gang's graffiti is untouched or unchallenged for a period of time, the gang's control in that area is reaffirmed.

Chicano gangs use a nonverbal communication that has existed for over 50 years. This method, called a *placa*, allows the Chicano gang member to express messages about himself, his gang, and other gangs and make direct challenges to others. A full *placa* expresses the gang's opinion of itself, its control of the area, and a warning that other gangs are helpless to do anything about it. Figure 8–2 is an example of a full placa, with the translation given at the right (the actual graffiti would be written in unique gang style). Figure 8–3 provides a guide to reading gang graffiti.

Hand signals, or "throwing signs," are made by forming letters or numbers with the hands and fingers depicting the gang symbol or initials. This allows the gang member to show which gang he belongs to and issues challenges to other gangs in the vicinity. Figure 8–4 shows some examples of hand signals as they are commonly used in the western United States.

Police agencies in jurisdictions experiencing gang problems must develop expertise in gang movements, activities, and all forms of nonverbal communication. Police have also developed intelligence files on known or suspected gang members.

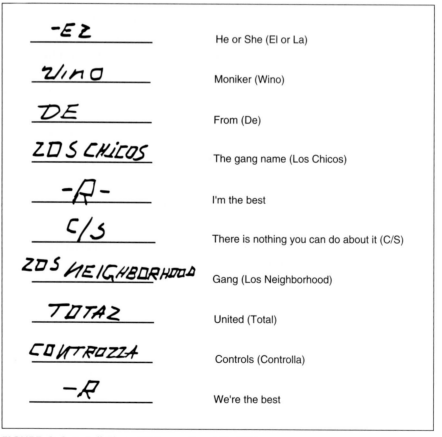

-EZ	He or She (El or La)
Wino	Moniker (Wino)
DE	From (De)
ZOS CHICOS	The gang name (Los Chicos)
-R-	I'm the best
C/S	There is nothing you can do about it (C/S)
ZOS NEIGHBORHOOD	Gang (Los Neighborhood)
TOTAZ	United (Total)
CONTROZZA	Controls (Controlla)
-R	We're the best

FIGURE 8–2 A Full *Placa.* (Chicano Gang Graffiti)

GUIDE TO READING GANG GRAFFITI

1. Step One
   Barrio or Varrio
   Meaning Neighborhood
   or Group/Clique

   B̲ H G R
   PQS
   -13-
   L' s

2. Step Two
   The 'HG' Meaning
   Hawaiian Gardens City
   and Gang/Clique

   B H̲ ̲G̲ R
   PQS
   -13-
   L' s

3. Step Three
   The Letter 'R' Meant to
   Be "RIFA" Meaning Rule,
   Reign, or Control

   B H G R̲
   PQS
   -13-
   L' s

4. Step Four
   The actual gang group
   abbreviation of 'PQS'
   "PEQUENOS" from Hawaiian
   Gardens. (Normally younger
   groups, i.e., Chicos,
   Midgets, or Tiny's.)

   B H G R
   P̲Q̲S̲
   -13-
   L' s

5. Step Five
   The Number '13' * stands
   for "SUR" meaning
   Southern California

   B H G R
   PQS
   -̲1̲3̲-̲
   L' s

* The number 13 is sometimes used by younger gang members to mean marijuana.

6. Step Six
   The letter 'L' or 'L's'
   is used to mean the Vato
   Locos or the Crazy Ones/
   Brave Ones. Not normally
   a separate gang or clique.

   B H G R
   PQS
   -13-
   L̲' ̲s̲

FIGURE 8–3 Guide to Reading Gang Graffiti.

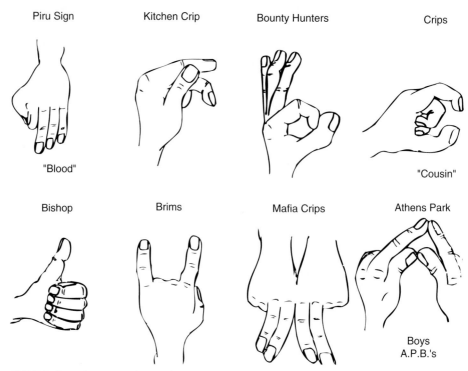

FIGURE 8–4 Gang Hand Signals.

SOCIETAL RESPONSES

Determining the best course of action for dealing with street gangs is not easy. Most experts describe programs in communities with gangs that would include some combination of the following:

- *Fundamental changes in the way schools operate.* Schools should broaden their scope of services and act as community centers involved in teaching, providing services, and serving as locations for activities before and after the school day.
- *Job skills development for youths and young adults accompanied by improvements in the labor market.* Many youths have dropped out of school and do not have the skills to find employment. Attention needs to be focused on ways to expand the labor market, including the development of indigenous businesses in these communities, and to provide job skills for those in and out of school.
- *Assistance to families.* A range of family services, including parent training, child care, health care, and crisis intervention, must be made available in communities with gangs.
- *Changes in the way the criminal justice system—particularly policing—responds generally to problems in these communities and specifically to gang problems.* Many people believe that police agencies need to increase their commitment to understanding the

communities they serve and to solving problems. This shift is currently being addressed through the proactive community oriented policing and problem solving approach (COPPS, discussed in Chapter 6). COPPS brings the community and police together in a partnership to work toward reducing neighborhood disorder and fear of crime.[40] Case studies about successes with the COPPS approach to gangs are becoming more commonplace.

- *Intervention and control of known gang members.* Illegal gang activity must be controlled by diverting peripheral members from gang involvement and criminal activity. Achieving control may mean making a clear statement—by arresting and incapacitating hard-core gang members—that communities will not tolerate intimidating, violent, and/or criminal gang activity.[41]

Obviously something more is needed than police work alone to break the cycle of gang delinquency. In too many communities, gang violence is tolerated as long as gang members victimize each other and do not bother the rest of society.[42] Without community support, the contemporary cycle of youth-gang activities will continue; even gang members who are imprisoned join branches of their gang behind bars, while replacements are found to take their place on the street.[43]

Across the country, many cities have responded to their gang problems by forming some type of special unit—often suppression oriented—as an initial response to major episodes of gang violence.[44] Today, about two-thirds of all large cities in the United States with a population of more than 200,000 have a specialized gang unit, while about half of all other cities have such a unit.[45] These gang intelligence units (GIUs) identify the core members of the gangs and target them for enforcement. Intelligence is collected by questioning suspected gang members who are arrested and talking with rival gang members and with residents in gang neighborhoods. At the same time, officers collect information on gang activities, monitor graffiti, collect information on assaults and homicides of gang members by rival gangs, and observe disputes involving drug sales. They have also enhanced collaboration with agencies external to the police agency (schools, local code enforcement agencies, probation agencies, community groups, and so on).[46]

Over time, however, many police agencies have shifted from an emphasis on suppression to one with more emphasis on education. A well-known police response to gangs is the Gang Resistance Education and Training (GREAT) program, which originated in 1991 in Phoenix, Arizona. GREAT emphasizes the acquisition of information and skills needed by students to resist peer pressure and gang influences; the curriculum contains nine-hour lessons offered to middle school students, mostly seventh-graders.

Federal law enforcement initiatives are also working to suppress gang activities. The Immigration and Naturalization Service's Violent Gang Task Force engages in proactive, interagency, multijurisdictional operations against criminal alien gangs. The Bureau of Alcohol, Tobacco and Firearms has also expanded its focus to include criminal gangs, particularly violent gang groups. Today the ATF maintains five databases that are dedicated to tracking outlaw motorcycle organizations, Bloods and Crips, prison gangs, and Asian gangs.[47]

Policing in very small communities can differ in some respects from that in larger jurisdictions. But, like their counterparts in larger jurisdictions, these officers still have serious order-maintenance responsibilities. (*Courtesy Dan Peak*)

A fundamental question to ask is this: How do we as a society replace the gang's social importance and financial benefits (children in gangs earn literally hundreds of dollars a day in drug-related activities) with education or work programs that pay minimum wage? This is a complex issue with no simple answers.

POLICING IN SMALL AND RURAL JURISDICTIONS

America is largely rural; hence, there is a style of policing that has a nature all its own: that of the small town or rural area. The rural environment is a significant part of the country, and **rural policing** employs a significant proportion of law officers. Approximately half of the nation's 17,000 local police departments and sheriff's offices employ fewer than 10 sworn officers; 90 percent serve a population of under 25,000.[48]

Today's new breed of rural officer is often well trained, and many come from large departments to the more peaceful, less stressful small-town environment. However, with this setting come several crime problems and special policing issues and concerns as well.

CRIME PROBLEMS

The belief that crime is less frequent in rural areas than in urban areas is supported by the FBI's Uniform Crime Reports. As with urban areas, larceny is the most common crime; motor vehicle theft is the least common. The greatest difference between rural and urban crime is robbery, which occurs almost 54 times more often per 100,000 citizens in urban areas.[49]

Rural areas have experienced a disproportionately large increase in crime, growing as much as three or four times faster than the population. A decline in a sense of acquaintanceship among community members is largely responsible.[50] Biker gangs have a long history of criminal activity in rural settings.[51] Also, there are indications that patterns of urban drug use, including the use of crack, are spreading to rural areas.[52] Some reports suggest that rural areas may serve as production sites for many forms of narcotics.[53] Alcohol is also of particular concern in rural areas, and DUI is more common there as well.[54] There is also reason to believe that vice and organized crime are features of the rural environment. Prostitution is often found to exist for truck drivers, and moonshine and bootlegged items, stolen auto parts, and other illegal merchandise are often transported on rural routes. So-called hate groups are also found to exist with frequency in rural areas. Sharing a combination of anti-Semitism, racism, fundamentalist Christianity, and a deep suspicion of government, homosexuality, driver's licenses, the IRS, the liberal media, the Federal Reserve Bank (which, they believe, is controlled by Jews), these groups have beliefs that can lead to violence, as the earlier discussion on militias showed.

Agricultural crimes are also a rural problem. Average citizens fail to realize that cattle rustling and other Old West–type crimes still occur. With the help of a herding dog, rustlers can load 10 to 15 head of cattle on a trailer and be gone within minutes—a crime that amounts to millions of dollars in losses per year in California alone. A unique form of assistance is now available for rural police officers in the West: a Rural Crime School. Twice a year, the California Rural Crimes Task Force, in Gilroy, operates its 40-hour school for investigators covering cattle rustling, theft of horses, deer poaching, oil and irrigation field theft, and dog fighting, among other things. Since 1996, the school has trained more than 400 police officers from California and other western states.[55]

High rates of poverty and the availability of guns (gun ownership is much more prevalent in rural areas) have also been associated with high levels of rural crime. And, with smaller tax bases, rural police departments are more likely to be seriously understaffed and without important resources; some are even unable to tap into statewide systems for record checks or vehicle registrations.

UNIQUE DUTIES

Small town police officers and county deputies often have vastly different duties than their urban counterparts. First, rural officers perform more mundane, non-police chores; some examples would include locking and unlocking municipal parking lots, emptying parking meters, delivering meeting agendas to the homes of governing officials, picking up the daily receipts at the local swimming pool or golf course, or logging the status of the machinery at the electrical plant. Rural agencies usually have little, if any, specialization; the officers become jack-of-all-trades generalists, working criminal cases from beginning to end. The work also varies—a deputy sheriff may be working a burglary today and a livestock theft tomorrow or patrolling the most remote areas in a four-wheel-drive truck or on a horse. (The two latter aspects of rural policing may actually be advantages, however.)

Disadvantages to rural policing include a reduced salary and benefits package; limited mobility, promotional advancement, or transfers to specialized assignments; job stability that is more tenuous where there is no civil service protection; few available social services agencies available for referral of people with problems; backup officers—if there are any—may be a considerable distance away; and being "under a microscope" more than their urban counterparts, both professionally and socially.[56]

Some Concerns

Given the large number of small police agencies in America—many of which no doubt have adequately trained employees—there are nonetheless several concerns with this level of policing. These concerns center on the ability possessed by some officers and the type of "justice" that some of these officers mete out (such as a marshall who issues traffic citations for "offenses" that were not personally witnessed but on the basis of information provided by a citizen).

A related problem is the ability of small agencies—which are often suffering training and personnel shortages—to provide a full-service police agency, one that can assure the proper handling and analysis of evidence at major crime scenes. As an outgrowth of this concern, in 1973, the National Advisory Commission on Criminal Justice Standards and Goals recommended that "at a minimum, police agencies that employ fewer than 10 sworn employees should consolidate for improved efficiency and effectiveness."[57]

Because the training and abilities of a vast number of small or rural police agencies cannot keep pace with today's sophisticated criminal element, many observers believe that contract and consolidated policing, while perhaps less politically popular, would greatly assist small jurisdictions.

POLICING THE HOMELESS

Contributing Factors

"It's just not going to go away no matter what type of enforcement posture you have on it. You can't just sweep them up and move them out of town."[58] That

Contemporary police officers must often deal with problems and offenses of homeless persons. (*Courtesy Ron Glensor*)

statement by a Santa Barbara, California, police officer sums up the prevailing frustration of the police toward America's **homeless** "street people." The pessimistic attitude reflects a dilemma police officers face in many jurisdictions. There are strong pressures on the police to "do something" about this growing segment of our society, but officers are severely limited in what they can do.

The **deinstitutionalization** policies of the 1970s said, in effect, that people could not be involuntarily committed unless they posed a risk of harm to themselves or others; this freed thousands of people from institutions but virtually threw many of them into the streets, riverbanks, parks, and gutters of America. Many of the homeless cannot fend for themselves; they have become a major social problem in our cities.

Who are the homeless? First, as always, it seems, estimates of people who are homeless continue to range widely, from a low of 230,000 to as many as 750,000 people nationwide.[59] Several studies have reported that 25 to 45 percent of the people living on the streets are alcoholics. A multicity study by the United States Conference of Mayors found that an average of 29 percent of the homeless people suffered from severe mental disorders. Many more homeless people suffer from less severe psychological disorders that nevertheless prevent them from holding stable jobs. A surprising number of the homeless are military veterans; runaways comprise another sizeable category. Many homeless people, however, are not alcoholic or mentally ill; they are destitute. Increasingly, intact families and women with children are homeless as a result of economic dislocation.[60]

A NEW INTOLERANCE

It has been more than a decade since reducing homelessness was viewed as an urgent national priority, when celebrities and citizens joined in high-profile efforts such as Hands Across America, which raised $15 million for the cause. Instead of sympathy, the street dwellers now often attract hostility, as citizens are tired of being accosted by aggressive panhandlers, and vendors say the homeless are harming their business.[61]

According to a report by the National Law Center on Homelessness and Poverty, at least 24 cities have used police to forcibly remove the homeless from certain sections of town. As the new century commences, the homeless problem is particularly perplexing, since it persists while so many other social ills seemed to be on the mend: Violent crime was down; the numbers of those on the welfare rolls were at their lowest in years; pregnancies, sex, and drug use among teens were dropping; and unemployment rates were the lowest they had been in a quarter-century. Yet the homeless situation remained as bad or grew worse.[62]

What seems to have changed is the nation's tolerance level, and many cities are working hard to make the homeless "disappear." Even such bastions of liberalism as California's Berkeley and San Francisco took legal action against homeless people. Berkeley's city council passed an ordinance banning groups of more than two dogs at a time on city streets (the law was clearly aimed at young homeless people, who often have canine companions), and San Francisco supervisors voted to bar alcohol in several parks frequented by the homeless and allow shopping centers to use private security guards to roust indigents.[63]

Still, the problem persists: The demand for emergency shelter has not declined in the past 15 years, and increased federal funding has not helped the situation, despite the $5 billion spent on programs designed to combat homelessness since 1993.[64]

WHAT CAN THE POLICE DO?

In addition to using the kinds of ordinances and statutes just described, there are two common police approaches to the homeless problem. First, police can employ a strict enforcement policy, routinely arresting street people for crimes and generally maintaining pressure on them to keep moving. This approach has worked in many cities and has allowed businesses to thrive in areas formerly rife with street people. Second, the police can employ a policy of benign neglect by which street people are virtually ignored except during the winter when the temperature drops below 10 degrees. Then, mental health workers patrol the streets in teams with the police, offering shelter to the homeless. However, the police often find that detox and mental health shelters are either in short supply or they simply refuse to accept police referrals of the homeless—especially chronic alcoholics.

Research has found that one-half of all police agencies have no training programs dedicated to handling the homeless. Roll-call training was the most common method of informing officers about shelter availability. Surveys have also found a surprisingly limited store of police data on the homeless. This gives rise to several questions: Are police executives misinformed concerning the degree of homelessness in their jurisdictions? Is the magnitude of the problem on a national level less than that portrayed by the media? Several policy questions also surface: Should police agencies consider conducting a census to assess the degree of homelessness within their jurisdictions? Should they consider the actual needs of the homeless? Should detailed record-keeping procedures be put into place? Should police work with shelters and other agencies to develop formal policies regarding referrals and police transportation of street people?[65]

A well-planned response to homeless populations, with an emphasis on a COPPS approach, would allow police departments to manage significant social problems while making efficient use of police resources.

POLICING AMERICA'S BORDERS

NIGHTLY DRAMAS AT OUR PERIMETER

In Chapter 2, we briefly discussed the Immigration and Naturalization Service (INS), which oversees the **U.S. Border Patrol**. The work of the border policing is unique and difficult, however, and thus it is included here for further discussion.

Illegal aliens closely watch Border Patrol vehicles, waiting for an opportunity to cross the border. (*Courtesy U.S. Border Patrol*)

Each night a drama is played out in many American cities as a hide-and-seek game unfolds at the nation's borders—especially in California, Arizona, and Texas—as thousands of people gather to break onto American soil and grasp at fortune and a new way of life. About 3 million of the 8.5 million Mexican-born people living in the United States are believed to be living here illegally.[66]

Many people do not consider illegals to be a crime problem. In fact, because of their general clandestine lifestyle, they are most likely to avoid contacting the police when they are victimized. One crime that has been disproportionately associated with illegal aliens is drug trafficking. This, too, police officials say, is a false impression. While a few illegal aliens do carry drugs across the border, the typical alien generally comes from a rural area and wants only to work. Most aliens do not trust banks and carry large sums of money in their pockets; therefore, they are prime targets for criminals.[67]

STEPPED-UP POLICE EFFORTS

Several actions by the federal government since the mid-1990s have combined to change the methods practiced by both the illegal immigrants and federal law enforcement agencies:

- In 1994, the U.S. government launched an operation (known as Operation Hold the Line) that dramatically reduced the number of migrants coming through key points in Texas.
- Beginning in 1995, the U.S. policy concerning illegal immigration shifted to a focus on deterring unauthorized entries instead of trying to catch illegals once they had crossed into the United States. This new policy doubled the number of agents and increased the Border Patrol budget to $1.4 billion to stanch the flow.[68]
- In 2000, the Border Patrol in the San Diego Sector increased its personnel (with over 2,000 agents) and equipment, effectively closing that point of entry by employing a fleet of helicopters; infrared systems; Zodiac boat teams patrolling river and ocean approaches; special units using vans, bicycles, all-terrain vehicles, and horses; and night-vision scopes and digitally scanned fingerprints (to speed identification of smugglers and migrants). Vegetation stress analysis is being tested in San Diego as well; software analyzes infrared aerial photography to determine differences between healthy vegetation and vegetation stressed by the passage of people and the waste they leave behind.[69]

As the United States has made it more difficult to cross the border, the price and the profits have shot up for the people smugglers, or "coyotes." A highly organized multimillion-dollar business was born, and smugglers even offered illegals a guarantee: You don't pay until you get to Phoenix; but then the fee rises to a steep $800. The coyotes—who realize an estimated $900 million per year in profits—use several types of employees: A *vendepollo*, or "chicken seller," works the streets in Mexico looking for new clients; a *brincador*, or "fence jumper," guides migrants across the border; a driver delivers them to a house in Phoenix, where they stay until the payment is wired from a relative in the United States.[70]

A Border Patrol infrared scope truck, capable of sensing body heat, is deployed for nighttime operations along the border. (*Courtesy U.S. Border Patrol*)

A New Hot Spot: Arizona

The result of these efforts has been a funneling of migrants into Arizona, now the most popular crossing point in the United States. About 450,000 illegals were caught there in 1999. The Border Patrol has begun fortifying the Arizona border, using agents, floodlights, infrared cameras, and seismic sensors that detect footsteps. Illegals who are captured are loaded into vans and sent back to Mexico—which represents the beauty of people smuggling: Once migrants are caught, the game merely starts over.[71]

Yet hundreds of human trails still weave through the desolate scrub near the Arizona-Mexico border. But seeking the American dream via the Arizona desert is fraught with danger. Indeed, in one incident 14 illegals died from exposure in the desert near Yuma, Arizona, in June 2001. They were part of 27 border crossers who were abandoned by their smuggler; they stumbled through harsh terrain for four days without sufficient water and in 115-degree heat. Four hundred Mexicans died during 2000 while taking more dangerous routes north to circumvent the tighter security; at least 140 died between October 2000 and June 2001. Critics blame the deaths on the U.S. policy of immigration prevention; agents block traditional crossing routes and therefore push migrants into more remote, deadly areas.[72] In July 2001, the Border Patrol expanded its special search-and-rescue team—the Border Search, Trauma, and Rescue (BORSTAR) team—in the southwest in response to the rising number of deaths in those sectors. During the first half of 2001, BORSTAR rescued about 300 immigrants—more than twice the number rescued in all of 2000.[73]

Drug smugglers are also busily trying to get into this country through Arizona. The amount of drugs captured in Arizona during fiscal year 1999 shows how large the problem has become in that area: over 11,000 pounds of cocaine, 168,000 pounds of marijuana, and $13.2 million in currency (more than five times the amount of the year before). Two-thirds of the cocaine headed to the United States now crosses the southwest border. Smugglers know the agents' houses and routines. But the Border Patrol is having some successes, owing to the use of Native American trackers. Such trackers travel light; they carry a submachine gun, a radio, and a global positioning system so they can call in backup Blackhawk helicopters. The smugglers' trails are distinctive because they carry heavy loads, digging their feet deep into the sandy ground.[74]

Exacerbating the problem of drug smuggling into Arizona is the fact that smugglers have been paying off some employees of the Immigration and Naturalization Service. While most immigration agents are honest, according to one FBI chief, arrests of federal agents constitutes a "national disgrace."[75] One arrested agent, who was earning $35,000 per year, had about $300,000 in cash in his home; two others were paid nearly $800,000 for allowing about 20 tons of cocaine to cross the border (which had a street value of $1.6 billion). Corrupt agents use such methods as waving through loads at preappointed times, driving the drugs through inspection points themselves, providing cartels with sensitive intelligence, and using a cellular phone and beeper to inform a smuggler when and where he would be working.[76]

Arrests of Mexicans crossing illegally into Arizona dropped 22 percent in early 2001; this reduction was believed to be due to continued beefed-up patrols and Mexico's improving economy. In Douglas, Arizona (where Border Patrol agents were catching an average of 2,100 illegals per day in early 2000), 12-foot metal fences block old crossing routes, powerful lights bathe miles of desert, and tower-mounted infrared cameras peer into the distance.

The election of Mexican president Vicente Fox in 2000 has also slowed the flow of migrants into the United States; a much more robust economy has provided a greater willingness in people to stay in Mexico. (Fox immediately made it clear to the United States, however, that he supports Mexican workers eventually being able to move freely across the border.)[77] Also, the United States became more willing to increase the number of Mexicans allowed to cross the border to work temporarily in this country. But the situation could sour in a hurry; experts predict that if the Mexican economy turns bad, "all the infrared cameras the Border Patrol can buy and all of Fox's charm will not be enough to stop those desperate enough from climbing, funneling, or hiking toward a brighter future in the United States."[78]

SUMMARY

Although police work generally tends to be very much the same from region to region and jurisdiction to jurisdiction, this chapter has demonstrated several functions—some of them unusually dangerous in nature—that require special methods. The police, to borrow a phrase, must put on different faces for different people.

As society and its problems become more diverse, the police will continue to face new challenges. It is important that police officers receive training and education that allows them to be prepared for impending exigencies and to ensure that they maintain a professional approach to any such emergencies.

ITEMS FOR REVIEW

1. Describe the cause and extent of the nation's gang problem and some options used by the police to deal with gangs.
2. Review how police methods differ with respect to investigating and prosecuting the Mafia and the three tactics the police have been using with great success to identify, arrest, and convict mafiosi.
3. Discuss some of the unique features of police work in rural jurisdictions. Include some of the unique crime problems in these jurisdictions. List some advantages and disadvantages of working as a police officer in a rural environment.
4. What are some of the social, economic, legal, and political problems encountered in attempting to police the homeless? The borders? What can be done by the police to control these problems?
5. Describe how hate crimes differ from other crimes and why the police must be vigilant in enforcing laws against such offenses.
6. Explain the three categories of terrorists and the kinds of political and law enforcement actions that might be taken to address the problem of terrorism.

NOTES

1. John F. Lewis Jr., "Fighting Terrorism in the 21st Century," *FBI Law Enforcement Bulletin*, March 1999, p. 3.
2. *Ibid.*
3. Chitra Ragavan, "FBI, Inc.: How the World's Premier Police Corporation Totally Hit the Skids," *U.S. News and World Report*, June 18, 2001, p. 14.
4. Danny Westneat, "Terrorists Go Green," *U.S. News and World Report*, June 4, 2001, p. 28.
5. Kit R. Roane, "It's Hardly Terror Inc.," *U.S. News and World Report*, June 11, 2001, p. 31.

6. Lewis, "Fighting Terrorism in the 21st Century," p. 4.
7. Roane, "It's Hardly Terror Inc.," p. 31.
8. Lewis, "Fighting Terrorism in the 21st Century," p. 6.
9. *Ibid.*, p. 5.
10. Walter Pincus and Dan Eggen, "Ashcroft Seeks New Surveillance Powers," *The Washington Post*, September 16, 2001, p. 1A.
11. Jim Morris, "Task Forces Formed to Recommend Means to Improve Aviation Security," *The Dallas Morning News*, September 16, 2001, p. 1A.
12. *Ibid.*, pp. 26–27.
13. Quoted in Joel Carlson, "Critical Incident Management in the Ultimate Crisis," *FBI Law Enforcement Bulletin*, March 1999, p. 19.
14. Southern Poverty Law Center, "Over 200 Militias and Support Groups Operate Nationwide," *Klan Watch Intelligence Report, #78* (Washington, D.C.: Author, June 1995).
15. J. Garofalo and S. Martin, "The Law Enforcement Response to Bias-Motivated Crimes," in *Bias Crime: American Law Enforcement and Legal Responses*, ed. R. Kelly (Chicago: University of Illinois at Chicago, 1998).
16. Larry K. Gaines, Victor E. Kappeler, and Joseph B. Vaughn, *Policing in America*, 3d ed. (Cincinnati: Anderson, 1999), p. 273.
17. David E. Kaplan and Lucian Kim, "Nazism's New Global Threat," *U.S. News and World Report*, September 25, 2000, p. 34.
18. *Ibid.*
19. *Ibid.*, p. 35.
20. *Ibid.*
21. Tom Morganthau, "The View from the Far Right," *Newsweek*, May 1, 1995, pp. 36–37.
22. Marc Cooper, "Montana's Mother of All Militias," *The Nation*, May 22, 1995, p. 716.
23. Elizabeth Gleick, "Who Are They?," *Time*, May 1, 1995, p. 50.
24. Marc Cooper, "Montana's Mother of All Militias," pp. 714, 716.
25. Elizabeth Gleick, "Who Are They?," p. 49.
26. James N. Gilbert, "Organized Crime on the Western Frontier" (paper presented at an annual conference of the Western Social Science Association, April 27, 1995, Oakland, California), p. 1.
27. David E. Kaplan, "Getting It Right: The FBI and the Mob," *U.S. News and World Report*, June 18, 2001, p. 20.
28. *Ibid.*
29. *Ibid.*
30. *Ibid.*
31. Carolyn R. Block and Richard Block, *Street Gang Crime in Chicago* (Washington, D.C.: National Institute of Justice, Research in Brief, 1993), p. 2.
32. Frederic M. Thrasher, *The Gang* (Chicago: University of Chicago Press, 1927), pp. 33, 339, 346.
33. See G. David Curry, Richard A. Ball, and Scott H. Decker, *Estimating the National Scope of Gang Crime from Law Enforcement Data* (Washington, D.C.: National Institute of Justice, Research in Brief, 1996).
34. Scott H. Decker and G. David Curry, "Responding to Gangs: Comparing Gang Member, Police, and Task Force Perspectives," *Journal of Criminal Justice* 28 (2000): 129–137.
35. Ronald C. Huff, *Comparing the Criminal Behavior of Youth Gangs and At-Risk Youths* (Washington, D.C.: National Institute of Justice, Research in Brief, 1998).
36. S. Ventakesh, "The Financial Activity of a Modern American Street Gang," in *Looking At Crime from the Street Level: Plenary Papers of the 1999 Conference on Criminal Justice Research and Evaluation—Enhancing Policing and Practice Through Research, Volume 1* (Washington, D.C.: U.S. Department of Justice, Office of Justice Programs, National Institute of Justice, 1999).
37. *Ibid.*
38. *Ibid.*
39. Gregory Vistica, "'Gangstas' in the Ranks," *Newsweek*, July 24, 1995, p. 48.
40. See, for example, Kenneth J. Peak and Ronald W. Glensor, *Community Policing and Problem Solving: Strategies and Practices,* 3d ed. (Upper Saddle River, NJ: Prentice Hall, 2002); Herman Goldstein, *Problem-Oriented Policing* (New York: McGraw-Hill, 1990); and Robert Trojanowicz, Robert Bucqueroux, and Bonnie Bucqueroux, *Community Policing: A Contemporary Perspective* (Cincinnati: Anderson, 1990).
41. Catherine H. Conly, Patricia Kelly, Paul Mahanna, and Lynn Warner, *Street Gangs: Current Knowledge and Strategies* (Washington, D.C.: U.S. Department of Justice, National Institute of Justice, 1993), pp. 65–66.

42. Ruth Horowitz, "Community Tolerance of Gang Violence," *Social Problems* 34 (December 1987): 437–450.
43. James B. Jacobs, *New Perspectives on Prisons and Imprisonment* (Ithaca, NY: Cornell University Press, 1983).
44. D. L. Weisel and E. Painter, *The Police Response to Gangs: Case Studies of Five Cities* (Washington, D.C.: Police Executive Research Forum, 1997).
45. *Ibid.*
46. *Ibid.*
47. Stephen Higgins, "Interjurisdictional Coordination of Major Gang Investigations," *The Police Chief* (June 1993): 46–47.
48. Brian A. Reaves, *State and Local Police Departments, 1992* (Washington, D.C.: National Institute of Justice), pp. 1–13.
49. Ralph A. Weisheit, David N. Falcone, and L. Edward Wells, *Rural Crime and Rural Policing* (Washington, D.C.: National Institute of Justice, October 1994), p. 2.
50. W. R. Freudenburg and R. E. Jones, "Criminal Behavior and Rapid Community Growth: Examining the Evidence," *Rural Sociology* 56 (1991): 619–645.
51. Howard Abadinsky, *Drug Abuse: An Introduction* (Chicago: Nelson-Hall, 1989).
52. See J. B. Treaster, "Study Finds Drug Use Isn't Just Urban Problem," *The New York Times*, October 1, 1991, A16.
53. Weisheit, Falcone, and Wells, "Rural Crime and Rural Policing," p. 4.
54. See J. B. Jacobs, *Drunk Driving: An American Dilemma* (Chicago: The University of Chicago Press, 1989). See also Ralph Weisheit and James M. Klofas, "The Social Status of DUI Offenders in Jail," *The International Journal of the Addictions* 27 (1992): 791–814.
55. "Poaching, Rustling, Horse Thievery—Very Real Problems for Rural PDs," *Law Enforcement News*, April 30, 2000, p. 7.
56. Weisheit, Falcone, and Wells, "Rural Crime and Rural Policing," p. 5.
57. National Advisory Commission on Criminal Justice Standards and Goals, *Police* (Washington, D.C.: U.S. Government Printing Office, 1973), p. 108.
58. Peter Finn, "Street People," United States Department of Justice, National Institute of Justice, Crime File Study Guide (Washington, D.C.: USGPO, 1988), p. 1.
59. Warren Cohen and Mike Tharp, "Fed-Up Cities Turn to Evicting the Homeless," *U.S. News and World Report*, January 11, 1999, p. 28.
60. *Ibid.*, p. 1.
61. Cohen and Tharp, "Fed-Up Cities Turn to Evicting the Homeless," p. 28.
62. *Ibid.*
63. *Ibid.*, p. 28.
64. *Ibid.*, p. 30.
65. Peter E. Finn and Monique Sullivan, *Police Response to Special Populations: Handling the Mentally Ill, Public Inebriate, and the Homeless* (Washington, D.C.: U.S. Department of Justice, National Institute of Justice, 1988), pp. 6–7.
66. Reed Karaim, "South of the Border: Illegal Crossings Dip," *U.S. News and World Report*, February 26, 2001, p. 28.
67. *Ibid.*, p. 8.
68. Karaim, "South of the Border," p. 28.
69. Bill McGarigle, "High-Tech Borders," *Government Technology*, May 2001, p. 56.
70. Alan Zarembo, "People Smugglers Inc.," *Newsweek*, September 13, 1999, p. 36.
71. *Ibid.*
72. Reed Karaim, "'Ruthless Callousness': Deaths on the Border," *U.S. News and World Report*, June 4, 2001, p. 28.
73. Associated Press, "Border Patrol to Expand Search, Rescue Teams," *Reno Gazette Journal*, July 24, 2001, p. 6c.
74. Donatella Lorch, "Trail Spotting," *Newsweek*, March 6, 2000, p. 42.
75. Dan McGraw, "The Corrupting Allure of Dirty Drug Money," *U.S. News and World Report*, March 8, 1999, p. 28.
76. *Ibid.*
77. Traci Carl, "Fox Seeks Change North of the Border," Associated Press, December 1, 2000.
78. Tom Fullerton, quoted in Karaim, "South of the Border," p. 28.

The Rule of Law

Key Terms and Concepts

Affidavit	Interrogation
Crime Control Model	Lineup
Due Process Model	*Mapp v. Ohio*
Entrapment	*Miranda v. Arizona*
Exclusionary rule	*Parens patriae*
Exigent circumstances	Probable cause
Fifth Amendment	Searches and seizures
Fourth Amendment	Sixth Amendment
Gideon v. Wainwright	*Terry v. Ohio*

The great can protect themselves, but the poor and humble require the arm and shield of the law.—*Andrew Jackson*

We must never forget that it is a constitution we are expounding.
—*John Marshall, in* McCulloch v. Maryland

INTRODUCTION

The Bill of Rights to the U.S. Constitution was passed largely to protect all citizens from excessive governmental power. The police are expected to control crime within the framework of these rights; they must conduct themselves in a manner that conforms to the law as set forth in the U.S. Constitution, state constitutions, statutes passed by state legislatures, and the precedent of prior interpretations by the courts.

In other words, under the rule of law of the United States, the means are more important than the ends. A nation's democratic form of government would be of little value if the police could arrest, search, and seize the persons and property of its citizens at will. If one considers the policing systems of some other non-democratic countries (such as those discussed in Chapter 13), it is readily apparent that from the citizen's standpoint, legal curbs on those who enforce the law are absolutely necessary.

This chapter examines three constitutional amendments that regulate the police and prevent abuses of power: the Fourth Amendment (probable cause, the exclusionary rule, arrest, search and seizure, electronic surveillance, and line-ups), the Fifth Amendment (confessions, interrogation, and entrapment), and the Sixth Amendment (right to counsel and interrogation). To avoid overwhelming the reader with case titles, only better-known court cases—such as *Miranda v. Arizona*—are included in the body of the chapter; others are cited in the notes section. To conclude, we discuss a related, yet in some ways very different, area of law and procedure—the law pertaining to juvenile offenders.

Because of space limitations, many Supreme Court decisions affecting police powers in this country have necessarily been omitted from this discussion. Many publications cover those decisions, however. A partial listing would include such monthly police periodicals as the *FBI Law Enforcement Bulletin* and *The Police Chief* magazine; other sources include the *Criminal Law Reporter, U.S. Law Week*, and the *Supreme Court Bulletin*.

THE RULE OF LAW

THE FOURTH AMENDMENT

> The right of the people to be secure in their persons, papers, and effects, against unreasonable searches and seizures, shall not be violated, and no Warrants shall issue, but upon probable cause, supported by Oath or affirmation, and particularly describing the place to be searched, and the persons or things to be seized.

The **Fourth Amendment** is intended to limit zealous behavior by the police. Its primary protection consists in requiring that that weight be provided by a neutral, detached magistrate rather than by the police officer. Crime, though a major concern to society, is balanced by the concern of officers who might thrust themselves

Officers have a responsibility to testify in court. (*Courtesy Washoe County, Nevada, Sheriff's Department*)

into our homes. The Fourth Amendment requires that the necessity for a person's right of privacy to yield to the society's right to search is best decided by a neutral judicial officer, not by an agent of the police.[1]

Probable Cause

The standard for a legal arrest is **probable cause**. This important concept is elusive at best and is often quite difficult for professors to explain and even more difficult for students to understand. One way to define probable cause is to say that for an officer to make an arrest, he or she must have more than a mere hunch yet less than actual knowledge that a crime was committed by the arrestee. I often use the following example to better explain the concept.

At midnight, a 55-year-old woman, having spent several hours at a city bar, wished to leave the bar and go to a nightclub in a rural part of the county. A man offered her a ride, but rather than driving directly to the nightclub, he drove to a remote place and parked the car. There he raped the woman and forced her to sodomize him. She fought him and later told the police she thought she had broken the temples of his black glasses. After the act, he drove her back to town; when she got out of the car, she saw the license plate number and thought that the hood of the car was colored red. Her account of the crime and her physical description of the rapist immediately prompted a photograph lineup; a known rape/sodomy suspect's picture was shown to her, along with those of several other men with a similar description. She tentatively identified the suspect in the mug shot but could not be certain; the suspect's mug shot had been taken several years previous to this act.

With this preliminary information, two police officers (one of whom was the author)—without a warrant—hurried to the suspect's home to question him. Upon entering the suspect's driveway the officers observed a beige car with a red hood; probable cause was beginning to build. Next the officers noted that the vehicle's license number matched that given by the victim—probable cause was now growing by leaps and bounds. Then the suspect exited the house and walked toward his car; the officers observed that his eyeglasses frame was black and the temples were gold, indicating the black temples had probably been broken and replaced by spare gold temples. We now have, by any standard, adequate probable cause to lead a "reasonable and prudent" person to believe this suspect is the culprit; failure to arrest him would be a gross miscarriage of justice. As a final stroke, the suspect was arrested and placed in an actual lineup, where he was identified by the victim. This was one of those rare cases where the evidence was so compelling that the defendant pled guilty at his initial appearance and threw himself at the judge's mercy.

Of course, the facts of each case and the probable cause present are different; the court will examine the type and amount of probable cause that the officer had at the time of the arrest. It is important to note that a police officer cannot add to the probable cause used to make the arrest after effecting the arrest; the court will determine whether or not there existed sufficient probable cause to arrest the individual, based on the officer's knowledge of facts at the time of taking the suspect into custody.

The Supreme Court has upheld convictions when probable cause was provided by a reliable informant,[2] when it came in an anonymous letter,[3] and when a suspect fit a Drug Enforcement Administration profile of a drug courier.[4] The Court has also held that police officers who "reasonably but mistakenly conclude that probable cause is present" are granted qualified immunity from civil action.[5]

The Exclusionary Rule

The Fourth Amendment recognizes the right to privacy, but its application raises some perplexing questions. First of all, not all searches are prohibited, only those that are unreasonable. Another issue has to do with how to handle evidence that is illegally obtained. Should murderers be released, Justice Cardoza asked, simply because "the constable blundered"?[6] The Fourth Amendment says nothing about how it is to be enforced; this is a problem that has stirred a good amount of debate for a number of years. Most of this debate has focused upon the wisdom of, and the constitutional necessity for, the so-called **exclusionary rule**, which provides that all evidence obtained in violation of the Fourth Amendment shall be excluded from government's use in a criminal trial.

The exclusionary rule has an interesting history. It first appeared in the federal criminal justice system in 1914, when the Supreme Court ruled in *Weeks v. United States* that all illegally obtained evidence was barred from use in federal prosecutions.[7] The practice has not existed in the state systems for nearly as long, however. Relatively few states employed the rule until 1950. Indeed, as of 1949 only 17 states followed the *Weeks* doctrine. There was considerable objection both within and outside the Supreme Court about the states being able to do what was forbidden to the federal government. The ludicrous nature of this situation becomes evident in that until 1960, any evidence illegally seized by state or local police officers could be turned over to and used by federal officers in federal courts; this was known as the "silver platter doctrine."[8]

In 1926, a police officer arrested a man in a hallway for stealing an over-coat—a misdemeanor. The officer then entered the defendant's room and searched it, discovering a bag containing a blackjack. Appealing his arrest for possessing the weapon, the defendant made a motion to suppress the blackjack as evidence, as it had been obtained through search without a warrant. Justice Cardoza delivered a classic objection to the rule and considered its far-reaching effects:

> A room is searched against the law, and the body of a murdered man is found. The privacy of the home has been infringed, and the murderer goes free. On the one side is the social need that crime shall be repressed. On the other hand, the social need that the law shall not be flouted by the insolence of office. There are dangers in any choice.[9]

Justice Oliver Wendell Holmes took an opposite view of the rule in his dissenting opinion in a wiretapping case.[10] Holmes said we must consider two desirable objects—that criminals should be detected and all available evidence used and that the government not itself commit other crimes while gathering evidence. Holmes was repulsed by the manner in which government "dirtied" itself by engaging in snooping and wiretapping activities. In *Olmstead*, Holmes made his classic statement, "For my part, I think it is a less evil that some criminals should escape than that the government should play an ignoble part."

The 1961 Supreme Court decision in ***Mapp v. Ohio***[11] put an end to confusion over the admissibility of illegally seized evidence in the state courts. (See next page.) But the Court's decision in *Mapp* did not end the controversy surrounding the exclusionary rule. Opponents of the rule are still left with an unhappy feeling that the rule is invoked only by someone—usually a guilty person—who does not want evidence of his or her crimes to be used at trial; furthermore, they believe that the suspect's behavior has been much more reprehensible than that of the police.[12]

The Supreme Court has objected to police behavior when it "shocks the conscience," excluding evidence, for example, that was obtained by forcibly extracting evidence (by stomach pump) from a man who had swallowed two morphine capsules in their presence.[13]

Modifications of the Exclusionary Rule

Three major decisions during the 1983–1984 term of the Supreme Court also served to modify the exclusionary rule. Justice William Rehnquist established a "public safety exception" to the doctrine. In that case, the defendant was charged with criminal possession of a firearm after a rape victim described him to the police. The officers located him in a supermarket and, upon questioning him

Court Closeup

Mapp v. Ohio, 367 U.S. 643 (1961)

In May 1957, three Cleveland police officers went to the home of Miss Dolree Mapp to follow up an informant's tip that a suspect in a recent bombing was hiding there. They also had information that a large amount of materials for operating a numbers game would be found. Upon arrival at the house, officers knocked on the door and demanded entrance, but Mapp, after telephoning her lawyer, refused them entry without a search warrant.

Three hours later, the officers again attempted to enter Mapp's home, and again she refused them entry. They then forcibly entered the home. Mapp confronted the officers, demanding to see a search warrant; an officer waved a piece of paper at her, which she grabbed and placed in her bosom. The officers struggled with Mapp to retrieve the piece of paper, at which time Mapp's attorney arrived at the scene. The attorney was not allowed to enter the house or see his client. Mapp was forcibly taken upstairs to her bedroom, where her belongings were searched. One officer found a brown paper bag containing books that he deemed to be obscene.

Mapp was charged with possession of obscene, lewd, or lascivious materials. At the trial, the prosecution attempted to prove that the materials belonged to Mapp; the defense contended that the books were the property of a former boarder who had left his belongings behind. The jury convicted Mapp, and she was sentenced to an indefinite term in prison.

In May 1959, Mapp appealed to the Ohio Supreme Court, claiming that the obscene materials were not in her possession and that the evidence was seized illegally. The court disagreed, finding the evidence admissible. In June 1961, the U.S. Supreme Court overturned the conviction, holding that the Fourth Amendment's prohibition against unreasonable search and seizure had been violated and that as

> *the right to be secure against rude invasions of privacy by state officers is . . .*
> *constitutional in origin, we can no longer permit that right to remain an empty*
> *promise. We can no longer permit it to be revocable at the whim of any police officer*
> *who, in the name of law enforcement itself, chooses to suspend its enjoyment.*

about the weapon's whereabouts (without giving him the Miranda warning), they found it located behind some cartons. Rehnquist said that the case presented a situation in which concern for public safety outweighed a literal adherence to the rules. The police were justified in questioning the defendant on the ground of "immediate necessity."[14]

Another 1984 decision announced the "inevitability of discovery exception." A 10-year-old girl was murdered and her body hidden in 1977. While transporting the suspect, detectives—who had promised the suspect's attorney they would not question him while in transit—appealed to his sensitivities by saying it would be proper to find the body in order that the girl's parents could give her a Christian burial. (This became known as the "Christian Burial Speech.") The suspect directed them to the body, and in 1977 the Supreme Court ruled that the detectives had violated his rights by inducing him to incriminate himself without the presence of counsel. But the Court left open the possibility that the state could introduce evidence that the body would have been found even without the suspect's testimony. Using this "inevitability of discovery" opening, the Iowa courts found the defendant guilty, and in 1984 the Supreme Court upheld his conviction.[15]

Also in 1984, the Court ruled that evidence can be used even if obtained under a search warrant later found to be invalid. The Court held that evidence obtained by police officers acting in good faith on a reasonable reliance on a search warrant issued by a neutral magistrate but found lacking in probable cause could be used at trial. This decision prompted a strong dissenting opinion by three justices, including William Brennan Jr., who said "It now appears that the Court's victory over the Fourth Amendment is complete."[16]

Another favorable ruling to the police was handed down in 1988. Federal agents, observing suspicious behavior in and around a warehouse, illegally entered the building (with force and without a warrant) and observed marijuana in plain view. They left and obtained a search warrant for the building, then returned and arrested the defendant for conspiracy to deliver illegal drugs. The Court allowed the evidence to be admitted at trial, saying it ought not to have been excluded simply because of unrelated illegal conduct by the police. If probable cause could be established apart from their illegal activity, the Court said, evidence obtained from the search should be admitted.[17]

Views Toward the Exclusionary Rule

Today there are basically two schools of thought or models regarding the use of illegally seized evidence. The **Crime Control Model** holds that the police are going to make mistakes occasionally. The victim is entitled to sue and have his or her complaint considered by a jury. Most important, this model holds, there is no reason why the evidence should not be used against the suspect at trial, even if obtained illegally. Proponents of the **Due Process Model**, conversely, believe the only way to deal with illegally obtained evidence is to suppress it before trial. When such evidence is allowed for use at trial, convictions based wholly or in part on it should be reversed. The victim of such seizures usually is in no financial position to sue, and departmental discipline of officers is totally ineffective as a deterrent. Thus, the only way to control the illegal collection of evidence is to take the profit out of it.[18]

The current trend is clearly toward the Crime Control Model, which weakens the impact of the exclusionary rule. Not only do many court watchers expect the Supreme Court to continue this trend, but a large number of law-and-order citizens also favor the rule's decline. Indeed, Chief Justice Rehnquist has openly called for its removal. Acknowledging that the primary reason for the rule is to control police misconduct, Rehnquist noted that since *Mapp* was decided, redress is more easily obtainable by a defendant whose constitutional rights have been violated, including the *Bivens* decision and the long-dormant Section 1983 actions. (See Chapter 11.) Further, Rehnquist believed that modern juries can be trusted to return fair awards to injured parties, saying that "I feel morally certain that the United States is the only nation in the world in which the most relevant, most competent evidence as to the guilt or innocence of the accused is mechanically excluded because of the manner in which it may have been obtained"[19]

In summary, since the Warren Court expanded the rights of criminal defendants in the 1960s, a surge of cases to the Supreme Court has raised further questions concerning the exclusionary rule. Many observers have expected the Court to overturn *Mapp*. Yet the Court has not done so, apparently believing that without *Mapp*, flagrant abuses that occurred prior to it could resurface.

Arrests With and Without a Warrant

The restriction on the right of the police to arrest is the hallmark of a free society. A basic condition of freedom is that one cannot be legally seized in an arbitrary and capricious manner at the discretion or whim of any government official. It is customary to refer to the writ of habeas corpus—the "Great Writ"—as the primary guarantee of personal freedom in a democracy. Habeas corpus is defined simply as a writ requiring an incarcerated person to be brought before a judge for an investigation of the restraint of that person's liberty. It should be noted that habeas corpus is the means of remedying wrongful arrest or other detention that has already occurred and that may have been illegal. The constitutional or statutory provisions for making an arrest are of crucial importance because they prevent police action that could be very harmful to the individual.[20]

It is always best for a police officer to effect an arrest with a warrant. In fact, in 1980, the Supreme Court required police officers to obtain warrants when making felony arrests, should there be time to do so—that is, when there are no **exigent circumstances** present.[21] To obtain an arrest warrant, the officer or a citizen swears in an **affidavit** (as an "affiant") that he or she possesses certain knowledge that a particular person has committed an offense. This could be, for example, a private citizen who tells police or the district attorney that he or she attended a party at a residence where drugs or stolen articles were present. Or, as is often the case, a detective gathers physical evidence and/or interviews witnesses or victims and determines that probable cause exists to believe that a particular person committed a specific crime. In any case, a neutral magistrate, if agreeing that probable cause exists, will issue the arrest warrant. Officers will execute the warrant, taking the suspect into custody to answer the charges.

An arrest without a warrant requires exigent circumstances and that the officer possess probable cause (as explained in the sodomy case on page 240). Street officers rarely have the time or opportunity to effect an arrest with warrant in hand. Although the following actual scenario involves a search preceding an arrest, it will make the point: One afternoon a police officer is sent to the residence of several college students. They report that four men left their party and soon thereafter another guest discovered that a stereo had been taken from a car parked in the yard. A description of the men and their vehicle is given to the officer, who shortly thereafter observes a vehicle and four men matching the description. The men are stopped in their vehicle, and the officer calls for backup.

The law does not require that the officer ask the subjects to stay put while he speeds off to the courthouse to attempt to secure a search warrant. The doctrine of probable cause allows the officer to search the vehicle and arrest the occupants if stolen or contraband items are found (as in this case, where the stolen merchandise was found under the driver's seat). These kinds of situations are encountered by police officers thousands of times each day. Such searches and arrests without benefit of a warrant are legally permissible, provided the officer had probable cause (which can later be explained to a judge) for his or her actions.

In 1979, the Supreme Court rendered two decisions relating to arrests. Police, the Court said, must have probable cause to take a person into custody and on to the police station for interrogation.[22] Police may not randomly stop a single vehicle to check the driver's license and registration; there must be probable cause for stopping automobile operators.[23] However, in 1990, the Court ruled that the stopping of all vehicles passing through sobriety checkpoints—a form of seizure—did not violate the Constitution, although singling out individual vehicles for random stops without probable cause is not authorized).[24] Several days later, it ruled that police were not required to give drunk-driving suspects a Miranda warning and could videotape their responses.[25]

Since 1975, police practice has been to ensure that a person arrested without a warrant is entitled to a "prompt" initial appearance for a probable cause determination to see if the police were justified in arresting and holding the detainee. In its 1990–1991 term, the Supreme Court said that "prompt" does not mean "immediate," and that 48 hours is generally soon enough.[26]

Search and Seizure in General

Because of the serious nature of police invasion of private property, the Supreme Court has had to examine several issues related to **searches and seizures**. In 1995, the Court affirmed without decision an opinion of the Pennsylvania Supreme Court that the police violated the Fourth Amendment when they broke down the door of a residence only one or two seconds after they knocked and announced their presence and that they had a warrant. There were no exigent circumstances present.[27] Furthermore, in *Wilson v. Arkansas*,[28] 1995, the Court found a search invalid when police in Arkansas, armed with a search warrant after receiving an informant's tip that drugs were being sold at the defendant's home, identified themselves as they entered the residence and subsequently found drugs and paraphernalia. The Court believed that the Fourth Amendment requires officers to knock and announce their presence prior to entry unless doing so would increase the threat of physical harm to themselves or others, increase the likelihood of escape, or increase the risk that evidence would be destroyed.

However, the Court has upheld a search (with a warrant) of a third party's property when police have probable cause to believe it contains fruits or instrumentalities of a crime (for example, a newspaper office containing photographs of a disturbance),[29] a search of a wrong apartment conducted with a warrant but with a mistaken belief that the address was correct,[30] and a warrantless search and seizure of garbage in bags outside the defendant's home.[31]

The Court has also attempted to define when a person is considered "seized," an important issue because seizure involves Fourth Amendment protections. Is a person "seized" while police are pursuing him or her? Basically, there is no rule that determines a point of seizure in all situations. The standard is whether a suspect believes his or her liberty is restrained. This is ultimately a question for a judge or jury to decide.[32] In a recent roadblock case, the Court did provide some guidance, however. Where a police roadblock resulted in the death of a speeder, the Court said roadblocks involve a "governmental termination of freedom of movement," that the victim was therefore seized under the Fourth Amendment, and so the police were liable for damages.[33]

Two decisions in the 1990–1991 Supreme Court term expanded police practices. The Court looked at a police drug-fighting technique known as "working the buses." Police board a bus at a regular stopping place, approach seated passengers, and ask permission to search their luggage for drugs. Justice O'Connor, writing for the majority, said that such a situation should be evaluated in terms of whether or not a person in the passenger's position would have felt free to decline the officer's request or to otherwise terminate the encounter; it was held that such police conduct does not constitute a search.[34] It also decided that no "seizure" occurs when a police officer seeks to apprehend a person through a show of authority but applies no physical force (such as in a foot pursuit). In this case, a juvenile being chased by an officer threw down an object, later determined to be crack cocaine. The Supreme Court, however, found no seizure or actual restraint in this situation and reversed the appellate court.[35] Also, it should be noted that no individualized suspicion of misconduct is required in either of these cases, the Court held.

Supreme Court decisions have also authorized a warrantless seizure of blood from a defendant to obtain evidence. (This was a DUI case, the drawing of blood was done by medical personnel in a hospital, and there were exigent circumstances—the evidence would have been lost by dissipation in the body).[36] However, when police compelled a robbery suspect to be subjected to surgery to remove a bullet, the Court believed such an intrusion to seize evidence was unreasonable; this case said there are limits to what police can do to solve a crime.[37]

Searches and Seizures, With and Without a Warrant

As is the case with making an arrest, the best means by which the police can search a person or premises is with a search warrant issued by a neutral magistrate. Such a magistrate has determined, after receiving information from a sworn affiant, that probable cause exists to believe that the fruits or instrumentalities of a crime(s) are possessed by a person or are present at a particular location. Again, as with an arrest with a warrant, the "luxury" of searching and seizing with a warrant is most often held by investigative personnel, who can interview victims and witnesses and gather other available evidence, then request the warrant. Street officers rarely have the opportunity to perform such a search, as the flow of events (for example, seeing an individual in a vehicle reported to have been involved in a robbery) normally requires quick action to prevent escape and to prevent the evidence from being destroyed or hidden.

Five types of searches may be conducted without a warrant: (1) searches incidental to lawful arrest, (2) searches during field interrogation (stop-and-frisk), (3) searches of automobiles under special conditions, (4) seizures of evidence that is in "plain view," and (5) searches when consent is given.

Searches Incidental to Lawful Arrest

In *United States v. Robinson*, 1973, the defendant was arrested and taken to the police station for driving without a permit—an offense for which a full-scale arrest could be made. Robinson was taken to jail and searched, and heroin was found. He tried to suppress the evidence on the ground that the full-scale arrest and custodial search were unreasonable for a driver's-license infraction. The Supreme Court disagreed, saying the arrest was legal and when police assumed custody of him, they needed total control and therefore could perform a detailed inventory of his possessions: "It is the fact of the lawful arrest that establishes the authority to search and we hold that in the case of lawful custodial arrest a full search of the person is not only an exception to the warrant requirement of the Fourth Amendment, but is also a 'reasonable' search under that Amendment."[38]

The rationale for this decision was in part the possibility that the suspect might destroy evidence unless swift action was taken. But in *Chimel v. California*, 1969, when officers without a warrant arrested an individual in one room of his house and then proceeded to search the entire three-bedroom house, including the garage, attic, and workshop, the Supreme Court said that searches incidental to lawful arrest are limited to the area within the arrestee's immediate control, or that area from within which he or she might obtain a weapon. Thus, if the police are holding a person in one room of the house, they are not authorized to search and seize property in another part of the house, away from the arrestee's immediate physical presence.[39]

The Court approved the warrantless seizure of a lawfully arrested suspect's clothes even after a substantial time period had elapsed between the arrest and the search.[40] Another advantage given the police was the Court's allowing a warrantless, in-home "protective sweep" of the area in which a suspect is arrested to determine if other persons who would pose a danger to them might be there. Such a search, if justified by the circumstances, is not a full search of the premises and may only include a cursory inspection of those spaces where a person may be found.[41]

Stop and Frisk: Lesser Intrusions During Field Interrogation

In 1968, the U.S. Supreme Court heard a case challenging the constitutionality of on-the-spot searches and questioning by the police. This case was *Terry v. Ohio*, involving a suspect who was stopped and searched while apparently "casing" a store for robbery.

The Court's dilemma in this case was whether to rule that in some circumstances the police did not need probable cause to stop and search people and thus appear to invalidate *Mapp v. Ohio*, or to insist on such a high standard for action by the police that they could not function on the streets.[42] The Court held that an on-the-spot stop for brief questioning, accompanied by a superficial search ("patdown") of external clothing for weapons was something less than a full-scale

---

 ### Court Closeup

Terry v. Ohio, 319 U.S. 1 (1968)

Cleveland Detective McFadden, a veteran of 19 years of police service, first noticed Terry and another man at about 2:30 P.M. on the afternoon of the arrest in October 1963. McFadden testified that it appeared the men were "casing" a retail store. He observed the suspects making several trips down the street, stopping at a store window, walking about a half block, turning around and walking back again, pausing to look inside the same store window. At one point they were joined by a third party, who conversed with them and then moved on. McFadden claimed that he followed them because he believed it was his duty as a police officer to investigate the matter further.

Soon the two rejoined the third man; at that point McFadden decided the situation demanded direct action. The officer approached the subjects and identified himself, then requested that the men identify themselves as well. When Terry said something inaudible, McFadden "spun him around so that they were facing the other two, with Terry between McFadden and the others, and patted down the outside of his clothing." In a breast pocket of Terry's overcoat, the officer felt a pistol. Subsequent to that finding McFadden found a pistol on the other man's person as well. Both Terry and his partner were arrested and ultimately convicted for concealing deadly weapons. Terry appealed, on the ground that the search was illegal and that the evidence should have been suppressed at trial.

The U.S. Supreme Court disagreed with Terry, holding that the police have the authority to detain a person briefly for questioning even without probable cause if they believe that the person has committed a crime. Such detention does not constitute an arrest. The officer may also frisk a person if the officer reasonably suspects that he or she is in danger.

search and therefore could be performed with less than the traditional amount of probable cause. This case instantly became, and remains, a major tool for the police.

While *Terry* said the stop-and-frisk was legal under the Fourth Amendment in cases involving direct police observation, other cases have said that such a stop was legal when based on information provided by an informant[43] and when an individual is the subject of a "wanted" flier from another jurisdiction.[44] In summary, police officers are justified, both for their own safety and in order to detect past or future crimes, in stopping and questioning people. A person may be frisked for a weapon if an officer fears for his or her life, and the officer may go through the individual's clothing if the frisk indicates the presence of a weapon. Regardless of the rationale for the stop-and-frisk, there will always be some argument about whether this search is being used frivolously or to harass individuals. However, in balancing the public's need for safety against individual rights, the Court was willing to tip the scales in favor of community protection, especially when the safety of the officer was concerned.[45]

An important expansion of the *Terry* doctrine was handed down in 1993 in *Minnesota v. Dickerson*.[46] There, a police officer observed a man leave a notorious crack house and then try to evade the officer. The man was eventually stopped and patted down, during which time the officer felt a small lump in the man's front pocket that was suspected to be drugs. After manipulating and squeezing the lump, the officer removed it from the man's pocket; the object was crack cocaine wrapped in a cellophane container. Although the defendant's arrest and conviction were later thrown out (the Supreme Court reasoning that the search was illegal because it was beyond the limited frisk for weapons as permitted by *Terry*), the Court also allowed such seizures in the future when officers' probable cause is established by the sense of touch.

Another case extending *Terry*, *Illinois v. Wardlow*,[47] was decided in January 2000. The Court held that a citizen's running away from the police—under certain conditions—supports reasonable suspicion to justify a search. Two Illinois police officers investigating drug transactions in an area of heavy drug activity observed Wardlow holding a bag. Upon seeing the two officers, Wardlow fled, but he was soon stopped. The officers conducted a protective pat down and then squeezed the bag; they felt a gun and arrested Wardlow. The Court reasoned that, taken together, several factors (i.e., the stop occurred in a high crime area; the suspect acted in a nervous, evasive manner; and the suspect engaged in unprovoked flight upon noticing the police[48]) were sufficient to establish reasonable suspicion.

Another important Supreme Court decision in February 1997 took officer safety into account. In *Maryland v. Wilson*,[49] the Court held that police may order passengers out of vehicles they stop, regardless of any suspicion of wrongdoing or threat to the officer's safety. Chief Justice Rehnquist cited statistics showing officer assaults and murders during traffic stops and noted that the "weighty interest" in officer safety is present whether a vehicle occupant is a driver or a passenger. (Here, a Maryland state trooper initiated a traffic stop and ordered an apparently nervous passenger, Wilson, to exit from the vehicle. While doing so, Wilson dropped a quantity of crack cocaine, for which he was arrested and convicted.)

Searches and Pursuits
of Automobiles

The third general circumstance allowing a warrantless search is when an officer has probable cause to believe that an automobile contains criminal evidence. The Supreme Court has traditionally distinguished searches of automobiles from searches of homes on the ground that a car involved in a crime can be rapidly moved and its evidence irretrievably lost. The Court first established this doctrine in *Carroll v. United States*, 1925. In this case, officers searched the vehicle of a known bootlegger without a warrant but with probable cause, finding 68 bottles of illegal booze. On appeal, the Court ruled that the seizure was justified. However, *Carroll* established two rules: First, to invoke the *Carroll* doctrine, the police must have enough probable cause that if there had been enough time, a search warrant would have been issued; second, urgent circumstances must exist that require immediate action.[50]

Extending the creation of the *Carroll* doctrine, however, two new questions confronted the justices: whether impounded vehicles were subject to warrantless search and whether searches could be made of vehicles stopped in routine traffic inspections. In *Preston v. United States* (1964), however, the Court ruled that once the police had made a lawful arrest and then towed the suspect's car to a different location, they could not conduct an incidental search of the vehicle. The Court

A police officer engages in a vehicle search. (*Courtesy Reno, Nevada, Police Department*)

reasoned that because such a search was remote in time and place from the point of arrest, it was not incidental and therefore was unreasonable.[51]

Harris v. United States, 1968, upheld the right of police to enter an impounded vehicle following a lawful arrest in order to inventory its contents.[52] Building on this decision, the Court later upheld a warrantless search of a vehicle in custody, saying that because the police had probable cause to believe it contained evidence of a crime and could be easily moved, it made little difference whether a warrant was sought or an immediate search conducted.[53]

By 1974, the expectation of citizens to privacy in their vehicles was further diminished; the Court said an automobile has "little capacity for escaping public scrutiny [as] it travels public thoroughfares where both its occupants and its contents are in plain view."[54] This view was reinforced in 1976, where the Court said that a validly impounded car may be searched without probable cause or warrant as it is reasonable for an inventory of its contents to be made as a protection against theft or charges of theft while the car is in police custody.[55]

An automobile may be searched following the lawful search of its driver or another occupant. Following the rationale of *Chimel*, the Court ruled that the entire interior of the car, including containers, may be examined even if the items are not within the driver's reach.[56] The Court went on to say that a warrantless search of an automobile incidental to a lawful arrest, including its trunk and any packages or luggage, is permissible if there is probable cause to believe it contains evidence of a crime.[57] The Court also authorized a protective pat-down for weapons (similar to that of persons in *Terry v. Ohio*) of vehicle passenger compartments after a valid stop and when officers have a reasonable belief they may be in danger.[58] Finally, it was recently decided that evidence seized by opening a closed container during a warrantless inventory search of a vehicle incidental to lawful arrest is admissible.[59]

During its 1990–1991 term, the Supreme Court extended the long arm of the law with respect to automobiles. In a May 1991 decision, the Court declared that a person's general consent to a search of the interior of an automobile justifies a search of any closed container found inside the car that might reasonably hold the object of the search; thus, an officer, after obtaining a general consent, does not need to ask permission to look inside each closed container.[60] One week later, the Court ruled that probable cause to believe that a container within a car holds contraband or evidence allows a warrantless search of that item under the automobile exception, even in the absence of probable cause extending to the entire vehicle.[61] This decision clarified the *Carroll* doctrine.

Finally, during the 1998–1999 term, the Court held that when an officer has probable cause to search a vehicle, the officer may search objects belonging to a passenger in the vehicle, provided the item(s) the officer is looking for could reasonably be in the passenger's belongings.[62] (Here the officer was searching an automobile for contraband and searched a passenger's purse, finding drug paraphernalia there.)

Plain-View and Open-Field Searches

The police do not have to search for items that are seen in plain view. If such items are believed to be fruits or instrumentalities of a crime and the police are lawfully on the premises, they may seize them. For example, if an officer has been admitted into a home with an arrest or search warrant and sees drugs and paraphernalia on a living-room table, he or she may arrest the occupants on those charges as well as the earlier ones. If an officer performs a traffic stop for an offense and observes drugs in the back seat of the car, he may arrest for that as well. Providing that the officer was lawfully in a particular place and the plain-view discovery was inadvertent, the law does not require the officer to ignore contraband or other evidence of a crime that is in plain view.

The Supreme Court has said that officers are not required to immediately recognize object in plain view as contraband before it may be seized. (For instance, an officer may see a balloon in a glove box with a white powdery substance on its tip and later determine the powder to be heroin.)[63] Furthermore, fences and the posting of "No Trespassing" signs afford no expectation of privacy and do not prevent officers from viewing open fields without a search warrant.[64] Nor are police prevented from making a naked-eye aerial observation of a suspect's backyard or other curtilage (the grounds around a house or building).[65]

Two recent decisions have further defined the plain-view doctrine. In one case, an officer found a gun under a car seat while looking for the vehicle identification number; the Court upheld the search and resulting arrest as being a plain-view discovery.[66] However, in another similar situation, the Court disallowed an arrest when an officer, during a legal search for weapons, moved a stereo system to locate its serial number, saying that constituted an unreasonable search and seizure.[67]

Consent to Search

Another permissible warrantless search involves citizens' waiving their Fourth Amendment rights and consenting to a search of their persons or effects. It must be established at trial, however, that the defendant's consent was given voluntarily. In some circumstances, as with metal detectors at airports, an agent's right to search is implied.

In the leading case on consent searches, *Schneckloth v. Bustamonte*, 1973, a police officer stopped a car for a burned-out headlight. Two other backup officers joined him. When asked if his car could be searched, the driver consented. The officers found several stolen checks in the trunk. Driver and passenger were arrested and convicted. On appeal, the defendants argued that the evidence should have been suppressed, as they did not know they had the right to refuse the officers' request to search the car. The Supreme Court upheld their convictions, reasoning that the individuals, although poor, uneducated, and alone

Police officers must often engage in warrantless searches of persons. (*Courtesy Reno, Nevada, Police Department*)

with three officers, could reasonably be considered capable of knowing and exercising their right to deny officers permission to search their car.[68]

However, police cannot deceive persons into believing they have a search warrant when they in fact do not. For example, the police, looking for a rape suspect, announced falsely to the suspect's grandmother that they had a search warrant for her home and were told, "Go ahead"; this evidence was invalid.[69] Nor can a hotel clerk give a valid consent to a warrantless search of the room of one of the occupants; hotel guests have a reasonable expectation of privacy and that right cannot be waived by hotel management.[70]

Electronic Surveillance

It was the original view of the Supreme Court, in *Olmstead v. United States*, 1928, that wiretaps were not searches and seizures and did not violate the Fourth Amendment; this represented the old rule on wiretaps. [71] However, that decision was overruled in 1967 in *Katz v. United States*, which held that any form of electronic surveillance, including wiretapping, is a search and violates a reasonable expectation of privacy.[72] Katz involved a public telephone booth, deemed by the Court to be a constitutionally protected area where the user has a reasonable expectation of privacy. This decision expressed the view that the Constitution protects people, not places. Thus, the Court has required that warrants for electronic surveillance be based on probable cause, describe the conversations that are to be overheard, be for a limited period of time, name subjects to be overheard, and be terminated when the desired information is obtained.[73]

However, the Supreme Court has held that while electronic eavesdropping (that is, an informant wearing a "bug," or hidden microphone) did not violate the Fourth Amendment (a person assumes the risk that whatever he or she says may be transmitted to the police),[74] the warrantless monitoring of an electronic beeper in a private residence violated the suspect's right to privacy. A federal drug agent had placed a beeper inside a can of ether, which was being used to extract cocaine from clothing imported into the United States, and monitored its movements.[75]

Lineups and Other Pretrial Identification Procedures

A police **lineup** or other face-to-face confrontation after the accused has been arrested is considered a critical stage of criminal proceedings; therefore, the accused has a right to have an attorney present. If counsel is not present, the evidence obtained is inadmissible.[76] However, the suspect is not entitled to the presence and advice of a lawyer before being formally charged.[77]

Lineups that are so suggestive as to make the result inevitable violate the suspect's right to due process. (In one case, the suspect was much taller than the other two people in the lineup and he was the only suspect wearing a leather jacket similar to that worn by the robber. In a second lineup, the suspect was the only person who had participated in the first lineup.)[78] In short, lineups must be fair to suspects; a fair lineup guarantees no bias against the suspect.

The Supreme Court has held that a suspect may be compelled to appear before a grand jury and give voice exemplars for comparison with an actual voice recording. Appearance before a grand jury is not a search, and giving of a voice sample is not a seizure that is protected by the Fourth Amendment.[79]

THE FIFTH AMENDMENT

No person shall be held to answer for a capital, or otherwise infamous crime, unless on a presentment or indictment of a Grand Jury, except in cases arising in the land or naval forces, or in the Militia, when in actual service in time of war or public danger; nor shall any person be subject for the same offense to be twice put in jeopardy of life or limb; nor shall be compelled in any criminal case to be a witness against himself, nor be deprived of life, liberty, or property, without due process of law; nor shall private property be taken for public use, without just compensation.

A major tool used in religious persecutions in England during the 1500s was the oath. Ministers were called before the Court of Star Chamber (courts that during much of the 1500s and 1600s, without a jury, enforced unpopular political policies and meted out severe punishment, including whipping, branding, and mutilation) and questioned about their beliefs. Being men of God, they were compelled to tell the truth and admitted to their nonconformist views; for this, they were often severely punished or even executed.[80] In the 1630s, the Star Chamber and similar bodies of cruelty were disbanded by Parliament. People had become repulsed by compulsory self-incrimination; the privilege against self-incrimination was recognized in all courts when claimed by defendants or witnesses. Today the **Fifth Amendment** clause applies not only to criminal defendants but also to any witness in a civil or criminal case and anyone testifying before an administrative body, a grand jury, or a congressional committee. However, the privilege does not extend to blood samples, handwriting exemplars, and other such items not considered to be testimony.[81]

The right against self-incrimination is one of the most significant provisions in the Bill of Rights. Basically it states that no criminal defendant shall be compelled to take the witness stand and give evidence against himself or herself. No one can be compelled to answer any question if his or her answer can later be used to implicate or convict him or her. Some people view the defendant's "taking the Fifth" as an indication of guilt; others view this as a basic right in a democracy, wherein defendants do not have to contribute to their own conviction. In either case, the impact of this amendment is felt daily by the criminal justice system.

Voluntary Confessions and Decisions Supporting Miranda

Traditionally, the U.S. Supreme Court has excluded physically coerced confessions on the ground that such confessions might very well be untrustworthy or unreliable in view of the duress surrounding them.

As the quality of police work has improved, so has police use of physical means to obtain confessions diminished. Some cases involved psychological rather than physical pressure on the defendant to confess. One such case involved an accused who was questioned for eight hours by six police officers in relays and was told falsely that the job and welfare of a friend who was a rookie cop depended on his confession. He was also refused contact with his lawyer. The Court reversed his conviction, not so much on the ground that the confession was unreliable, but on the ground that it was obtained unfairly.[82]

In the 1960s, the Supreme Court ruled in *Escobedo v. Illinois*, 1964,[83] discussed on page 259, and in *Miranda v. Arizona*, 1966,[84] that confessions made by suspects who have not been notified of their constitutional rights cannot be admitted into evidence. In these cases, the Court emphasized the importance of a defendant having the "guiding hand of counsel" present during the interrogation process.

Once a suspect has been placed under arrest, the Miranda warnings must be given before interrogation for any offense, be it a felony or misdemeanor. An

Court Closeup

Miranda v. Arizona, 384 U.S. 436 (1966)

While walking to a Phoenix, Arizona, bus stop on the night of March 2, 1963, 18-year-old Barbara Ann Johnson was accosted by a man who shoved her into his car, tied her hands and ankles, and drove her to the edge of the city, where he raped her. He then drove Johnson to a street near her home, letting her out of the car and asking that she pray for him.

The Phoenix police subsequently picked up Ernesto Miranda for investigation of Johnson's rape and included him in a lineup at the police station. Miranda was identified by several women; one identified him as the man who had robbed her at knifepoint a few months earlier, and Johnson thought he was the rapist.

Miranda was a 23-year-old eighth-grade dropout with a police record dating back to age 14. He had numerous problems with the police and had been given an undesirable discharge by the army. He had also served time in prison for driving a stolen car across a state line. During questioning, the police told Miranda that he had been identified by the women; Miranda then made a statement in writing that described the incident. He also noted that he was making the confession voluntarily and with full knowledge of his legal rights. He was soon charged with rape, kidnapping, and robbery.

At trial, Miranda's court-appointed attorney got the officers to admit that during the interrogation the defendant was not informed of his right to have counsel present and that no counsel was present. Nonetheless, Miranda's confession was admitted into evidence. He was convicted and sentenced to serve 20 to 30 years for kidnapping and rape.

On appeal, the U.S. Supreme Court held that

the current practice of incommunicado interrogation is at odds with one of our Nation's most cherished principles—that the individual may not be compelled to incriminate himself. Unless adequate protective devices are employed to dispel the compulsion inherent in custodial surroundings, no statement obtained from the defendant can truly be the product of free choice.

exception is the brief, routine traffic stop; however, a custodial interrogation of a DUI (driving under the influence) suspect requires the Miranda warning.[85] Moreover, after an accused has invoked the right to counsel, the police may not interrogate the same suspect about a different crime.[86] Once a "Mirandized" suspect invokes his or her right to silence, interrogation must cease. The police may not readminister Miranda and interrogate the suspect later unless the suspect's attorney is present. If, however, the suspect initiates further conversation, any confession he or she provides is admissible.[87]

Decisions Eroding Miranda

Miranda, along with *Escobedo* and *Mapp*, combined to represent the centerpiece of the Warren Court's "due process revolution" of the 1960s. However, several decisions, including many by the Burger Court, have dealt severe blows to *Miranda*.

It has been held that a second interrogation session held after the suspect had initially refused to make a statement did not violate *Miranda*.[88] If a suspect waives his or her Miranda rights and makes voluntary statements while irrational (allegedly "following the advice of God"), those statements too are admissible.[89] The Court also decided that when a suspect waives his or her Miranda rights, believing the interrogation will focus on minor crimes, but the police shift their questioning to a more serious crime, the confession is valid—there was no police deception or misrepresentation.[90] And when a suspect invoked his or her right to assistance of counsel and refused to make written statements, then voluntarily gave oral statements to police, the statements were admissible (defendants have "the right to choose between speech and silence").[91] Finally, a suspect need not be given the Miranda warning in the exact form as it was outlined in *Miranda v. Arizona*. In one case, the waiver form said the suspect would have an attorney appointed "if and when you go to court." The Court held that as long as the warnings on the form reasonably convey the suspect's rights, they need not be given verbatim.[92]

In 1994, the Supreme Court ruled that after police officers obtain a valid Miranda waiver from a suspect, they may continue questioning him or her when he or she makes an ambiguous or equivocal request for counsel during questioning. In this case,[93] the defendant stated during an interview and after waiving his rights, "Maybe I should talk to a lawyer." The officers inquired about this statement and determined that he did not want a lawyer and continued their questioning. When a suspect unequivocally requests counsel, all questioning must cease. However, here the Court held that when the suspect mentions an attorney, the officers need not interrupt the flow of the questioning to clarify the reference but may continue questioning until there is a clear assertion of the right to counsel, such as "I want a lawyer."

Entrapment

The due process clause of the Fifth Amendment requires "fundamental fairness"; government agents may not act in a way that is "shocking to the universal sense of justice." Thus, the police may not induce or encourage a person to commit a crime that he or she would otherwise not have attempted.[94] This is the current test used by many courts to evaluate police behavior. Some states take a broader view than others as to what constitutes **entrapment**. For example, a police department in a western state had police officers impersonate homeless people. The decoys pretended to be asleep or passed out from intoxication on a public bench, and paper money visibly protruded from their pockets. Several passersby helped themselves to the money and were arrested on the spot. On appeal, the prosecution argued that a thief is a thief, the people had the intent to commit theft, and the decoy operation simply provided an opportunity for dishonest people to get caught. The state supreme court disagreed, calling the operation entrapment, adding that the situation could cause even honest people to be overcome by temptation.

However, the Supreme Court has approved an undercover drug agent's provision of an essential chemical for the manufacture of illegal drugs. (The defendant, the majority said, was an "unwary criminal" who was already "predisposed" to committing the offense).[95] Nor is it entrapment when a drug agent sells drugs to a suspect, who then sells it to government agents. Government conduct in this case is shocking to civil libertarians, but the focus here is the conduct of the defendant, not the government. As long as government's conduct is not outrageous and the defendant was predisposed to crime, the arrest is valid.[96]

The Supreme Court has held that police officers "may not originate a criminal design, implant in an innocent person's mind the disposition to commit a criminal act, and then induce commission of the crime."[97]

THE SIXTH AMENDMENT

> In all criminal prosecutions the accused shall enjoy the right to a speedy and public trial, by an impartial jury of the State and district wherein the crime shall have been committed, which district shall have been previously ascertained by law, and to be informed of the nature and cause of the accusation; to be confronted with the witnesses against him; to have compulsory process for obtaining witnesses in his favor, and to have the assistance of counsel for his defense.

Right to Counsel

Many people believe that the **Sixth Amendment** right of the accused to have the assistance of counsel before and at trial is the greatest right we enjoy in a democracy. Indeed, a close reading of the cases mentioned here would reveal the negative outcomes that are possible when a person, rich or poor, illiterate or intelligent, has no legal representation.

Over a half-century ago, in *Powell v. Alabama*, 1932, it was established that in a capital case, when the accused is poor and illiterate, he or she enjoys the right to assistance of counsel for his or her defense and due process.[98] In ***Gideon v. Wainwright***, 1963, the Supreme Court mandated that all indigent persons charged with felonies in state courts be provided counsel.[99]

Note that *Gideon* applied only to felony defendants. In 1973, *Argersinger v. Hamlin* extended the right to counsel to indigent persons charged with misdemeanor crimes if the accused faces the possibility of incarceration (however short the incarceration may be).[100]

Another landmark decision concerning the right to counsel is *Escobedo v. Illinois*, 1964.[101] Danny Escobedo's brother-in-law was fatally shot in 1960; Escobedo was arrested without a warrant and questioned, but he made no statement to the police. He was released after 14 hours of interrogation. Following police questioning of another suspect, Escobedo was again arrested and questioned at police headquarters. Escobedo's request to confer with his lawyer was denied, even after the lawyer arrived and requested to see his client. This questioning of Escobedo lasted several hours, during which time he was handcuffed and left standing. Eventually he admitted being an accomplice to murder; under Illinois law, he was as guilty as the person firing the fatal bullet. At no point was Escobedo advised of his rights to remain silent or confer with his attorney.

Escobedo's conviction was ultimately reversed by the Supreme Court, based on a violation of Escobedo's Sixth Amendment right to counsel. However, the real thrust of the decision was his Fifth Amendment right not to incriminate himself; when a defendant is scared, flustered, ignorant, alone, and bewildered, he or she is often unable to effectively make use of protections granted under the Fifth Amendment without the advice of an attorney.[102] The *Miranda* decision set down two years later simply established the guidelines for the police to inform suspects of all of these rights.

What Constitutes an Interrogation?

The Supreme Court has stated that an **interrogation** takes place not only when police officers ask direct questions of a defendant but also when the police make remarks designed to appeal to a defendant's sympathy, religious interest, and so forth. (See the discussion of the "Christian Burial Speech" case on page 243.) This has been deemed soliciting information through trickery and deceit. The Christian Burial Speech case and *Escobedo* demonstrated that even before and certainly after a suspect has been formally charged, a suspect in police custody should not be interrogated without an attorney present unless he or she has waived the right to counsel.

However, the Supreme Court upheld a conviction when two police officers, in a suspect's presence, discussed the possible whereabouts of the shotgun used in a robbery and expressed concern that nearby school children might be endangered by it. Hearing this conversation, the suspect led officers to the shotgun, thereby implicating himself. On appeal, the Court said interrogation includes words and actions intended to elicit an incriminating response from the defendant and that no such interrogation occurred here; this was a mere dialogue between officers, and the evidence was admissible.[103]

If the police were present at and recorded a conversation between a husband and wife (this tape was later used against the husband at trial, where he claimed insanity in the killing of his son), an interrogation did not occur. The Court believed that the police merely arranged a situation in which it was likely the suspect would make incriminating statements, so anything recorded could be used against him in court.[104]

Two cases on police interrogations were heard during the 1990–1991 Supreme Court term. First, the Court held that a defendant who is in custody and has been given the Miranda warning may be questioned later on a separate, as-yet-uncharged offense. Here, the defendant appeared with an attorney at a bail hearing on robbery charges. Later, while he was still in custody, the police, after reading him his rights, questioned him about a murder; the defendant agreed to discuss the murder without counsel and made incriminating statements that were used to convict him.[105] Second, in the one victory for the defense, the Court held that once a criminal suspect has asked for and consults with a lawyer, interrogators may not later question him without his lawyer being present.[106]

JUVENILE RIGHTS

The criminal justice philosophy toward juveniles is very different from its philosophy toward adults. Consequently, police officers, who are constantly dealing with juvenile offenders, must know and apply a different standard of treatment in these situations. The approach toward juvenile offenders is generally that society, through poor parenting, poverty, and so forth, is primarily responsible for their criminal behavior.

The prevailing doctrine that guides our treatment of juveniles is *parens patriae*, meaning the "state is the ultimate parent" of the child. In effect, this means that as long as we adequately care for and provide at least the basic amenities for our children as required under the law, they are ours to keep. But when our children are physically or emotionally neglected or abused by their parents or guardians, the juvenile court and police may intervene and remove the child from that environment. Then, the doctrine of *in loco parentis* takes hold, meaning that the state will act in place of the parent. The author can state from experience that there is probably no more overwhelming or awe-inspiring duty for a police officer than having to testify in juvenile court that a woman is an unfit mother and that parental ties should be legally severed. However, when a person chooses to be a negligent and/or abusive parent, it is clearly in everyone's best interest for the state to assume care and custody of the child(ren).

The juvenile justice system, working through and with the police, seeks to protect the child. It seeks to rehabilitate, not punish; its procedure is generally amiable, not adversarial. That is why the term *In Re*, meaning "concerning" or "in the matter of," is commonly used in many juvenile case titles—for example, a case would be called *In Re Smith* rather than the adversarial and more formal *State v. Smith*. Juvenile court proceedings are generally shrouded in privacy, heard before a judge only. However, when a juvenile commits an act that is so heinous that the protective and helpful juvenile court philosophy will not work, the child may be remanded to the custody of the adult court to be tried as an adult.

Juvenile delinquency (an ambiguous term that has no widespread agreed-upon meaning but has a multitude of definitions under state statutes)[107] became recognized as a national problem in the 1950s. As a result, the period of 1960 to 1970 witnessed several important decisions by the Supreme Court that addressed the rights of juveniles. *Kent v. United States*,[108] 1966, involved a 16-year-old male who was arrested in the District of Columbia for robbery, rape, and burglary. The juvenile court, without holding a formal hearing, waived the matter to a criminal court, and Kent was tried and convicted as an adult. Kent appealed, arguing that the waiver without a hearing violated his right to due process. The Supreme Court agreed.

Another landmark case extending due process to juveniles was *In Re Gault*, 1967.[109] Gerald Gault was a 15-year-old who resided in Arizona and allegedly made obscene telephone calls. When a neighbor complained to police, Gault was arrested and eventually sent to a youth home (a previous crime, stealing a wallet, was also taken into account), to remain there until he either turned 21 or was paroled. Prior to his hearing, Gault did not receive a timely notice of charges; at his hearing, Gault had no attorney present, nor was his accusor present; no transcript was made of the proceedings; and Gault was not read his rights or told he could remain silent. Gault appealed on the grounds that all of these due process rights should have been provided. The Supreme Court reversed his conviction, declaring that these Fourteenth Amendment protections applied to juveniles as well as adults. This case remains the most significant juvenile-rights decision ever rendered.

In 1970, the Supreme Court decided *In Re Winship*, involving a 12-year-old boy convicted in New York of larceny.[110] At trial, the court relied upon the "preponderance of the evidence" standard of proof against him rather than the more demanding "beyond a reasonable doubt" standard used in adult courts. At that time, juvenile courts could actually apply any of three standards of proof that existed (the third was "clear and convincing evidence"). The Court reversed Winship's conviction on the ground that the "beyond a reasonable doubt" standard had not been used.

Other precedent-setting juvenile cases followed. In *McKeiver v. Pennsylvania*, 1971, the Supreme Court said juveniles do not have an absolute right to trial by jury; whether or not a juvenile receives a trial by jury is left to the discretion of state and local authorities.[111] Finally, in *Breed v. Jones*, 1975, the Court concluded that the Fifth Amendment protected juveniles from double jeopardy, or being tried twice for the same offense (Breed had been tried in both California juvenile court and later in Superior Court for the same offenses).[112]

SUMMARY

Our society places great importance on individual freedom, and the power of government has traditionally been feared; therefore, our Constitution, courts, and legislatures have seen fit to rein in the power of government agents through what is commonly referred to as the "rule of law." This necessary aspect associated with having police in a democracy carries with it a responsibility for police practitioners to understand the law and—more important, perhaps—keep abreast of the legal changes our society is constantly undergoing.

The law is dynamic; that is, it is constantly changed by the Supreme Court and other federal courts and by state courts and legislatures. It is imperative that police agencies have a formal mechanism for imparting these legal changes to their officers.

The number of successful criminal and civil lawsuits against police officers today demonstrates that the police have not always done their homework and simply do not apply the law in the manner in which the federal courts intended. Officers must understand and enforce the law properly. In this grave business of adult cops and robbers, the means are in many respects more important than the ends. The courts and the criminal justice system should expect and allow nothing less.

ITEMS FOR REVIEW

1. Outline the protections afforded citizens by the Fourth, Fifth, and Sixth Amendments.
2. Define and give an example of probable cause.
3. Discuss, from both the police and community perspectives, the ramifications of having an exclusionary rule. What would be the ramifications for the police and the public if the rule were discontinued?
4. Distinguish between arrests and searches and seizures with and without a warrant. Which form is best? Why? Cite examples of each.
5. In what significant way(s) has the *Miranda* decision been eroded? What is its long-term outlook, given the shifting composition of judges on the Supreme Court?
6. Delineate the major rights of juveniles. What are the major differences in philosophy and treatment between juvenile and adult offenders? How are these differences manifested toward the offender in court?

NOTES

1. David Neubauer, *America's Courts and the Criminal Justice System*, 7th ed. (Belmont, CA: Wadsworth, 2002), pp. 294–300.
2. *Draper v. United States*, 358 U.S. 307 (1959).
3. *Illinois v. Gates*, 462 U.S. 213 (1983).
4. *United States v. Sokolow*, 109 S.Ct. 1581 (1989).
5. *Hunter v. Bryant*, 112 S.Ct. 534 (1991).
6. *People v. Defore*, 242 N.Y. 214, 150 N.E. 585 (1926).
7. 232 U.S. 383 (1914).
8. This doctrine was overruled by the Supreme Court in *Elkins v. United States*, 364 U.S. 206 (1960).
9. *People v. Defore* (1926).
10. *Olmstead v. United States*, 277 U.S. 438, 48 S.Ct. 564 (1928).
11. 367 U.S. 643 (1961).
12. John Kaplan, Jerome H. Skolnick, and Malcolm M. Feeley, *Criminal Justice: Introductory Cases and Materials*, 5th ed. (Westbury, NY: The Foundation Press, 1991), pp. 258–259, 269.
13. *Rochin v. California*, 342 U.S. 165 (1952).
14. In *New York v. Quarles*, 467 U.S. 649 (1984).
15. *Nix v. Williams*, 52 LW 4732 (1984). This case began as *Brewer v. Williams*, 430 U.S. 387 (1977).
16. *United States v. Leon*, 82 L.Ed.2d 677 (1984).

17. *Murray v. United States*, 487 U.S. 533 (1988).
18. Kaplan, Skolnick, and Feeley, *Criminal Justice*, p. 269.
19. *California v. Minjares*, 443 U.S. 916 (1979).
20. Alexander B. Smith and Harriet Pollack, *Criminal Justice: An Overview* (New York: Holt, Rinehart and Winston, 1980), pp. 154–155.
21. *Payton v. New York*, 445 U.S. 573 (1980).
22. *Dunaway v. New York*, 442 U.S. 200 (1979).
23. *Delaware v. Prouse*, 440 U.S. 648 (1979).
24. *Michigan Department of State Police v. Sitz*, 110 S.Ct. 2481, 110 L.Ed.2d 412 (1990).
25. *Pennsylvania v. Muniz*, 110 S.Ct. 2638, 110 L.Ed.2d 528 (1990).
26. *Riverside County, Calif. v. McLaughlin*, 59 LW 4413 (May 13, 1991).
27. *Pennsylvania v. Bull*, 63 LW 3695 (1995).
28. 115 S.Ct. 1914 (1995).
29. *Zurcher v. Stanford Daily*, 436 U.S. 547 (1978).
30. *Maryland v. Garrison*, 480 U.S. 79 (1987).
31. California v. Greenwood, 486 U.S. 35 (1988).
32. Rolando V. Del Carmen and Jeffrey T. Walker, *Briefs of 100 Leading Cases in Law Enforcement* (Cincinnati: Anderson, 1991), p. 49.
33. *Brower v. County of Inyo*, 109 U.S. 1378 (1989).
34. *Florida v. Bostick*, 59 LW 4708 (June 20, 1991).
35. *California v. Hodari D.*, 59 LW 4335 (April 23, 1991).
36. *Schmerber v. California*, 384 U.S. 757 (1966).
37. *Winston v. Lee*, 470 U.S. 753 (1985).
38. 414 U.S. 218 (1973).
39. 395 U.S. 752 (1969).
40. *United States v. Edwards*, 415 U.S. 800 (1974).
41. *Maryland v. Buie*, 58 LW 4281 (1990).
42. Smith and Pollack, *Criminal Justice*, p. 161.
43. *Adams v. Williams*, 407 U.S. 143 (1972).
44. *United States v. Hensley*, 469 U.S. 221 (1985).
45. Smith and Pollack, *Criminal Justice*, p. 162.
46. *Minnesota v. Dickerson*, 113 S.Ct. 2130 (1993).
47. 120 S.Ct. 673 (2000).
48. *Ibid.*, at 673.
49. *Maryland v. Wilson*, 117 S.Ct. 882 (1997).
50. 267 U.S. 132 (1925).
51. 376 U.S. 364 (1964).
52. 390 U.S. 234 (1968).
53. *Chambers v. Maroney*, 399 U.S. 42 (1970).
54. *Cardwell v. Lewis*, 417 U.S. 583 (1974).
55. *South Dakota v. Opperman*, 428 U.S. 364 (1976).
56. *New York v. Belton*, 453 U.S. 454 (1981).
57. *United States v. Ross*, 456 U.S. 798 (1982).
58. *Michigan v. Long*, 463 U.S. 1032 (1983).
59. *Colorado v. Bertine*, 479 U.S. 367 (1987).
60. *Florida v. Jimeno*, 59 LW 4471 (May 23, 1991).
61. *California v. Acevedo*, 59 L.W. 4559 (May 30, 1991).
62. *Wyoming v. Houghton*, 119 S.Ct. 1297 (1999).
63. *Texas v. Brown*, 460 U.S. 730 (1983).
64. *Oliver v. United States*, 466 U.S. 170 (1984).
65. *California v. Ciraolo*, 476 U.S. 207 (1986).
66. *New York v. Class*, 54 LW 4178 (1986).
67. *Arizona v. Hicks*, 55 LW 4258 (1987).
68. 412 U.S. 218 (1973).
69. *Bumper v. North Carolina*, 391 U.S. 543 (1968).
70. *Stoner v. California*, 376 U.S. 483 (1964).
71. 277 U.S. 438 (1928).
72. 389 U.S. 347 (1967).
73. *Berger v. New York*, 388 U.S. 41 (1967).

74. *On Lee v. United States*, 343 U.S. 747 (1952).
75. *United States v. Karo*, 468 U.S. 705 (1984).
76. *United States v. Wade*, 388 U.S. 218 (1967).
77. *Kirby v. Illinois*, 406 U.S. 682 (1972).
78. *Foster v. California*, 394 U.S. 440 (1969).
79. *United States v. Dionisio*, 410 U.S. 1 (1973).
80. Kaplan, Skolnick, and Feeley, *Criminal Justice*, pp. 219–220.
81. *Ibid.*, pp. 220–221.
82. *Spano v. New York*, 360 U.S. 315 (1959).
83. 378 U.S. 478 (1964).
84. 384 U.S. 436 (1966).
85. *Berkemer v. McCarty*, 468 U.S. 420 (1984).
86. *Arizona v. Roberson*, 486 U.S. 675 (1988).
87. *Edwards v. Arizona*, 451 U.S. 477 (1981).
88. *Michigan v. Mosley*, 423 U.S. 93 (1975).
89. *Colorado v. Connelly*, 479 U.S. 157 (1986).
90. *Colorado v. Spring*, 479 U.S. 564 (1987).
91. *Connecticut v. Barrett*, 479 U.S. 523 (1987).
92. *Duckworth v. Eagan*, 109 S.Ct. 2875 (1989).
93. *Davis v. United States*, 114 S.Ct. 2350 (1994).
94. *Sherman v. United States*, 356 U.S. 369 (1958).
95. *United States v. Russell*, 411 U.S. 423 (1973).
96. *Hampton v. United States*, 425 U.S. 484 (1976).
97. *Ibid.*, at p. 1540.
98. 287 U.S. 45 (1932).
99. 372 U.S. 335 (1963).
100. 407 U.S. 25 (1973).
101. 378 U.S. 478 (1964).
102. Smith and Pollack, *Criminal Justice*, p. 177.
103. *Rhode Island v. Innis*, 446 U.S. 291 (1980).
104. *Arizona v. Mauro*, 481 U.S. 520 (1987).
105. *McNeil v. Wisconsin*, 59 LW 4636 (June 13, 1991).
106. *Minnick v. Mississippi*, 59 LW 4037 (1990).
107. Arnold Binder, Gilbert Geis, and Dickson Bruce, *Juvenile Delinquency: Historical, Cultural, Legal Perspectives* (New York: Macmillan Publishing Company, 1988), pp. 6–9.
108. 383 U.S. 541 (1966).
109. 387 U.S. 9 (1967).
110. 397 U.S. 358 (1970).
111. 403 U.S. 528 (1971).
112. 421 U.S. 519 (1975).

Accountability

Ethics,
Force and Corruption,
and Discipline

Key Terms and Concepts

Code of Silence
Complaints
Ethics
Garrity v. *New Jersey*
Lautenberg Amendment
Limitations on officers'
    constitutional rights

Police brutality
Police corruption
Police firearms regulations
Police use of force
Racial profiling
Slippery slope

This is your duty, to act well the part that is given to you.—*Epictetus*

"Learn what is true in order to do what is right" is the summing up of the whole duty of man.—*T. H. Huxley*

INTRODUCTION

The new millennium certainly came in like a lion with respect to controversial police practices and public outcry; we can hope in this regard its first decade will go out more like a lamb. Citizens in some communities took their anger against police to the streets, engaging in violent civil unrest (see, for example, the description of recent actions in Cincinnati, discussed later). Perhaps no aspect of policing carries more controversy, problems, and questions than how to maintain police accountability; this topic of policing includes ethical conduct, police use of force and corruption, and meting out officer discipline. This chapter addresses that triad.

First is an examination of the complicated concept of police ethics, including types of and potential problems under community policing. Next we review the use of violence and force against the public, including police brutality in general and the use and control of lethal force; included is the contemporary hot-button issue of racial profiling and other field tactics. Then we consider police corruption: types and causes, problems posed by the police code of silence, and how to deal with it. Next is an overview of areas in which the federal courts have placed limitations on police behaviors by virtue of their unique role: freedom of speech and religion, sexual misconduct, search and seizure, self-incrimination, residency, moonlighting, misuse of firearms, and alcohol and drug abuse. The chapter concludes with an examination of disciplinary policies and practices, including dealing with citizens' complaints and doling out sanctions.

This chapter necessarily lays bare several sordid aspects of police behavior, especially the use of force. We should all keep in mind the positive findings of a recent federal study, however: Of nearly 44 million contacts the police had with the public in a year's time, less than 1 percent resulted in force being used or threatened. And in more than half of the cases in which force was used by the police, citizens acknowledged having argued with, disobeyed, or resisted police during the interaction or were drunk or high on drugs at the time.[1]

IN THE BEGINNING: PROBLEMS GREET THE NEW MILLENNIUM

TROUBLES IN CITIES LARGE AND SMALL

There was no dearth of high-profile investigations of, and outcomes involving, metropolitan police problems during the early part of the new millennium:

- Seventy officers of the Los Angeles Police Department's (LAPD) Rampart Division came under scrutiny in early 2000 after a disgraced former officer (who was eventually sentenced to five years in prison for stealing cocaine) told investigators that he and other anti-gang-unit officers framed suspects and falsified evidence. Nearly 100 people were framed by "rogue" officers over a three-year period by planting drugs

The police have always been criticized for a variety of
reasons, as shown in this 1874 caricature of police as pigs.

on them, and a few unarmed citizens were shot. Dozens of cases brought by the unit
were thrown out, and dozens of officers were convicted, fired, suspended, or quit.[2] In
early 2000, the Los Angeles police chief, Bernard C. Parks, issued a stinging indictment of his own department, saying that lax oversight and poor adherence to departmental policies helped corruption to flourish in the LAPD. Parks said he needed at
least $9 million and hundreds of new positions (primarily in the internal affairs divi-
sion) to fix the problems.[3]

- Also in early 2000, the verdict was rendered in the New York police killing of an unarmed West African, Amadou Diallo; officers, mistaking Diallo's wallet for a handgun, fired 41 shots at Diallo, striking him 19 times, in February 1999; four NYPD officers who were involved were acquitted.

While policing problems have plagued the cities of Los Angeles and New
York for 40 years, big-city policing abuses are also seeping into smaller towns and
suburbs:

- Following some kind of confrontation, a police detective in Prince George's County,
Maryland, killed a black 25-year-old university student in September 2000, shooting
him in the back five times; he was one of 12 people shot there in 13 months, five
fatally.

- In Riverside, California, an unconscious 19-year-old woman in a parked car died after being shot 12 times by local police in 1998.
- And in mid-2000, the U.S. Justice Department began probing racial profiling in Eastpointe, Michigan, and Orange County, Florida.[4]

Possible explanations for such problems in small-town and suburban police agencies abound, with a *Washington Post* columnist offering that "it is the police culture, more than race, that is at the crux of the problem . . . a mentality of brutality."[5] (See Exhibit 10–1, which discusses methods of measuring a police department's "culture of integrity.")

A New Tool: Federal Investigations

The federal government—specifically, the U.S. Justice Department—has a new legal means to investigate allegations of racial bias in police departments. The law authorizing such investigations, passed in 1994 after the Rodney King beating in Los Angeles, has gotten results while generating a lot of controversy. It compels police agencies to initiate safeguards against excessive force and racial bias— for instance, computer systems to track complaints and disciplinary actions. The Fraternal Order of Police filed a lawsuit against the law in 2001, arguing that the Justice Department's use of the law was unconstitutional.

An example of the law's use occurred with the April 2001 fatal police shooting in Cincinnati of Timothy Thomas, an unarmed 19-year-old black man. Thomas's killing—the 15th young black man killed by Cincinnati police in six years—ignited a storm of violence and looting and resulted in the city's mayor requesting U.S. Attorney General John Ashcroft to review the "practices, procedures, and training" of the police department. At the same time, the Justice Department was investigating 12 other local police departments (in Buffalo, New York; Charleston, West Virginia; Cleveland; Detroit; Eastpointe, Michigan; New Orleans; New York City; Orange County, Florida; Prince George's County, Maryland; Riverside, California; Tulsa, Oklahoma; and Washington, D.C.) to determine whether they engaged in "a pattern and practice" of racial discrimination or brutality.[6]

POLICE ETHICS

A Scenario

Assume that the police have strong suspicions that Jones is a serial rapist, but they have not secured enough probable cause to obtain a search warrant for Jones's car and home, where evidence might be present. Officer Brown feels frustrated and, early one morning, uses a razor blade to remove the current registration decal from the license plate on Jones's car. The next day he stops Jones for operating his vehicle with an expired registration; he impounds and inventories the vehicle and finds evidence of several sexual assaults, which ultimately leads

---

EXHIBIT 10-1

Measuring a Police Department's "Culture of Integrity"

Researchers believe they have found a quantitative method that allows police executives to assess their agency's level of resistance to corruption. A national survey of 3,235 officers in 30 police departments asked them to examine 11 common scenarios of police misconduct. The study was based on the premise that organizational and occupational culture can create an atmosphere in which corruption is not tolerated. Survey questions were designed to indicate whether officers knew the rules governing misconduct, how strongly they supported those guidelines, if they knew the disciplinary penalties for breaking those rules and believed them to be fair, and if they were willing to report misconduct. Respondents found some types of transgressions to be significantly less serious than others, and the more serious the transgression was perceived to be, the more willing officers were to report a colleague and to believe severe discipline appropriate. Four scenarios that were not considered major transgressions by officers included the off-duty operation of an outside security business, receiving free meals, accepting free holiday gifts, and the cover-up of a police drunk-driving accident.

Indeed, a majority of respondents said they would not report a fellow officer for accepting free gifts, meals, or discounts or for having a minor traffic accident while under the influence of alcohol. The intermediate levels of misconduct included the use of excessive force on a car thief following a foot pursuit, a supervisor who offers time off during holidays in exchange for a tune-up on his personal vehicle, and accepting free drinks in return for ignoring a late bar closing. Very serious forms of misconduct, as perceived by the respondents, included accepting a cash bribe, stealing money from a found wallet, and stealing a watch from a crime scene.

Source: Adapted from "How Do You Rate? The Secret to Measuring a Department's 'Culture of Integrity.'" Reprinted with permission from *Law Enforcement News*, October 15, 2000, pp. 1, 6. John Jay College of Criminal Justice (CUNY) 555 West 57th St., New York, NY 10019.

---

to Jones's conviction on 10 counts of forcible rape and possession of burglary tools and stolen property. Brown receives accolades for the apprehension. Was Officer Brown's act legal regarding the vehicle registration? Should Brown's actions, even if improper or illegal, be condoned for "serving the greater public good"? Did Brown use the law properly?

This hypothetical sequence of events and accompanying questions should be kept in mind as we consider the problem of police ethics.

EXHIBIT **10–2**

Law Enforcement Code of Ethics

All law enforcement officers must be fully aware of the ethical responsibilities of their position and must strive constantly to live up to the highest possible standards of professional policing.

The International Association of Chiefs of Police believes it is important that police officers have clear advice and counsel available to assist them in performing their duties consistent with these standards, and has adopted the following ethical mandates as guidelines to meet these ends.

PRIMARY RESPONSIBILITIES OF A POLICE OFFICER

A police officer acts as an official representative of government who is required and trusted to work within the law. The officer's powers and duties are conferred by statute. The fundamental duties of a police officer include serving the community; safeguarding lives and property; protecting the innocent; keeping the peace; and ensuring the rights of all to liberty, equality and justice.

PERFORMANCE OF THE DUTIES OF A POLICE OFFICER

A police officer shall perform all duties impartially, without favor or affection or ill will and without regard to status, sex, race, religion, political belief or aspiration. All citizens will be treated equally with courtesy, consideration and dignity.

Officers will never allow personal feelings, animosities or friendships to influence official conduct. Laws will be enforced appropriately and courteously and, in carrying out their responsibilities, officers will strive to obtain maximum cooperation from the public. They will conduct themselves in appearance and deportment in such a manner as to inspire confidence and respect for the position of public trust they hold.

DISCRETION

A police officer will use responsibly the discretion vested in the position and exercise it within the law. The principle of reasonableness will guide the officer's determinations and the officer will consider all surrounding circumstances in determining whether any legal action shall be taken.

Consistent and wise use of discretion, based on professional policing competence, will do much to preserve good relationships and retain the confidence of the public. There can be difficulty in choosing between conflicting courses of action. It is important to remember that a timely word of advice rather than arrest—which may be correct in appropriate circumstances—can be a more effective means of achieving a desired end.

USE OF FORCE

A police officer will never employ unnecessary force or violence and will use only such force in the discharge of duty as is reasonable in all circumstances.

Force should be used only with the greatest restraint and only after discussion, negotiation and persuasion have been found to be inappropriate or ineffective. While the use of force is occasionally unavoidable, every police officer will refrain from applying the unnecessary infliction of pain or suffering and will never engage in cruel, degrading or inhuman treatment of any person.

CONFIDENTIALITY

Whatever a police officer sees, hears or learns which is of a confidential nature will be kept secret unless the performance of duty or legal provision requires otherwise.

Members of the public have a right to security and privacy, and information obtained about them must not be improperly divulged.

INTEGRITY

A police officer will not engage in acts of corruption or bribery, nor will an officer condone such acts by other police officers.

The public demands that the integrity of police officers be above reproach. Police officers must, therefore, avoid any conduct that might compromise integrity and thus undercut the public confidence in a law enforcement agency. Officers will refuse to accept any gifts, presents, subscriptions, favors, gratuities or promises that could be interpreted as seeking to cause the officer to refrain from performing official responsibilities honestly and within the law. Police officers must not receive private or special advantage from their official status. Respect from the public cannot be bought; it can only be earned and cultivated.

COOPERATION WITH OTHER OFFICERS AND AGENCIES

Police officers will cooperate with all legally authorized agencies and their representatives in the pursuit of justice.

An officer or agency may be one among many organizations that may provide law enforcement services to a jurisdiction. It is imperative that a police officer assist colleagues fully and completely with respect and consideration at all times.

Personal/Professional Capabilities

Police officers will be responsible for their own standard of professional performance and will take every reasonable opportunity to enhance and improve their level of knowledge and competence.

Through study and experience, a police officer can acquire the high level of knowledge and competence that is essential for the efficient and effective performance of duty. The acquisition of knowledge is a never-ending process of personal and professional development that should be pursued constantly.

Private Life

Police officers will behave in a manner that does not bring discredit to their agencies or themselves.

A police officer's character and conduct while off duty must always be exemplary, thus maintaining a position of respect in the community in which he or she lives and serves. The officer's personal behavior must be beyond reproach.

Source: Adopted by the Executive Committee of the International Association of Chiefs of Police on October 17, 1989. Used with permission.

DEFINITION AND TYPES

Proper ethical behavior has always been the cornerstone of policing and is what the public expects of its public servants. **Ethics** usually involves standards of moral conduct and what we call conscience, the ability to recognize right from wrong, and acting in ways that are good and proper. One definition for ethics is "the discipline dealing with what is good and bad and with moral duty and obligation."[7]

There are both absolute and relative ethics. *Absolute* ethics is a concept wherein an issue only has two sides: Something is either good or bad, black or white. The original interest in police ethics focused on such unethical behaviors as bribery, extortion, excessive force, and perjury. Few communities can tolerate the absolute unethical behavior of rogue officers; for instance, anyone would have a hard time trying to rationally defend a police officer's stealing.

Relative ethics is more complicated and can have a multitude of sides and varying shades of gray. What is considered ethical behavior by one person may be deemed highly unethical by someone else. However, not all police ethical issues are clear cut. For example, communities seem willing at times to tolerate extra-legal behavior by the police if there is a greater public good, especially in dealing with such problems as gangs and the homeless—and with such offenders as that in the serial rapist scenario. As one police captain put it, "When you're shoveling society's garbage, you gotta be indulged a little bit."[8]

Another type of ethical dilemma faced by police officers involves *situational* ethics, which is related to relativism. For instance, consider a situation in which a police officer stops his father for driving while intoxicated. The officer must decide whether to do his or her duty—make an arrest—or allow the offender to go away free.

ETHICS AND COMMUNITY POLICING

With the shift to community oriented policing and problem solving (COPPS, discussed in Chapter 6), some concerns have been raised about the increased potential for ethical dilemmas to confront the police because they have greater discretion and more public interaction.

An example of an ethical problem that can arise with more frequency under COPPS involves gratuities—free gifts that are supposedly given to the police without obligation by the giver. Whether the police should receive such minor gratuities as free coffee and meals is a longstanding and controversial issue, one for which there will probably never be widespread consensus of opinion. Proponents argue that police deserve such perks of the job and that minor gratuities are the building blocks of positive social relationships. Harmless gratuities, it is maintained, may create good feelings in the community toward police officers, and vice versa. Opponents believe, however, that the receipt of gratuities can lead to future deviance—the **slippery slope** perspective, which proposes that the acceptance of minor gratuities begins a process wherein the recipient's integrity is gradually subverted and eventually leads to more serious unethical conduct.[9] Given that judges, educators, and other professionals neither expect nor receive such gifts, some people (and police agencies) conclude that gratuities are unethical. As an example, after firing an officer, the Bradenton, Florida, Police Department recently established a policy prohibiting sworn personnel from accepting any discounted meals offered by restaurant owners. The officer was believed to have "outright stolen" items from merchants, taking and never paying for cigars, sandwiches, magazines, or other goods.[10] Following is a scenario involving COPPS and gratuities:

> The sheriff's department has a long-standing policy concerning the solicitation and acceptance of gifts (written like that in Figure 10–1, p. 289). A deputy has been working a problem-solving project in a strip mall area with juvenile loitering, drug use, prostitution, and vandalism after hours in the parking lot. The mall manager, Mr. Chang, believes it is his moral duty to show his appreciation to the deputy and has made arrangements for the deputy and family to receive a 15 percent discount when

shopping at any store in the mall. Knowing that the agency policy requires that such offers be declined, the deputy is also aware that Chang will be very hurt or upset if the proffered gift is refused.[11]

A particularly strong consideration in this scenario is that the mall manager is Asian American and might be extremely hurt if his gift were rejected. Some policy issues are also presented in this scenario. For example, in developing a rapport with a mall restaurant manager, is an officer who was formerly prohibited from accepting a free meal now free to do so? Assume that other deputies learn of Chang's new discount arrangement and several of them go to the mall expecting to be treated similarly, resulting in complaints to the sheriff by several business owners. Certainly some people would hold this action to be unethical, given their intent—motivation for personal gain and exploitation of the situation.

In sum, the subject of police ethics is not simplistic in nature. We all know that officers should do right, not wrong, but the existence and use of relative ethics makes this a complicated issue at best. What can be said is that police officers must be recruited and trained with ethics in mind, because they will be given much freedom to become more involved in their community and given wider discretion to make important decisions when addressing neighborhood disorder.

USE OF VIOLENCE AND FORCE

A TRADITION OF PROBLEMS

Throughout its history, policing has experienced allegations of brutality and corruption. In the late 1800s, New York police sergeant Alexander "Clubber" Williams epitomized police brutality; he spoke openly of using his nightstick to knock a man unconscious, batter him to pieces, or even kill him. Williams supposedly coined the term "tenderloin," when he commented "I've had nothing but chuck steaks for a long time, and now I'm going to have me a little tenderloin."[12] Williams was referring to opportunities for graft in an area in downtown New York that was the heart of vice and nightlife, often termed Satan's Circus. This was Williams's beat, where his reputation for brutality and corruption became legendary.[13]

Although police brutality and corruption are no longer openly tolerated, a number of events throughout history have demonstrated that the problem still exists and requires the attention of police officials. Several of these events, such as the so-called police riot in 1968 during the Democratic National Convention in 1968, were discussed in Chapter 1.

THE PREROGATIVE TO USE FORCE

Our society recognizes three legitimate and responsive forms of force: the right of self-defense, including the valid taking of another person's life in order to protect

"Clubber" Williams. (*Courtesy NYPD Photo Unit*)

ourselves from harm; the power to control persons for whom some responsibility is granted an authority figure for care and custody (such as a prison guard or attendant in a mental hospital); and the relatively unrestricted authority of police to use force as required. Police work is dangerous; a routine arrest may result in a violent confrontation, sometimes triggered by drugs, alcohol, or mental illness. To cope, police officers are given the unique right to use force, even deadly force, against their fellows. There are, of course, limitations as to when the officer may exercise deadly force, which are discussed later in this chapter.

Egon Bittner defined **police use of force** as the "distribution of non-negotiably coercive remedies,"[14] and asserted that the duty of police intervention in matters of societal disorder

> means above all making use of the authority to overpower resistance. This feature of police work is uppermost in the minds of people who solicit police aid. Every conceivable police intervention projects the message that force may . . . have to be used to achieve a desired objective.[15]

Exhibit 10-3

Good News, Better News
on Use of Force

First, the good news: Of more than 43 million contacts law enforcement had with the public during 1999, less than 1 percent resulted in force being used or threatened. Now the better news: In more than half of those cases, the subjects acknowledged having argued with, disobeyed or resisted police during the interaction, or had been drunk or high on drugs at the time.

The Bureau of Justice Statistics reported this month that an estimated 43.8 million people age 16 or older had face-to-face contact with police in 1999, with about half of those involving a vehicle stop. Of the nation's 18.1 million black licensed drivers, 12.3 percent were pulled over at least once, compared with 10.4 percent of the 143 million licensed white drivers.

The study, "Contacts Between Police and the Public—Findings from the 1999 National Survey," was based on data from BJS's Police-Public Contact Survey, which was conducted during the last six months of that year. Some 80,000 participants were used to develop the representative sample.

In addition to a larger percentage of blacks being subject to traffic stops, researchers also found that both blacks and Hispanics proportionately accounted for more searches, as well. Some 1.3 million drivers or vehicles were searched in 1999, said the study. Eleven percent of African Americans and 11.3 percent of Hispanics were searched by police, compared with 5.4 percent of whites. In nearly 90 percent of the searches conducted without the express consent of drivers, no drugs, alcohol, illegal weapons or evidence of criminal wrongdoing were found.

"What I think is important is when we asked drivers whether they felt it was a legitimate stop, about 84 percent of white drivers said yes, and about 74 percent of black drivers said yes," said Lawrence Greenfeld, a BJS statistician and co-author of the study. "There is some disparity there, but for the vast majority of black and white drivers who said yes, there was a legitimate reason."

The study also found that the race of the officer involved in the stop played little or no role in whether the motorists felt it was justified, Greenfeld told Law Enforcement News. "It suggests that the perception of whether the stop was valid or not has nothing to do with the officer's race," he said. . . .

Of those who had contact with police during 1999, the report found that 422,000 contacts resulted in force or the threat of force. An estimated 20 percent of those incidents involved only the threat of force. Among blacks and Hispanics, 2 percent said they experienced the threat or use of force by police, compared to 1 percent of whites. . . .

An estimated 49 percent of those involved in a force situation had charges filed against them, ranging from a traffic offense to assaulting an officer, said the study. Whites were as likely as blacks in such incidents to describe the force used against them as excessive and the vast majority, 92 percent, said police had acted improperly.

Some 90 percent of respondents overall, however, said police had acted properly during the encounter. An estimated 51 percent of drivers said they had been speeding, 24 percent cited reckless driving, an illegal turn or some other traffic violation.

Source: Reprinted with permission from *Law Enforcement News,* March 15, 2001, p. 1, John Jay College of Criminal Justice (CCNY) 555 West 57th St., New York, NY 10019.

The exercise of force by police can take several forms, escalating from a simple verbal command to a light touch on the arm to encourage someone to move along or comply with an order, the use of the baton or Mace to control an individual, and the use of the carotid restraint (the so-called sleeper hold) to the use of lethal force.

POLICE BRUTALITY

Many people contend that there are actually three means by which the police can be "brutal." There is the *literal* sense of the term, which involves the physical abuse of others. There is the *verbal* abuse of citizens, exemplified by slurs or epithets. Finally, for many who feel downtrodden the police *symbolize* brutality because the officers represent the establishment's law, which serves to keep the minority groups in their place. It is perhaps the latter form of **police brutality** that is of the greatest concern for those persons interested in improving community relations. Because it is a philosophy or frame of mind, it is probably the most difficult to overcome.

Citizen use of the term "police brutality" encompasses a wide range of practices, from the use of profane and abusive language to the actual use of physical force or violence.[16] Some would deny that police brutality exists or that there is little, if any, police brutality in today's enlightened police agency. Others acknowledge that police brutality exists today, but that *"brutality is the prerogative of the police state. To tolerate any of it is to differ from the police state only in degree"* (emphasis in original).[17]

While no one can deny that brutal practices are used by some police officers, it is impossible to know with any degree of accuracy how often and to what extent these incidents occur. They are low-visibility acts, and many victims decline to report them. Although it is widely believed that brutality is a racial matter primarily involving white police and black victims, Albert Reiss found that lower-class white men were as likely to be brutalized by the police as lower-class black men.[18] What is most disturbing is that 37 percent of the instances of excessive force occur in settings controlled by the police—station houses and patrol cars. In half the situations, a police officer did not participate but did not restrain his or her colleague, indicating that the informal police culture did not disapprove of the behavior.[19]

It is impossible to realistically expect that police brutality will at some point disappear forever. There are always going to be, in the words of A. C. Germann, Frank Day, and Robert Gallati, "Neanderthals" who enjoy their absolute control over people and become tyrannical in their arbitrary application of power.[20] Obviously, the existence of a formal citizen complaint review procedure within police agencies for investigating allegations of brutality and excessive use of force is necessary.

USE AND CONTROL OF LETHAL FORCE

The lethal use of force by a police officer is one of the most tragic and unfortunate circumstances that can occur in our society. It is tragic for the victim, the victim's

Scenes from the *Walker Report* of the 1968 Chicago Democratic National Convention.

family, the officer, and, most likely, the officer's family. Estimates of the number of citizens killed annually by the police range from 300 to 600; about 1,500 more are wounded.[21] This number appears to be declining from the high level set in the 1970s.[22] However, it remains a notable fact that the typical victim of deadly force has been a young African American male.[23] For this reason, deadly force has become one of the major sources of tension between the police and minority communities.

When the colonists came to this country from England, they brought with them a common law principle that authorized the use of deadly force to apprehend any and all fleeing felony suspects. In such circumstances, it was reasoned, deadly force was appropriate and not disproportionate to the punishments arrestees received. As our laws and society evolved, however, it became possible for police to use deadly force with firearms against people who were at great distances from them, often for nonviolent property crimes. The justification and necessity for the fleeing felon rule came into question.[24] Many states modified their criminal laws, narrowing the range of fleeing suspects police were authorized to shoot.

Then the Supreme Court's 1985 decision in *Tennessee v. Garner*[25] greatly curtailed the use of deadly force. There, the Court held that the use of deadly force to prevent the escape of all felony suspects was constitutionally unreasonable. It is not better, the Court reasoned, that all felony suspects die than that they escape. Where the suspect poses no immediate threat to the officer or to others, the harm resulting from failing to apprehend him does not justify the use of deadly force to do so. In short, the Fourth Amendment prohibits the seizure of an unarmed, nondangerous suspect by means of deadly force.

Only recently has any meaningful attempt been made to examine or control the use of deadly force by police officers. Prior to the 1970s, police officers had tremendous discretion regarding their use of firearms, and police departments often had poorly defined or nonexistent policies on this issue. Investigations of police shootings were sometimes conducted half-heartedly, and police agencies did not always keep records of all firearm discharges by officers.[26] This situation has changed dramatically. First, many states modified the old fleeing-felon doctrine that had allowed police to use deadly force to prevent a suspected felon from escaping. Second, the U.S. Supreme Court ruled that shootings of any unarmed nonviolent fleeing felony suspects violated the Fourth Amendment to the Constitution (discussed later). Third, almost all major urban police departments enacted restrictive policies regarding deadly force. And finally, the Supreme Court made it easier for a private citizen to successfully sue and collect damages as a result of a questionable police shooting.[27]

FURTHER SOURCES OF TENSION: RACIAL PROFILING AND OTHER FIELD TACTICS

Are the police biased in their treatment of minorities? A study in the early 1980s by the RAND Corporation found that although minorities are treated differently at a few points in the criminal justice system, there was no evidence that this results from widespread and consistent racial prejudice in the system.[28]

Many people remain convinced, however, that the justice system unfairly draws minorities into its web and that police methods are at the forefront of this practice. **Racial profiling**—also known as "Driving While Black or Brown" (DWBB)—may be defined as a police officer stopping a vehicle simply because the driver is of a certain race; the officer does so because he or she is acting on a personal bias.[29] This issue has driven a deep wedge between the police and minorities, many of whom claim to be victims of this practice. Indeed, the New Jersey state police superintendent was fired by that state's governor in March 1999 for statements concerning racial profiling that were perceived as racially insensitive.

Anecdotal evidence of racial profiling has been accumulating for years, and now many people and groups, such as the Police Practices Project of the American Civil Liberties Union (ACLU), believe that all "pretext" traffic stops are wrong, because the chance that racism and racial profiling will creep into such stops is very high. As one ACLU official argued, "Police can find an excuse to stop virtually anyone for anything. 'Out of place' stops are the most blatant example. It's also about officers who are a product of a society that is infected with conscious and unconscious racism."[30] Many states are now dealing with this problem (Exhibit 10–4).

For their part, many police executives defend such tactics as an effective way to focus their limited resources on likely lawbreakers; they argue that profiling is based not on *prejudice* but on *probabilities*—or the statistical reality that young minority men are disproportionately likely to commit crimes. (However, this argument provides little comfort to the many minority men, including many middle-class professionals, who have complained about police harassment based solely on their skin color.) As explained by Bernard Parks, African American police chief of Los Angeles,

> We have an issue of violent crime against jewelry salespeople. The predominant suspects are Colombians. We don't find Mexican-Americans, or blacks, or other immigrants. It's a collection of several hundred Colombians who commit this crime. If you see six in a car in front of the Jewelry Mart, and they're waiting and watching people with briefcases, should we play the percentages and follow them? It's common sense.[31]

Still, it is very difficult for the police to combat the public's perception that traffic stops of minorities simply on the basis of race are widespread and prejudicial in nature.

Some police agencies, including those in San Diego and San Jose, California, voluntarily collect data to determine if racial profiling exists. Many police executives, however, fear that the federal government will step in with legislation requiring that much more information be collected during traffic stops. Because of the often-contentious nature of traffic stops, police are concerned that having to collect additional information will only exacerbate an already tense situation. Whatever the approach that is taken, this is an area that is evolving. Traditional practices are being examined and redressed.

Exhibit 10–4

Racial Profiling:
The Beat Goes On

New Developments from the White House
to the Statehouses

According to a memo released by the White House on Feb. 28, U.S. Attorney General John Ashcroft has been directed by President Bush to "review the use by federal law enforcement authorities of race as a factor in conducting stops, searches and other investigative procedures," as a step toward reducing the practice of racial profiling.

The memo also directs Ashcroft to work with Congress and state and local law enforcement officials to collect data and "assess the extent and nature of any such practices." . . .

In other developments around the nation on the racial profiling issue:

Arkansas—On the grounds that data collection would be too costly and dangerous, state lawmakers in February pulled a bill that would have required state police to gather information on traffic stops as a means of tracking racial profiling. . . .

Maryland—A bill that would require officers to record racial data from all traffic stops garnered support last month from more than 20 witnesses who appeared before the House Commerce and Government Matters Committee, including elected officials and representatives of chiefs of police, sheriffs and the Fraternal Order of Police.

Minnesota—While Minneapolis and St. Paul police chiefs Robert Olson and William Finney pushed this month for a statewide, state-funded study that would require officers to collect racial data from traffic stops, Gov. Jesse Ventura told members of the state's police chiefs' and sheriffs' associations that he was not inclined to mandate the gathering of such information. Ventura's budget contains $280,000 for such a study, but the governor said he would be willing to redirect the money to training instead.

Missouri—Data from 4,480 traffic stops made by Columbia police showed that black drivers made up 20 percent of those who were pulled over, although they represent just 10 percent of the city's population, according to the 1990 census. The report, released in February, also found that 42 percent of cars searched by local police had black drivers.

Under Missouri's racial profiling law, the 712 police agencies in the state are required to hand in a report on the race and gender of drivers they stopped, and departments that fail to do so by March, or that refuse to, could face a loss of state funds. By February, 127 reports had been turned in, said a spokesman for the attorney general's office.

New Jersey—The four victims of a 1998 shooting by state police on the New Jersey Turnpike, which ignited the national furor over racial profiling, were awarded $12.9 million in a settlement with the state in February. . . .

Officials from the state troopers' union have pointed to a dramatic decline in the number of drug arrests made on the turnpike and Garden State Parkway as evidence of the chilling effect that fear of racial-profiling accusations has had on the force. The number of drug charges resulting from stops on the turnpike fell from 494 in 1999 to 370 last year, and from 783 to 350 on the parkway. . . .

Oregon—The initial results of a 10-month study of racial profiling by the Hillsboro Police Department, released in February, found that while officers probably do not target minorities, they might be more likely to search, arrest or ticket a black driver once a stop has occurred.

Pennsylvania—Computerized records maintained by the Philadelphia Police Department will now include not only the race, age and gender of people stopped, and the basis for the action, but will also note the same characteristics of the officer. The information will be available to the PPD's internal affairs unit and to civil rights lawyers.

South Dakota—Requiring police to collect racial data on the state's large Indian population during traffic stops would only serve to further alienate that group, said Representative Bill Napoli, a Rapid City Republican whose impassioned speech swayed lawmakers to kill the legislation by a vote of 11–2.

Utah—A bill that would have required drivers to list a racial category on their license, so that it could be recorded by police in the event of a traffic stop, died in the state Senate this month after a vote of 13–13. The bill was supported by law enforcement agencies, the state attorney general and the governor's office.

Washington—The creation of a Tacoma Police Department task force to examine racial profiling was approved in January by the City Council. Police found recently that nearly one-fourth of the 31,114 traffic tickets issued in 1999 went to black drivers.

Source: Reprinted with permission from *Law Enforcement News*, February 28, 2001, p. 6. John Jay College of Criminal Justice (CUNY) 555 West 57th St., New York, NY 10019.

Other police field practices are the source of tensions between minorities and police, such as the following:

1. *Delay in responding to calls for service.* Several studies of police work have found that patrol officers would often deliberately delay responding to calls for service, especially in cases of family disturbances.[32] Although this delay may be justified on grounds of officer safety (i.e., awaiting backup) and while these studies did not demonstrate any pattern of racial bias, it does not help public perceptions of the police.

2. *Verbal abuse, epithets, and other forms of disrespect.* Offensive labels for people are a regular aspect of the working language of some police officers. One study found that 75 percent of all officers used some racially offensive words, most of which were not

uttered in the presence of citizens; however, police openly ridicule and belittle citizens in only 5 percent of all encounters.[33] It has also been found that verbal expressions of disrespect did not necessarily translate into discriminatory behavior.[34] Derogatory terms are often an expression of the general alienation from the public that the police feel. In some situations, the police use the terms as a control technique, in an attempt to establish their authority.[35] Nonetheless, they should be avoided at all times.

3. *Excessive questioning and frisking of minority citizens.* Allegations of harassment by police are often raised by racial minorities who believe they have been unnecessarily subjected to field interrogations. Many officers, because they are trained to be suspicious and often confront individuals in questionable circumstances, regard such activities as legitimate and effective crime-fighting tactics.

4. *Discriminatory patterns of arrest and traffic citations.* African Americans are arrested more often than whites relative to their numbers in the population.[36] African American complainants request arrests more often than whites. Since most incidents are intraracial, this can result in more arrests of African Americans.[37] Like crime and punishment generally, police and minority relations cannot be properly analyzed apart from the broader social, political, and economic situations from which they emerge. Police have been found more likely to arrest both white and African American suspects in low-income areas. Insofar as African Americans are disproportionately represented among the poor, however, this factor is likely to result in a disproportionate rate of arrests of African Americans.[38]

5. *Excessive use of physical force.* Police have been found to use force in about 5 percent of all encounters involving offenders. In about two-thirds of the incidents involving force, its application is judged as reasonable. White and African-American officers used excessive force at nearly the same rate. It is known that "a sizable minority of citizens experience police misconduct at one time or another."[39] The result, of course, is that many racial minorities *perceive* that their race is being unduly brutalized. And, to them, perception is reality.

A Related Issue: Domestic Violence

Police firearm use has recently undergone a major change: A 1996 federal law, entitled the Domestic Violence Offender Gun Ban (popularly known as the **Lautenberg Amendment**), bars anyone—including police and military personnel—from carrying firearms if they have a misdemeanor conviction for domestic violence. Although no figures are available regarding loss of jobs, that has been the case for hundreds of police officers across the nation. In many cases, officers found to have past misdemeanor convictions have lost their jobs.[40]

No one denies that a police officer who beats a spouse or child should be fired. The ban's supporters maintain that the police must also be held accountable when they commit domestic violence and that their easy access to firearms can cause a domestic argument to escalate to homicide.[41] However, critics of the law, including many politicians, police associations, and unions, argue that the law is too broad.

Assume, for example, that a female police officer tells her 15-year-old son he cannot leave the house and hang out with some kids she knows to be using drugs.

He attempts to leave and calls her some names, so she grabs him by the arm and sits him down. Except for a bruise on his arm, he is not injured, but he calls the police, a report is filed, and she is convicted of misdemeanor assault in a trial by judge with no right to trial by jury. Because of the domestic-violence law, she loses her right to carry a gun and, thus, her career is ended.[42]

The issue is not whether abusive police officers should be fired, but whether the law as it is written is effective and legal. Several state lawsuits have challenged the constitutionality of the law, including those by the National Association of Police Officers (which argues that officers are being "sacrificed on the altar of political correctness"[43]), and in August 1998, the U.S. Court of Appeals for the District of Columbia Circuit exempted police and federal law enforcement agents within its jurisdiction from the law. Unless the law is amended, because police agencies typically have no unarmed positions, it in effect ends the careers of officers who are affected by it.

POLICE CORRUPTION

A Long-Standing "Plague" on Policing

"For as long as there have been police, there has been police corruption."[44] Thus observed Lawrence Sherman about the oldest and most persistent problem in American policing. To make the point, corruption has long plagued the New York City Police Department, as determined by the Knapp Commission, which investigated police corruption there in the early 1970s.[45] Knapp's 1973 report stated that there are two primary types of corrupt police officers: the "meat-eaters" and the "grass-eaters." Meat-eaters, who probably constitute a small percentage of police officers, spend a good deal of their working hours aggressively seeking out situations they can exploit for financial gain, including gambling, narcotics, and other lucrative enterprises. No change in attitude is likely to affect meat-eaters; their income is so large that the only way to deal with them is to get them off the force and, it is hoped, prosecute them. Grass-eaters constitute the overwhelming majority of those officers who accept payoffs. They are not aggressive but will accept gratuities from contractors, tow-truck operators, gamblers, and the like.

The Knapp Commission also identified several factors that influence how much graft police officers receive, the most important of these being the character of the individual officer. The branch of the department and type of assignment also affect opportunities for corruption. Typically, a plainclothes officer will have more and varied opportunities than the uniformed patrol officer, and uniformed officers located in beats with, say, several vice dens will have more opportunities for payoffs. Another factor is rank; the amount of payoffs received generally ascends proportionately with rank.

Police corruption can be defined broadly, from major forms of police wrongdoing to the pettiest forms of improper behavior. Another definition is "the misuse of authority by a police officer in a manner designed to produce personal gain for the officer or for others."[46] Police corruption is not limited to monetary gain, however. Gains may be made through the acceptance of services received, status, influence, prestige, or future support for the officer or someone else.[47]

Events such as those described in Los Angeles and other cities have focused attention on the broader issue of rogue cops in police departments across the country, especially in minority neighborhoods.[48] The brazenness and viciousness of today's corrupt police officers also troubles even their most staunch defenders.

TYPES AND CAUSES

Several factors contribute to police corruption, among them the rapid hiring of personnel, civil service and union protections that make it difficult to fire officers,[49] and temptations from money and sex.

Two theories—the "rotten apple" theory and the "environmental" theory— have been suggested to explain police corruption. The rotten apple theory holds that corruption is the result of having a few bad apples in the barrel who probably had character defects prior to employment. The environmental theory suggests that corruption is more the result of a widespread politically corrupt environment; politically corrupt cities create an environment in which police misconduct flourishes.[50]

Police corruption takes two basic forms: external and internal. The external form includes those activities (such as gratuities and payoffs) that occur from and through police contacts with the public. Internal corruption involves the relationships among police officers within the workings of the police department; this would include payments to join the police force, for better shifts or assignments, for promotions, and the like.[51]

Ellwyn Stoddard coined the term "blue-coat crime" and described several different forms of deviant practices among both police and citizens. In the following list, those coming first would probably elicit the least fear of prosecution and those at the end would probably invoke major legal ramifications:

1. *Mooching:* receiving free coffee, meals, liquor, groceries, laundry services, and so forth
2. *Chiseling:* demanding free admission to entertainment or price discounts
3. *Favoritism:* using license tabs, window stickers, or courtesy cards to gain immunity from traffic arrest
4. *Prejudice:* behaving less than impartially toward minority group members or others who are less likely to have influence in city hall
5. *Bribery:* receiving payments of cash or gifts for past or future assistance in avoiding prosecution, also includes political payoffs for favoritism in promotions; police officers accepting payoffs or protection money are said to be "on the pad"
6. *Shakedown:* appropriating expensive items for personal use and attributing the loss to criminal activity when investigating a burglary or unlocked door
7. *Perjury:* following the "code" that demands that officers lie to provide an alibi for fellow officers apprehended in unlawful activity

8. *Premeditated theft:* planned burglaries that involve the use of tools or keys to gain entry; any prearranged act of unlawful acquisition of property that cannot be explained as a "spur of the moment" theft[52]

The most common and extensive form of corruption involves the receipt by police officers of small gratuities or tips. Discounts and free services may be regarded by officers as relatively unimportant, while the payment of cash—bribery—is a very different matter.[53] Former NYPD Commissioner Patrick V. Murphy was one of those who "drew the short line," telling his officers that "except for your paycheck, there is no such thing as a clean buck."[54] Such police officials would argue that even the smallest gratuities can create an expectation of some patronage or favor in return. Professionalism always discourages gratuities for these occupations, and an argument could be made that the police are in no less sensitive a position.

Another serious form of police corruption is that which is drug related. Until the 1960s, most police corruption was associated with the protection of gambling operations, illegal liquor establishments, prostitution, and similar "victimless" activities. More recently, however, drug-related police corruption has probably surpassed those earlier forms of deviance. A typology of drug-related corruption has been developed by David Carter,[55] who believes that the numbers of such cases "have notably increased":

> *Type I Drug Corruption*—an officer seeks to use his or her position simply for personal gain. Includes the following: giving information to drug dealers about investigations, names of informants, planned raids, and so forth; accepting bribes from drug dealers in exchange for actions such as nonarrest, evidence tampering, perjury; theft of drugs from police property room for personal consumption; "seizure" of drugs for personal use without arresting the person possessing the drugs; taking profits of drug dealers' sales and/or their drugs for resale; extorting drug traffickers for money or property in exchange for nonarrest or nonseizure of drugs.
>
> *Type II Drug Corruption*—this form of corruption involves the officer's search for legitimate goals and may not even be universally perceived as being corrupt. "Gain" may involve organizational benefit—perhaps a form of "winning" or "revenge." Included are such actions as giving false statements to obtain arrest or search warrants against suspected or known drug dealers; perjury during hearings and trials of drug dealers; "planting," or creating evidence against "known" drug dealers; entrapment; and falsely spreading rumors that a dealer is a police informant in order to place that person's safety in jeopardy.

THE CODE OF SILENCE

Patrick V. Murphy wrote that "the most difficult element to overcome in the fight against corruption in the department was the code of silence."[56] This **code of silence**—keeping quiet in the face of misconduct by other officers—has been well documented. Evidence of the fraternal bond that exists in policing was found by William Westley as early as 1970, when more than 75 percent of the officers surveyed said that they would not report another officer for taking money from a prisoner or testify against an officer accused by a prisoner.[57] (In a related vein, see Exhibit 10–1.)

Officer Jack Smith finds himself in a moral dilemma. He knows of another officer's misconduct; he witnessed the officer putting expensive ink pens in his pocket while securing an unlocked office supply store on graveyard shift. If reported, the misconduct will ruin the officer involved, but if not reported, the behavior can eventually cause enormous harm. To outsiders, this is not a dilemma at all; the only proper path is for Smith to report the misconduct. To philosophers, the doctrines of utilitarianism (the ethic of good consequences) and deontology (the ethic of rights and duties) require that this officer work to eliminate corruption. But the outsiders and philosophers are not members of the close fraternity of police, nor do they have to depend upon other officers for their own safety.

There are several arguments for and against Officer Smith's informing on his partner. Reasons for informing would include the fact that the harm caused by a scandal would be outweighed by the public's knowing that the police department is free of corruption; also, individual episodes of corruption would be brought to a halt. The officer, moreover, has a sworn duty to uphold the law. Any employee has a right to be allowed to do his or her duty, including blowing the whistle on employers or colleagues. Reasons against Officer Smith's informing include the fact that a skilled police officer is a valuable asset whose social value far outweighs the damage done by moderate corruption. Also, discretion and secrecy are obligations assumed by joining and remaining within the police fraternity; dissenters should resign rather than inform. Furthermore, it would be unjust to inflict punishment of dismissal and disgrace on an otherwise decent officer.[58]

How does one reconcile these two varying points of view? Probably the first thing to do is to realize that each view is morally defensible. A person who is in charge of investigating police corruption would no doubt be warmer toward the punitive view, while at the other extreme would be the person who would overlook such behaviors at all times. The ideal position might be in the middle—to maintain a commitment to professionalism and ethics without overreacting by insisting that officers report on their fellows for every acceptance of a cup of coffee and so forth.

The good news is that a recent survey by the National Institute of Justice found that about 83 percent of all officers in the United States do not accept the code of silence as an essential part of the mutual trust necessary to good policing.[59]

INVESTIGATION AND PROSECUTION

Federal powers and jurisdiction for investigating and prosecuting police corruption were significantly expanded through the Hobbs Act in 1970. (See 18 U.S.C., Section 1955.) Two important elements of this federal statute that allow the investigation of police corruption are extortion and commerce. Whenever a police officer solicits a payoff from a legitimate businessowner to overlook law violations (for example, a tavern owner who was selling alcohol to minors), extortion (involving fear) occurs that inherently affects legitimate commerce. The Hobbs Act may be employed by the prosecutor when these two elements are present. The meaning of extortion has been expanded so it now covers most payoff arrangements that involve public officials.[60] The only area of police corruption that may be beyond the reach of the Hobbs Act is that which is purely internal in nature or isolated gratuities. The federal perjury statute (18 U.S.C. 1621) and the federal false sworn declaration statute (18 U.S.C. 1623), both enacted in 1970, have also become powerful weapons for prosecutors in investigating public corruption. Both statutes deal with false testimony under oath, and in an investigation of corruption they are pertinent at the grand jury stage.[61]

POSSIBLE SOLUTIONS

There are several other possible measures for overcoming the pernicious effects of police corruption. In addition to the obvious need for an honest and effective police administration, training of recruits on the need for a corruption-free department is also necessary. The creation and maintenance of an internal affairs unit and vigorous prosecutions of law-breaking police officers are also critical to the function of integrity. Finally, there should be some mechanism for rewarding the honest police officers, which should minimally include protection from retaliation when they inform on crooked cops. All police officers should be given written formal guidelines to the departmental policy on soliciting and accepting gifts or gratuities. This apprises officers of the administration's view of such behavior and assists the chief executive in maintaining integrity and disciplining officers accordingly. Figure 10–1 is an example of a good functional policy concerning gratuities.

Finally, computers can also assist with police corruption. Indeed, information that was uncovered about corruption within the Chicago Police Department

---

1. Without the express permission of the Sheriff, members shall not solicit or accept any gift, gratuity, loan, present, or fee where there is any direct or indirect connection between this solicitation or acceptance of such gift and their employment by this office.
2. Members shall not accept, either directly or indirectly, any gift, gratuity, loan, fee or thing of value, the acceptance of which might tend to improperly influence their actions, or that of any other member, in any matter of police business, or which might tend to cast an adverse reflection on the Sheriff's Office.
3. Any unauthorized gift, gratuity, loan, fee, reward or other thing falling into any of these categories coming into the possession of any member shall be forwarded to the member's commander, together with a written report explaining the circumstances connected therewith. The commander will decide the disposition of the gift.

FIGURE 10–1 Washoe County, Nevada, Sheriff's Office Gratuity Policy.
(Washoe County [Nevada] Sheriff's Office)

was obtained through an $850 software program known as "Brainmaker," an early-warning device intended to flag at-risk officers before they commit acts that could get them arrested or fired. With this approach, such officers can be provided with counseling before serious problems occur.[62]

LIMITATIONS ON OFFICERS' CONSTITUTIONAL RIGHTS

Police officers are generally afforded the same rights, privileges, and immunities outlined in the U.S. Constitution for all citizens. However, by virtue of their position, they may be compelled to give up certain rights in connection with an investigation of on-duty misbehavior or illegal acts. These rights are the basis for legislation such as the Peace Officers' Bill of Rights (discussed in Chapter 12), labor agreements, and civil service and departmental rules and regulations that guide an agency's disciplinary process.

Following is a brief overview of some areas in which the federal courts have placed **limitations on officers' constitutional rights** and hold sworn officers more accountable by virtue of the higher standard that exists in light of their occupation.

Free Speech

Although the right of freedom of speech is one of the most fundamental and cherished of all American rights, the Supreme Court has indicated that "the State has interests as an employer in regulating the speech of its employees that differ significantly from those it possesses in connection with regulation of the speech of the citizenry in general."[63] Thus, the state may impose restrictions on its employees that it would not be able to impose on the citizenry at large. However, these restrictions must be reasonable. For example, a department may not prohibit "any activity, conversation, deliberation, or discussion which is derogatory to the Department," as such a rule obviously prohibits all criticism of the agency by its officers, even in private conversation.[64]

Another First Amendment–related area is that of personal appearance. The Supreme Court has upheld several grooming standards (regarding hair, sideburn, and moustache length) for officers to make officers readily recognizable to the public and to maintain the esprit de corps within the department.[65]

Searches and Seizures

The Fourth Amendment to the U.S. Constitution protects "the right of the people to be secure in their persons, houses, papers, and effects, against unreasonable searches and seizures." The Fourth Amendment usually applies to police officers when they are at home or off duty in the same manner as it applies to all citizens.

However, because of the nature of their work, police officers can be compelled to cooperate with investigations of their behavior when ordinary citizens would not. Examples would include equipment and lockers provided by the department to the officers. There, the officers have no expectation of privacy that affords or merits protection.[66]

However, lower courts have established limitations where searches of employees themselves are concerned. The rights of prison authorities to search their employees arose in a 1985 Iowa case, when employees were forced to sign a consent form as a condition of hire. The court disagreed with such a broad policy, ruling that the consent form did not constitute a blanket waiver of all Fourth Amendment rights.[67] Police officers may also be forced to appear in a lineup, a clear "seizure" of his or her person.

SELF-INCRIMINATION

The Supreme Court has addressed questions concerning the Fifth Amendment as it applies to police officers who are under investigation. In *Garrity v. New Jersey*,[68] a police officer was ordered by the attorney general to answer questions or be discharged. The officer testified that information obtained as a result of his answers was later used to convict him of criminal charges. The Supreme Court held that the information obtained from the officers could not be used against him at his criminal trial, because the Fifth Amendment forbids the use of coerced confessions.

However, it is proper to fire a police officer who refuses to answer questions that are related directly to the performance of his or her duties, provided that the officer has been informed that any answers may not be used later in a criminal proceeding.[69]

RELIGIOUS PRACTICES

Police work requires that personnel are available and on duty 24 hours per day, seven days a week. Although it is not always convenient or pleasant, such shift configurations require that many officers work weekends, nights, and holidays. It is generally assumed that one who takes such a position agrees to work such hours and abide by other such conditions; it is usually the personnel with the least seniority on the job who must work the most undesirable shifts. However, there are occasions when one's religious beliefs are in direct conflict with the requirements of the job, such as the carrying of firearms or the ability to attend religious services or periods of religious observance that occur during one's shift assignments. The carrying of firearms may even conflict with one's beliefs. In these situations, the employee may be forced to choose between his or her job and religion.

Title VII of the Civil Rights Act of 1964 prohibits religious discrimination in employment. Thus, Title VII requires reasonable accommodation of religious beliefs, but not to the extent that the employee has complete freedom of religious expression.[70]

Sexual Misconduct

To be blunt, there is ample opportunity for police officers to become engaged in adulterous or extramarital affairs. It is likely that no other occupation or profession offers the opportunities for sexual misconduct that the police occupation does. Police officers frequently work alone, usually without direct supervision, in activities that involve frequent contact with citizens, usually in relative isolation. The problem seems to be pervasive in police departments from the smallest to the largest. Unfortunately, it is also an area of police behavior that is not easily quantified or understood.[71]

Allen Sapp[72] suggested several possible categories of sexually motivated behaviors by police officers (again, the extent to which each or all occur is unknown):

- *Sexually motivated nonsexual contacts:* Officers initiate contacts with female citizens without probable cause or any legal basis for the purpose of obtaining names and addresses for possible later contacts by the officer.
- *Voyeuristic contacts:* Officers seek opportunities to view unsuspecting women partially clad or nude, such as in parked cars on "lovers lane."
- *Contacts with crime victims:* A wide variety of behavior can occur, including unnecessary callbacks to homes of female victims, bodily contact with accident victims, and sexual harassment by officers.
- *Contacts with offenders:* Officers may also harass female offenders by conducting body searches and pat-downs or frisks.
- *Sexual shakedowns:* Officers may demand sexual services from prostitutes or other citizens.
- *Citizen-initiated sexual contacts:* "Police groupies"—often young females who are sexually attracted to the uniform, weapons, or power of the police officer—may seek to participate in sexual activities with officers. This category may also include offers of sexual favors in return for preferential treatment or calls to officers from lonely or mentally disturbed women seeking officers' attention.
- *Sex crimes by officers:* Officers may sexually assault jail inmates and rape citizens.

In a related vein, several federal courts have recently considered whether police agencies have a legitimate interest in the sexual activities of their officers when such activities affect job performance. In one such case, the court held that the dismissal of a married police officer for living with another man's wife was a violation of the officer's privacy and associational rights.[73]

Other courts, however, have found that off-duty sexual activity *can* affect job performance. Where a married city police officer was found having consensual sexual relations with unmarried women other than his wife, the department contended that the officer's conduct—which became public—severely damaged public confidence. Here, a Utah court held that adultery was not a fundamental right and refused to strike down a statute criminalizing adultery.[74] And in a Texas case, when a male officer's extramarital affair led to his being passed over for promotion, the city civil service commission, the Texas Supreme Court, and the U.S. Supreme Court upheld the denial. They concurred with the city police chief's argument that such a promotion would adversely affect the efficiency and morale of the department and would have been disruptive.[75]

RESIDENCY REQUIREMENTS

Many government agencies specify that all or certain members in their employ must live within the geographical limits of their employing jurisdiction. In other words, employees must reside within the county or city of employment. Such residency requirements have been justified on the grounds that officers should become familiar with and be visible in the jurisdiction of employment or that they should reside where they are paid by the taxpayers to work. Perhaps the strongest rationale given by employing agencies is that criminal justice employees must live within a certain proximity of their work in order to respond quickly in the event of an emergency.

MOONLIGHTING

Moonlighting means to hold a second job in addition to one's normal full-time occupation. The courts have traditionally supported the practice of police agencies placing limitations on the amount and kinds of outside work their employees can perform.[76] For example, police restrictions on moonlighting range from a complete ban on outside employment to permission to engage in certain forms of work, such as investigations, private security, teaching police science courses, and so on. The rationale for agency limitations is that "outside employment seriously interferes with keeping the [police and fire] departments fit and ready for action at all times."[77]

MISUSE OF FIREARMS

Police agencies typically attempt to restrain the use of firearms through written policies and frequent training of a "shoot/don't shoot" nature. Still, a broad range of potential and actual problems remains with respect to the use and possible misuse of firearms. Police agencies generally have policies regulating the use of handguns and other firearms by their officers, both on and off duty. The courts have held that such regulations need only be reasonable and that the burden rests with the police officer being disciplined to show that the regulation was arbitrary and unreasonable.[78]

Police firearms regulations tend to address three basic issues: (1) requirements for the safeguarding of the weapon, (2) guidelines for carrying the weapon while off duty, and (3) limitations on when the weapon may be fired.[79] Courts and juries are increasingly becoming more harsh in dealing with police officers who misuse their firearms. The current tendency is to investigate police shootings in order to determine if the officer acted negligently or the employing agency negligently trained and supervised the officer/employee.

ALCOHOL AND DRUG ABUSE

Alcoholism and drug abuse problems are much more acute when they involve police employees. It is obvious, given the law of most jurisdictions and the nature of their work, that police officers must not be walking time bombs; they must be able to perform their work with a clear head that is unbefuddled by alcohol or drugs.[80] Police departments typically specify in their policy manual that no alcoholic beverages will be consumed within a specified period prior to reporting for duty. Such regulations have been upheld uniformly as rational because of the hazards of the work.

Enforcing such regulations will occasionally mean that police employees are ordered to submit to drug or alcohol tests. In 1989, the U.S. Supreme Court issued a major decision on drug testing: *National Treasury Employees Union v. Von Raab*,[81] which dealt with drug-testing plans for U.S. customs workers. This decision addressed all three of the most controversial drug testing issues: whether testing should be permitted when there is no indication of a drug problem in the workplace, whether the testing methods are reliable, and whether a positive test proves there was on-the-job impairment.[82]

The Supreme Court held that although only a few U.S. customs employees tested positive, drug use is such a serious problem that the program was warranted. Second, the Court found nothing wrong with the testing protocol. And, finally, while tests may punish and stigmatize a person for extracurricular drug usage that may have no effect on the worker's on-the-job performance, the Court indicated that this dilemma is still no impediment to testing.

DISCIPLINARY POLICIES AND PRACTICES

MAINTAINING THE PUBLIC TRUST

Clearly, the public's trust and respect are precious commodities, quickly lost through improper behavior by police employees and the improper handling of an allegation of misconduct. Serving communities professionally and with integrity should be the goal of every police agency and its employees in order to ensure that trust and respect are maintained. The public expects that police agencies will make every effort to identify and correct problems and respond to citizens' complaints in an equally judicious, consistent, fair, and equitable manner.

Employee misconduct and violations of departmental policy are the two principal areas in which discipline is involved.[83] Employee misconduct includes those acts that harm the public, such as corruption, harassment, brutality, and violations of civil rights. Violations of policy may involve a broad range of issues, from substance abuse and insubordination to minor violations of dress or tardiness.

Due Process Requirements

There are well established, minimum due process requirements for discharging public employees. They must

1. Be afforded a public hearing
2. Be present during the presentation of evidence against them and have an opportunity to cross-examine their superiors
3. Have an opportunity to present their own witnesses and other evidence concerning their side of the controversy
4. Be permitted to be represented by counsel
5. Have an impartial referee or hearing officer presiding
6. Have an eventual decision based on the weight of the evidence introduced during the hearing

Such protections apply to any disciplinary action that can significantly affect a police employee's reputation and/or future chances for special assignment or promotion.[84]

At times, police administrators will determine that an employee must be disciplined or terminated. Grounds for discipline or discharge can vary widely from agency to agency. Certainly the agency's formal policies and procedures should specify and control what constitutes proper and improper behavior.

Dealing with Complaints

Origin of Complaint

A personnel **complaint** is an allegation of misconduct or illegal behavior against an employee by anyone inside or outside the organization. Internal complaints—those made from persons within the organization—may involve supervisors who observe officer misconduct, officers who complain about supervisors, supervisors who complain about supervisors, civilian personnel who complain about officers, and so on. External complaints originate from sources outside the organization and usually involve the public.

Every complaint, regardless of the source, must be accepted and investigated in accordance with established policies and procedures. Anonymous complaints are the most difficult to investigate because there is no opportunity to obtain further information or question the complainant about the allegation. Such complaints can have a negative impact on employee morale, as officers may view them as unjust and frivolous.

Types and Causes

Complaints may be handled informally or formally, depending on the seriousness of the allegation and preference of the complainant. A formal complaint occurs when a written and signed and/or tape-recorded statement of the allegation is made and the complainant requests to be informed of the investigation's disposition. Figure 10–2 provides an example of a complaint form used to initiate a personnel investigation.

An informal complaint is an allegation of minor misconduct made for informational purposes that can usually be resolved without the need for more formal processes. The supervisor may simply discuss the incident with the employee and resolve it through informal counseling as long as more serious problems are not discovered and there is no history of similar complaints.

The majority of complaints against officers fall under the general categories of verbal abuse, discourtesy, harassment, improper attitude, and ethnic slurs.[85] It is clear that the verbal behavior of officers generates a significant number of complaints. Finally, minority citizens and those with less power and fewer resources are more likely to file complaints of misconduct and to allege more serious forms of misconduct than persons with greater power and more resources.

Receipt and Referral

Administrators should have in place a process for receiving complaints that is clearly delineated by departmental policy and procedures. Generally, a complaint will be made at a police facility and referred to a senior officer in charge to determine its seriousness and whether or not there is a need for immediate intervention.

In most cases, the senior officer receiving the investigation will determine the nature of the complaint and the employees involved; the matter will be referred to the employee's supervisor to conduct an initial investigation. The supervisor completes the investigation, recommends any discipline, and sends the matter to the Internal Affairs Unit and the agency head to finalize the disciplinary process. This method of review ensures that consistent and fair standards of discipline are applied.

The Investigative Process

Generally, the employee's supervisor will conduct a preliminary inquiry of the complaint, commonly known as fact-finding. If it is determined that further investigation is necessary, the supervisor may question employees and witnesses, obtain written statements from those persons immediately involved in the incident, and gather any necessary evidence, including photographs. Care must be exercised that the accused employee's rights are not violated. The initial investigation would be sent to an appropriate division commander and forwarded to an IAU for review.

****

Control Number_____

Date & Time Reported   Location of Interview   Interview

_____   _____   _____Verbal _____Written _____Taped

| Type of complaint: | ____Force ____Procedural ____Conduct
 ____Other (Specify) |

Source of complaint: _____In Person ____Mail _____Telephone
                        _____Other (Specify)

Complaint originally   _____Supervisor _____On Duty Watch Commander ____Chief
Received by: ____IAU _____Other (Specify)

Notifications made: _____Division Commander   _____Chief of Police
Received by: _____On-Call Command Personnel
                       _____Watch Commander   _____Other (Specify)

Copy of formal personnel complaint given to complainant? ____Yes ____No

****

Complainant's name: Address:
                                                            Zip_____
Residence Phone: Business Phone:

DOB: Race: Sex: Occupation:

****

Location of Occurrence: Date & Time of Occurrence:

Member(s) Involved: Member(s) Involved:
(1) _____   (2)_____
(3) _____   (4)_____
Witness(es) Involved: Witness(es) Involved:
(1) _____   (2)_____
(3) _____   (4)_____

****

(1) _____ Complainant wishes to make a formal statement and has requested an investigation into the matter with a report back to him/her on the findings and actions.
(2) _____ Complainant wishes to advise the Police Department of a problem, understands that some type of action will be taken, but does not request a report back to him/her on the findings and actions.

****

CITIZEN ADVISEMENTS

(1) If you have not yet provided the department with a signed written statement or a tape-recorded statement, one may be required in order to pursue the investigation of this matter.
(2) The complainant(s) and/or witness(es) may be required to take a polygraph examination in order to determine the credibility concerning the allegations made.
(3) Should the allegations prove to be false, the complainant(s) and/or witness(es) may be liable for criminal and/or civil prosecution.

_____   _____
Signature of Complainant   Date & Time

_____
Signature of Member Receiving Complaint

FIGURE 10–2 Formal Complaint Form.

Making a Determination and Disposition

Once an investigation is completed, the supervisor or IAU officer must make a determination about the culpability of the accused employee and report that determination to the administrator. Following are the categories of dispositions that are commonly used:

- *Unfounded.* The alleged act(s) did not occur.
- *Exonerated.* The act occurred, but it is lawful, proper, justified, and/or in accordance with departmental policies, procedures, rules, and regulations.
- *Not sustained.* There is insufficient evidence to prove or disprove the allegations made.
- *Misconduct not based on the complaint.* Sustainable misconduct was determined but is not a part of the original complaint. For example, a supervisor investigating an allegation against an officer of excessive force may find the force used was within departmental policy but that the officer made an unlawful arrest.
- *Closed.* An investigation may be halted if the complainant fails to cooperate or it is determined that the action does not fall within the administrative jurisdiction of the police agency.
- *Sustained.* The act did occur and it was a violation of departmental rules and procedures. Sustained allegations include misconduct that falls within the broad outlines of the original allegation(s).

Once a determination of culpability has been made, the complainant should be notified of the department's findings. Details of the investigation or recommended punishment will not be included in the correspondence. As shown in Figure 10–3, the complainant will normally receive only information concerning the outcome of the complaint.

Appealing Disciplinary Measures

If an officer disagrees with a supervisor's recommendation for discipline, the first step of an appeal may involve a hearing before the division commander, who is usually of the rank of captain or deputy chief. The accused employee may be allowed labor representation or an attorney to assist in asking questions of the investigating supervisor, clarifying issues, and presenting new or mitigating evidence. If the employee is still not satisfied, an appeal hearing before the chief executive is granted. This is usually the final step in appeals within the agency. The chief or sheriff communicates a decision to the employee in writing. Depending upon labor agreements and civil service rules and regulations, some agencies extend their appeals of discipline beyond the department. For example, employees may bring their issue before the civil service commission or city or county manager for a final review. Employees may also have the right to an independent arbitrator's review of discipline.

Police Department
3300 Main Street
Downtown Plaza
Anywhere, U.S.A. 99999
June 20, 2002

Mr. John Doe
2200 Main Avenue
Anywhere, U.S.A.

Re: Internal affairs #000666-98
 Case Closure

Dear Mr. Doe,

Our investigation into your allegations against Officer Smith has been completed. It has been determined that your complaint is SUSTAINED and the appropriate disciplinary action has been taken.

Our department appreciates your bringing this matter to our attention. It is our position that when a problem is identified, it should be corrected as soon as possible. It is our goal to be responsive to the concerns expressed by citizens so as to provide more efficient and effective services.

Your information regarding this incident was helpful and of value in our efforts to attain that goal. Should you have any further questions about this matter, please contact Sergeant Jane Alexander, Internal Affairs, at 555-9999.

Sincerely,

I.M. Boss
Lieutenant
Internal Affairs Unit

FIGURE 10–3 Citizen's Notification of Discipline Letter.

DETERMINING THE LEVEL AND NATURE OF SANCTIONS

When an investigation against an employee is sustained, the sanctions and level of discipline must be decided. Management must be very careful when recommending and imposing discipline because of its impact on the overall morale of the agency's employees. If the recommended discipline is viewed by employees as too lenient, it may send the wrong message that the misconduct was insignificant. On the other hand, discipline that is viewed as too harsh may have a demoralizing effect on the officer(s) involved and other agency employees and result in allegations that the leadership is unfair.

Listed here are disciplinary actions commonly used by agencies in order of their severity:

Counseling. This is usually a conversation between the supervisor and employee about a specific aspect of the employee's performance or conduct. It is warranted when an employee has committed a relatively minor infraction or the nature of the offense is such that oral counseling is all that is necessary. No documentation or report is placed in the employee's personnel file.

Documented Oral Counseling. This is usually the first step in a progressive disciplinary process and is intended to address relatively minor infractions. This is provided when there are no previous reprimands or more severe disciplinary action of the same or similar nature.

Letters of Reprimand. These are formal written notices regarding significant misconduct, more serious performance violations, or repeated offenses. It is usually the second step in the disciplinary process and is intended to provide the employee and agency with a written record of the violation of behavior. It identifies what specific corrective action must be taken to avoid subsequent more serious disciplinary action.

Suspension. This is a severe disciplinary action that results in an employee being relieved of duty, often without pay. It is usually administered when an employee commits a serious violation of established rules or after written reprimands have been given and no change in behavior or performance has resulted.

Demotion. In this situation, an employee is placed in a position of lower responsibility and pay. It is normally used when an otherwise good employee is unable to meet the standards required for the higher position or when the employee has committed a serious act requiring that he or she be removed from a position of management or supervisor.

Termination. This is the most severe disciplinary action that can be taken. It usually occurs when previous serious discipline has been imposed and there has been inadequate or no improvement in behavior or performance. It may also occur when an employee commits an offense so serious that continued employment would be inappropriate.

Transfer. Many agencies use the disciplinary transfer to deal with problem officers; officers can be transferred to a different location or assignment, and this action is often seen as an effective disciplinary tool.[87]

SUMMARY

This chapter has examined the serious nature of police misbehavior and society's attempts to hold officers accountable. Police behavior is being closely scrutinized today, and the underlying theme of the chapter is that the work of policing carries with it a higher standard of expectations of and behavior by the police. The police are held to a much higher standard of behavior, especially since incidents of misprision of office often harm innocent or undeserving people and receive national attention.

ITEMS FOR REVIEW

1. Define ethics and delineate some of the unique ethical problems that can be posed under community policing.
2. What are some factors that contribute to police violence? What is the law of the U.S. Supreme Court regarding police use of deadly force?
3. What are the types of police brutality? Can all of its forms ever be totally eliminated?
4. Explain racial profiling and why it is a tinderbox in relations between minorities and police.
5. Review how and why police corruption began and what factors within both the community and policing seem to foster and maintain it.
6. Delineate the constitutional limitations that have been placed on officers' rights and behavior by federal courts.
7. Describe the general process that is used by police agencies to deal with citizen complaints.
8. Review some of the factors that enter into determining the level and nature of sanctions for officers who are to be disciplined.

NOTES

1. "Good News, Better News on Use of Force," *Law Enforcement News*, March 15, 2001, p. 7.
2. "L.A. Police Corruption Case Continues to Grow," *Washington Post*, February 13, 2000, p. 1A.
3. Scott Glover and Matt Lait, "LA Police Chief Issues Critical Review of Department," *Los Angeles Times*, February 5, 2000, p. 1A.
4. Angie Cannon, "Cop-Shop Woes in the Burbs," *U.S. News and World Report*, September 25, 2000, p. 26.
5. *Ibid.*
6. Kit R. Roane, "Policing the Police Is a Dicey Business: But the Feds Have a Plan to Root Out Racial Bias," *U.S. News and World Report*, April 30, 2001, p. 28.
7. *Webster's New World Dictionary* (New York: World Publishing, 1981), p. 389.
8. U.S. Department of Justice, National Institute of Justice, Office of Community Oriented Policing Services, *Police Integrity: Public Service with Honor* (Washington, D.C.: U.S. Government Printing Office, 1997), p. 62.
9. Brian Withrow, "When Strings Are Attached: Understanding the Role of Gratuities in Police Corruptibility," in eds. Quint Thurman and Jihong Zhao, *Contemporary Issues and Challenges in a Changing Police Environment* (Los Angeles: Roxbury, forthcoming).
10. "No Free (or Discounted) Lunch for Bradenton Cops," *Law Enforcement News*, October 15, 2000, p. 6.
11. Kenneth J. Peak, B. Grant Stitt, and Ronald W. Glensor, "Ethics in Community Policing and Problem Solving," *Police Quarterly* 1 (1998): 30–31.
12. L. Morris, *Incredible New York* (New York: Bonanza, 1951).
13. James A. Inciardi, *Criminal Justice*, 5th ed. (Orlando, Fla.: Harcourt Brace, 1996).
14. Egon Bittner, "The Functions of the Police in Modern Society," in *Policing: A View From the Street*, ed. Peter K. Manning and John Van Maanen (Santa Monica, CA: Goodyear Publishing Company, 1978), pp. 32–50.
15. *Ibid.*, p 36.
16. George F. Cole and Christopher E. Smith, *The American System of Criminal Justice*, 8th ed. (Belmont, CA: West/Wadsworth, 1998), p. 228.
17. A. C. Germann, Frank D. Day, and Robert R. J. Gallati, *Introduction to Law Enforcement and Criminal Justice* (Springfield, IL: Charles C. Thomas, 1976,) p. 225.
18. Albert J. Reiss Jr., "Police Brutality: Answers to Key Questions," *Transaction*, July–August 1968, pp. 10–19.
19. *Ibid.*

20. Germann, Day, and Gallati, *Introduction to Law Enforcement*, p. 225.
21. *Ibid.*; also see William Geller, "Deadly Force: What We Know," *Journal of Police Science and Administration* 10 (1982): 151–177.
22. Lawrence Sherman and Ellen G. Cohn, "Citizens Killed by Big City Police: 1970–84," unpublished manuscript, Crime Control Institute, Washington, D.C., October 1986.
23. James J. Fyfe, "Reducing the Use of Deadly Force: The New York Experience," in U.S. Department of Justice, *Police Use of Deadly Force* (Washington, D.C.: U.S. Government Printing Office, 1978), p. 28.
24. James J. Fyfe, Jack R. Greene, William F. Walsh, O. W. Wilson, and Roy Clinton McLaren, *Police Administration*, 5th ed. (New York: McGraw-Hill, 1997), pp. 1997–1998.
25. 471 U.S. 1, 105 S.Ct. 1694, 85 L.Ed.2d 1 (1985).
26. Mark Blumberg, "Controlling Police Use of Deadly Force: Assessing Two Decades of Progress," in *Critical Issues in Policing: Contemporary Readings*, ed. Roger G. Dunham and Geoffrey P. Alpert (Prospect Heights, IL: Waveland Press, 1989), pp. 442–464.
27. *Monell v. Department of Social Services*, 436 U.S. 658, 98 S.Ct. 2018 (1978).
28. Joan Petersilia, "Racial Disparities in the Criminal Justice System: Executive Summary of RAND Institute Study, 1983," in *The Criminal Justice System and Blacks*, ed. Daniel Georges-Abeyle (New York: Clark Boardman, 1984), pp. 231, 249.
29. Ron Neubauer, quoted in Keith W. Strandberg, "Racial Profiling," *Law Enforcement Technology*, June 1999, p. 62.
30. John Crew, quoted in *ibid.*
31. Randall Kennedy, "Suspect Policy," *The New Republic*, September 13, 1999, pp. 30–35.
32. Donald Black, *Manners and Customs of the Police* (New York: Academic Press, 1980), p. 117; Richard J. Lundman, "Domestic Police-Citizen Encounters," *Journal of Police Science and Administration* 2 (March 1974): 25.
33. Jerome Skolnick, *The Police and the Urban Ghetto* (Chicago: American Bar Foundation, 1968).
34. Albert J. Reiss Jr., *The Police and the Public* (New Haven, CT: Yale University Press, 1971), p. 142.
35. Walker, *The Police in America*, p. 234.
36. *Ibid.*, p. 235.
37. Robert Friedrich, "Racial Prejudice and Police Treatment of Blacks," in *Evaluating Alternative Law Enforcement Policies*, ed. Ralph Baker and Fred A. Meyers (Lexington, MA: Lexington Books, 1979), pp. 160–161.
38. Douglas A. Smith and Christy A. Visher, "Street-Level Justice: Situational Determinants of Police Arrest Decisions," *Social Problems* 29 (December 1981): 167–187.
39. Reiss, *The Police and the Public*, p. 151.
40. "Beat Your Spouse, Lose Your Job," *Law Enforcement News*, December 31, 1997, p. 9.
41. *Ibid.*
42. Adapted from Jerry Hoover, "Brady Bill Unfair in Broad Approach to Police Officers," *Reno Gazette Journal*, September 25, 1998, p. 11A.
43. "Denver Cop's Case Galvanizes Opponents of Lautenberg Gun Ban," *Law Enforcement News*, February 14, 1999, p. 1.
44. Lawrence W. Sherman, ed., *Police Corruption: A Sociological Perspective* (Garden City: Anchor Books, 1974), p. 1.
45. See Peter Maas, *Serpico* (New York: The Viking Press, 1973).
46. Herman Goldstein, *Policing a Free Society*, p. 188.
47. *Ibid.*
48. Gordon Witkin, "When the Bad Guys Are Cops," *Newsweek*, September 11, 1995, p. 20.
49. See *ibid.*, p. 22.
50. Lawrence W. Sherman, "Becoming Bent," in *Moral Issues in Police Work*, ed. F. A. Elliston and M. Feldberg (Totowa, NJ: Rowan and Allanheld, 1985), pp. 253–265.
51. Christian P. Potholm and Richard E. Morgan, *Focus on Police: Police in American Society* (New York: Schenkman Publishing Company, 1976), p. 140.
52. Ellwyn R. Stoddard, "Blue Coat Crime," in *Thinking About Police: Contemporary Readings*, ed. Carl B. Klockars (New York: McGraw-Hill Book Company, 1983), pp. 338–350.
53. Walker, *The Police in America: An Introduction*, p. 175.
54. "Police Aides Told to Rid Commands of All Dishonesty," *New York Times*, October 29, 1970.

55. David L. Carter, "Drug Use and Drug-Related Corruption of Police Officers," in *Policing Perspectives: An Anthology*, ed. Larry K. Gaines and Gary W. Cordner (Los Angeles: Roxbury, 1999), pp. 311–323.
56. Patrick V. Murphy and Thomas Plate, *Commissioner: A View from the Top of American Law Enforcement* (New York: Simon and Schuster, 1977), p. 226.
57. William A. Westley, *Violence and the Police* (Cambridge, MA: MIT Press, 1970), pp. 113–114.
58. Thomas E. Wren, "Whistle-Blowing and Loyalty to One's Friends," in *Police Ethics: Hard Choices in Law Enforcement*, ed. William C. Heffernan (New York: John Jay Press, 1985), pp. 25–43.
59. David Weisburd and Rosanne Greenspan, *Police Attitudes Toward Abuse of Authority: Findings from a National Study* (Washington, D.C.: U.S. Department of Justice, National Institute of Justice Research in Brief, May 2000), p. 5.
60. See, for example, *United States v. Hyde*, 448 F.2d 815 (5th Cir. 1971) cert. den., 404 U.S. 1058 (1972); *United States v. Addonizia*, 451 F.2d 49 (3d Cir.), cert. den., 405 U.S. 936 (1972); and *United States v. Kenney*, 462 F.2d 1205 (3d Cir.) as amended, 462 F.2d 1230 (3d Cir.), cert. den. 409 U.S. 914 (1972).
61. Herbert Beigel, "The Investigation and Prosecution of Police Corruption," in *Focus on Police: Police in American Society*, ed. C. P. Potholm and R. E. Morgan (New York: Schenkman Publishing Company, 1976), pp. 139–166.
62. "Artificial Intelligence Tackles a Very Real Problem—Police Misconduct Control," *Law Enforcement News*, September 30, 1994, p. 1.
63. *Pickering v. Board of Education*, 391 U.S. 563 (1968), at p. 568.
64. *Muller v. Conlisk*, 429 F.2d 901 (7th Cir. 1970).
65. *Kelley v. Johnston*, 425 U.S. 238 (1976).
66. See *People v. Tidwell*, 266 N.E.2d 787 (Ill. 1971).
67. *McDonell v. Hunter*, 611 F. Supp. 1122 (SD Iowa, 1985), affd. as mod., 809 F.2d 1302 (8th Cir. 1987).
68. *Garrity v. New Jersey*, 385 U.S. 483 (1967).
69. See *Gabrilowitz v. Newman*, 582 F.2d 100 (1st Cir. 1978).
70. *United States v. City of Albuquerque*, 12 EPD 11, 244 (10th Cir.); also see *Trans World Airlines v. Hardison*, 97 S.Ct. 2264 (1977).
71. Allen D. Sapp, "Police Officer Sexual Misconduct: A Field Research Study," in *Crime and Justice in America: Present Realities and Future Prospects*, ed. Paul F. Cromwell and Roger G. Dunham (Upper Saddle River, NJ: Prentice Hall, 1997), pp. 139–151.
72. *Ibid.*
73. See *Briggs v. North Muskegon Police Department*, 563 F. Supp 585 (W.D. Mich. 1983), aff'd. 746 F.2d 1475 (6th Cir. 1984).
74. *Oliverson v. West Valley City*, 875 F. Supp.1465 (D. Utah 1995).
75. *Henery v. City of Sherman*, 928 S.W.2d 464 (Sup. Ct. Texas), cert. denied, 17 S.Ct. 1098 (1997).
76. See, for example, *Cox v. McNamara*, 493 P.2d 54 (Ore. 1972); *Brenckle v. Township of Shaler*, 281 A.2d 920 (Pa. 1972); *Flood v. Kennedy*, 239 N.Y.S.2d 665 (1963); and *Hopwood v. City of Paducah*, 424 S.W.2d 134 (Ky. 1968).
77. Richard N. Williams, *Legal Aspects of Discipline by Police Administrators*, Traffic Institute Publication 2705 (Evanston, IL: Northwestern University, 1975), p. 4.
78. See *Lally v. Department of Police*, 306 So.2d 65 (La. 1974).
79. Charles R. Swanson, Leonard Territo, and Robert W. Taylor, *Police Administration*, 3d ed. (New York: Macmillan, 1993), p. 433.
80. See *Krolick v. Lowery*, 302 N.Y.S.2d 109 (1969), p. 115; *Hester v. Milledgeville*, 598 F.Supp. 1456, 1457 (M.D.Ga. 1984).
81. *National Treasury Employees Union v. Von Raab*, 489 U.S. 656 (1989).
82. Robert J. Aalberts and Harvey W. Rubin, "Court's Rulings on Testing Crack Down on Drug Abuse," *Risk Management* 38 (March 1991): 36–41.
83. V. McLaughlin and R. Bing, "Law Enforcement Personnel Selection," *Journal of Police Science and Administration* 15 (1987): 271–276.
84. *Ibid.*
85. A. E. Wagner and S. H. Decker, "Evaluating Citizen Complaints Against the Police," in *Critical Issues in Policing: Contemporary Readings*, ed. R. G. Dunham and G. P. Alpert (Prospect Heights, IL: Waveland, 1997).
86. Kim Michelle Lersch, "Police Misconduct and Malpractice: A Critical Analysis of Citizens' Complaints," *Policing* 21 (1998): 80–96.
87. Ronald W. Glensor, Kenneth J. Peak, and Larry K. Gaines, *Police Supervision* (New York: McGraw-Hill, 1999), pp. 213–215.

Civil Liability

Failing the Public Trust

Key Terms and Concepts

Bivens suit
Negligence
Qualified immunity
Respondeat superior
Section 1983
Section 242

Sovereign immunity
Stare decisis
Tort liability
Vicarious liability
Wrongful death

All men are liable to error; and most men are . . . under temptation to it.
—*John Locke*

Where laws end, tyranny begins.—*William Pitt*

INTRODUCTION

Policing is a challenging occupation. The police must enforce the laws, perform welfare tasks, protect the innocent, and attempt to prevent crime. They see people at their worst and participate in searches and seizures, major incidents (such as hostage situations), and high-speed pursuits. They make split-second decisions. They are custodians of offenders in local jails. Legal actions can arise against the police in the course of these and many other kinds of actions. Perhaps no other occupation, with the exception of medicine, is so vulnerable to legal attack for the actions of its practitioners. Some observers believe that community oriented policing and problem solving (COPPS, discussed in Chapter 6) could lead to an increase in civil liability filings because of the greater degree of involvement of police in the lives of citizens.[1]

This chapter discusses the current and developing laws that serve to make criminal justice practitioners legally accountable, both civilly and criminally, for acts of misconduct and negligence. For many years this has been a rapidly growing area of litigation and body of legal precedent; lawyers are becoming very knowledgeable about the statutes and court decisions in this regard and are more experienced and willing to seek redress in the federal court system as they protect the rights of their clients.

It cannot be overstated how important it is for students of criminal justice and police personnel to know and understand the area of liability. It is far better to learn which police actions are blameworthy—and potentially costly—through education and training than to learn it as a defendant in a lawsuit. Today's police executives must be proactive and follow appropriate laws and guidelines in order to avoid legal difficulty; failure to do so will place themselves and their jurisdiction at serious financial risk.

This chapter analyzes the legal history of federal civil liability suits, the kinds of police actions that promote liability suits, the liability of supervisors who fail to control their personnel, and some cases in which police officers have been criminally punished for misconduct. The chapter concludes with a discussion of three expanding areas of potential liability: vehicle pursuits, handling computer evidence, and disseminating public information.

A LEGAL FOUNDATION

Laws are enacted in three ways: by legislation, by regulation, and by court decision. Statutes and ordinances are laws passed by legislative bodies, such as the U.S. Congress, state legislatures, county commissions, and city councils. These bodies sometimes create a general outline of the laws they enact, leaving to a particular governmental agency the authority to fill in the details of the law through rules and regulations; these administrative rules and regulations, especially at the federal level, constitute one of the fastest growing bodies of new law in the past two decades.

When the solution to a legal dispute cannot be found in the existing body of law—statutes, rules, or regulations—judges must rely on prior decisions that their own or other courts have already made on similar issues. These judicial decisions are known as **stare decisis**, meaning "let the decision stand." They are also relying on precedent. Of course, prior court decisions can be overruled or modified by a higher court or by the passage of new legislation. Furthermore, judges sometimes create their own tests to fairly resolve an issue. Statutes, judicial decisions, and tests may differ greatly from state to state; therefore, it is important for lawyers or criminal justice practitioners to read and understand the laws as they apply in their own jurisdictions.

It is also important to have a basic understanding of **tort liability**. A tort is the infliction of some injury upon one person by another. Three categories of torts generally cover most of the lawsuits filed against criminal justice practitioners: negligence, intentional torts, and constitutional torts.

Negligence arises when a police officer's conduct creates a danger to others. In other words, the officer did not conduct his or her affairs in a manner so as to avoid subjecting others to a risk of harm. The officer will be held liable for the injuries caused to others through his or her negligent acts. Various levels or degrees of negligence are recognized in the law: simple, gross, and willful or criminal negligence. Simple negligence involves a reasonable act performed by a reasonable officer in the scope of employment but performed without due care; the result is usually a charge of mental pain and anguish, for which an employer or insurance company will pay damages. Gross negligence involves an unreasonable act, for which damages for mental pain and anguish will also be paid by either an employer (if the officer's acts were within the scope of employment) or the individual officer. Willful or criminal negligence actually involves an intentional act rather than negligence; the plaintiff will receive actual damages, plus mental pain and anguish damages, plus punitive damages. The damages will be paid by the individual officer only; neither the employer nor insurance company will be compelled to pay.[2]

Intentional torts occur when an officer engages in a voluntary act that had a substantial likelihood of resulting in injury to another; examples are assault and battery, false arrest and imprisonment, malicious prosecution, and abuse of process. Constitutional torts involve police officers' duty to recognize and uphold the constitutional rights, privileges, and immunities of others; violations of these guarantees may subject the officer to civil suit, most frequently brought in federal court under 42 U.S.C. Section 1983 (discussed later).[3]

Assault, battery, false imprisonment, false arrest, invasion of privacy, negligence, defamation, and malicious prosecution are examples of torts that are commonly brought against police officers.[4] False arrest is the arrest of a person without probable cause—an arrest that is made even though an ordinarily prudent person would not have concluded that a crime had been committed or that the person arrested had committed it. False imprisonment is the intentional illegal detention of a person, not only jailing but any confinement to a specified area.

Most false arrest suits result in a false imprisonment charge as well, but a false imprisonment charge sometimes can follow a valid arrest. For example, the police may fail to release an arrested person after a proper bail or bond has been posted, they can delay the arraignment of an arrested person unreasonably, or authorities can fail to release a prisoner after they no longer have authority to hold him or her. "Brutality" is not a legal tort action per se; rather, it must be brought as a civil assault and/or battery.[5]

A single act may also be a crime as well as a tort. For example, if Officer Smith, in an unprovoked attack, injures Citizen Jones, the state will attempt to punish Smith in a *criminal* action by sending him to prison or fining him or both. The state would have the burden of proof at criminal trial, having to prove Smith guilty "beyond a reasonable doubt." Furthermore, Jones may sue Smith for money damages in a *civil* action for the personal injury he suffered. Jones would argue that Smith failed to carry out his duty to act reasonably and prudently, and this failure resulted in Jones's injury. This legal wrong, of course, is a tort; Jones would have the burden of proving Smith's acts were tortious by a "preponderance of the evidence"—a lower standard and thus easier to satisfy in civil court.

Our system of government has both federal and state courts. Federal courts are intended to have somewhat limited jurisdiction, however, and tend not to hear cases involving private (as opposed to public) controversies unless federal law is involved or both parties agree to have their dispute settled there. Thus, most tort suits have been filed in state courts. There are two means by which a federal court may acquire jurisdiction of police misconduct suits. The first means is the predominant source of our later discussions, referred to as a "1983" suit. This name is derived from the fact that the suits are brought under the provisions of Title 42, Section 1983 of the U.S. Code. The significant part of this statute and its legislative history follow.

The second means by which a federal court may assume jurisdiction over a police misconduct suit is to allege what some legal commentators call a *Bivens* tort. This name derives from a 1971 case, *Bivens v. Six Unknown Named Agents of the Federal Bureau of Narcotics*.[6] The U.S. Supreme Court held that a civil suit based directly on the Fourth Amendment could be filed. In *Bivens*, federal narcotics agents conducted an illegal search, arrest, and interrogation, but a suit by the plaintiffs could not be filed under Section 1983 because that section covers only police agents acting under *state* law. Civil suits to recover damages for violations of constitutional rights by *federal* officers have thus become known as **Bivens suits**.

A preliminary question is "Who can sue whom?" Under common law, police officers were held personally liable for damage they caused to someone while exceeding the boundaries of permissible behavior. This law applied even when the officer may have been ignorant of the law. Today this rule is the basis for one of the major risks of policing.[7]

A suit may also be filed against an employer under the doctrine of **respondeat superior**, an old legal maxim meaning "let the master answer." This doctrine is also termed **vicarious liability**. In sum, an employer is liable in certain instances for wrongful acts of the employee. It is generally inapplicable if a jury determines that the employee's negligent or malicious acts were outside the legitimate scope of the employer's authority. Although U.S. courts have expanded the extent to which employers can be sued for the torts of their employees, they are still reluctant to extend this doctrine to police supervisors (sergeants and lieutenants) and administrators. The courts realize that, first of all, police supervisors have little discretion in hiring decisions. Second, the duties of police officers are largely established by the governmental authority that hired them rather than by their supervisors. However, if a supervisor has abused his or her authority, was present when misconduct occurred and did nothing to stop it, or otherwise participated in the conduct, he or she can be held liable for the tortious behavior of his or her officers.[8] This issue is discussed at greater length later in this chapter.

Another difficult issue is whether the police department and the employing governmental unit can be sued for damages caused by police misconduct. It is clear that under common law, the government could not be sued, because the king could do no wrong. This doctrine, known as **sovereign immunity**, was initially adopted by the U.S. court system; however, recent court decisions and legislative enactments have modified it extensively. The courts began to erode the doctrine in the mid-1900s, on the ground that modern times no longer seemed to justify it. Davis reported that 18 states had "abolished chunks of sovereign immunity" by 1970, that 29 had done so by 1975, and 33 had done so by 1978.[9] Indeed, by 1978, only two states still adhered fully to the traditional common law approach that government was totally immune from liability for torts occurring in the exercise of governmental functions.[10]

HISTORY AND GROWTH OF SECTION 1983 LITIGATION

In the years following the Civil War, Congress, in reaction to the states' inability to control the Ku Klux Klan's lawlessness, enacted the Ku Klux Klan Act of 1871. This was later codified as Title 42, U.S. Code **Section 1983**. Its statutory language is as follows:

> Every person who, under color of any statute, ordinance, regulation, custom, or usage of any State or Territory, subjects, or causes to be subjected, any citizen of the United States or any other person within the jurisdiction thereof to the deprivation of any rights, privileges, or immunities secured by the Constitution and laws, shall be liable to the party injured in an action at law, suit in equity, or other proper proceeding for redress.

This legislation was intended to provide civil rights protection to all "persons" protected under the act, when a defendant acted "under color of law" (misused power of office). It is also meant to provide an avenue to the federal courts for relief of alleged civil rights violations.

The original intent of the law did not include police misconduct litigation. In fact, the law was virtually ignored for 90 years, until the U.S. Supreme Court's decision in *Monroe v. Pape* in 1961.[11] There, 13 members of the Chicago Police Department broke into a home without a warrant, forced the family out of bed at gunpoint, made them stand naked while the officers ransacked the house, and subjected the family to verbal and physical abuse. Monroe was then taken to the police station, where he was held incommunicado and questioned for 10 hours before being released without charges. The plaintiffs (Monroe and his family) claimed that the officers acted "under color of law" as set forth in Section 1983, thus violating their constitutional rights. The U.S. Supreme Court agreed, holding the officers liable; however, the Court also held the City of Chicago immune from liability under the statute.

In the five-year period following *Monroe*, only a few police misconduct suits were filed in the federal courts, where this federal statute is used. As an example, the New York City Police Department led the country in lawsuits during the five-year period from 1967 to 1972, but only 5 percent of those cases were brought to the federal forum.[12] There was a virtual boom of Section 1983 suits from 1967 through 1976, however. Studies conducted by a national group, Americans for Effective Law Enforcement, found that police misconduct suits in *state* courts rose from 1,556 in 1967 to 8,007 in 1976 (an increase of 415 percent). During the same period, however, *federal* civil rights actions alleging police misconduct rose from 167 to 2,226 (or an increase of 1,233 percent). Combining state and federal jurisdictions, misconduct suits grew from 5.5 per 1,000 officers to 19.6 per 1,000 in this 10-year period. And from 1972 through 1976, 32 percent of all legal actions were arrest related and 32 percent alleged excessive nondeadly force.[13]

Several factors contributed to this surge in Section 1983 actions. First, some lawyers believe that clients receive more competent judges and juries in the federal forum than in state courts. Federal judges, who are appointed for life, can be less concerned about the political ramifications of their decisions than locally elected judges often are. Federal prosecutors can be more aggressive in arguing to jurors from a multicounty area; local prosecutors must argue to jurors who elected them and may know the defendant-officer. Furthermore, federal rules of pleading and evidence are uniform; federal procedures of discovery are more liberal; and lawyers have easier access to published case law in assisting them to prepare a federal suit.[14]

Just as important, Congress passed Section 1988 of the Civil Rights Act in 1976, which allows attorney's fees to the prevailing party over and above the award for compensatory and punitive damages. This provision in the law did as much to spur the use of Section 1983 as any other factor. Attorneys usually accept personal injury cases on a contingency basis. If they win, they receive one-third of the amount awarded. Thus, a plaintiff's verdict in a police shooting case can be quite profitable; a lawsuit that may not have been worth the time and effort is now, thanks to Sections 1983 and 1988, potentially quite lucrative.

Initially, the Supreme Court concluded that Congress did not intend Section 1983 to apply to municipalities; the cities believed they had absolute immunity.

However, in 1978, upon examining this question in *Monell v. Department of Social Services*,[15] the Court modified its earlier thinking. This decision has been another major factor in the increased use of Section 1983. In *Monell*, a class of female employees filed a Section 1983 suit alleging that their employers had, as a matter of official policy, forced female employees to take unpaid leaves of absence before such leaves were required for medical reasons. The Supreme Court overruled its previous decisions in *Monroe* and other cases, stating that Congress, in the Civil Rights Act of 1871, *did* intend that municipalities and other local governments be included as "persons" to whom Section 1983 applies. Local governing bodies and corporate "persons," therefore, can be sued for damages under Section 1983. *Monell* made it clear that an individual had legal recourse when he or she had been deprived of a right or privilege guaranteed under the Constitution and laws if such deprivation was the direct result of an official policy or custom of a local unit of government.

Defenses and immunities against Section 1983 suits exist, however. The states themselves, for example, are granted absolute immunity from Section 1983 suits,[16] as are judges, prosecutors, legislators, and federal officials. Federal officials usually act under color of federal law, as opposed to state law, specified in the act. Police officers are granted **qualified immunity**, meaning that as long as they acted in "good faith" and their conduct is reasonable, they have a defense. Over the years, the courts have struggled to develop a test for "good faith." In 1975, the Supreme Court developed a test that considered both the officer's state of mind at the time of the act in question (the subjective element) and whether or not the officer's act violated clearly established legal rights (the objective element).[17] Overzealous conduct that is not undertaken in good faith and without regard for the rights of citizens can and will result in a finding of liability.

POLICE ACTIONS PROMOTING SECTION 1983 LIABILITY

As indicated in the introduction, no group of workers in the private or public sectors (with the exception of physicians) is more susceptible to litigation and liability as an outgrowth of their work than police officers. The police are frequently cast into confrontational situations and, given the complex nature of their work, will from time to time act in a manner that evokes public scrutiny and complaint. Compounding this problem is the fact that some officers are overzealous in the pursuit of their work; intentionally or not, some officers will violate the rights of the citizens they are sworn to protect. We now look at some police behaviors that have resulted in officer and/or supervisor liability in the areas of brutality, off-duty activities, wrongful deaths, false arrest, illegal search and seizure, and negligence.

Section 1983 is an appropriate legal tool for citizens who believe they have been victims of *police brutality*. Such cases abound. Following are a few of them that demonstrate the police behaviors that have resulted in officer liability: In *Jennings v. City of Detroit*,[18] a 22-year-old single African American man was permanently paralyzed following a beating at a police station. Jury award: $8 million (settled for $3.5 million). In *Gilliam v. Falbo*,[19] the U.S. District Court for Ohio awarded $72,000 to a young man beaten by two officers. And in *Haygood v. City of Detroit*,[20] a 35-year-old plaintiff was awarded $2.5 million punitive and $500,000 compensatory damages after being subjected to racial slurs, beaten, and chained to a bed for 12 hours without charges ever being filed.

Even off-duty activities may get police officers into serious difficulty for acting "under color of law." Their part-time work as security guards often opens the door to legal problems. In *Carmelo v. Miller*,[21] two off-duty officers were working security at a baseball game. They received information that someone was displaying a gun and stopped a man who fit the description. The officers searched, arrested, beat, and kicked the suspect and his companion; no gun was found in the area. One of the beaten men required medical treatment. The officers were found liable. In *Stengel v. Belcher*,[22] an off-duty officer entered a bar carrying a .32 handgun (which he was required to carry off duty at all times) and a can of Mace. An altercation broke out and, without identifying himself, the officer got involved, killing two men and seriously wounding another. The plaintiffs recovered $800,000 in compensatory damages.

Clearly, the use of off-duty weapons and policies requiring that they be carried pose a risk of liability. In *Bonsignore v. City of New York*, a mentally unstable 23-year veteran police officer shot his wife five times and then killed himself, using a .32 caliber pistol that departmental policy required that he carry when off duty. Evidence produced at trial demonstrated that Officer Bonsignore's unsuitability for police duties was well known by the department—it had even provided him a limited-duty assignment as station house janitor—yet the police code of silence protected him. The jury awarded Mrs. Bonsignore nearly a half-million dollars.

Wrongful death suits are also becoming more frequent. The following cases illustrate how the law applies in this regard. In *Prior v. Woods*,[23] a 24-year-old man was killed outside his home by police officers who mistook him for a burglar. The jury awarded his estate $5.75 million. In *Burkholder v. City of Los Angeles*,[24] an LAPD officer killed a man in his early twenties who, while naked and under the influence of drugs, was climbing a light pole. The man had seized the officer's club but had not struck the officer. The jury awarded $450,000 in damages and $150,000 in attorney's fees. In *Webster v. City of Houston*, an officer killed and attempted to cover up the killing (by using a "throw down" weapon) of an unarmed 17-year-old boy who was stopped in a stolen vehicle. The jury awarded his estate nearly $1.5 million in punitive and compensatory damages.

Generally, police officers are not liable for damages under Section 1983 merely for *arresting* someone. However, that protective shroud vanishes if the plaintiff proves the officer was negligent or violated an established law or right (false arrest). As an illustration, in *Murray v. City of Chicago*,[25] the plaintiff's purse and checkbook were stolen; she reported the theft to the police. Later, some of the stolen checks were cashed and the plaintiff was arrested; she appeared in court and cleared up the matter, all charges being dropped. Several months later, she was arrested again at her home by Chicago officers who used an invalid arrest warrant that was related to the earlier mix-up. Murray was taken to the police station, was strip searched by male officers, and was detained for six hours before being released. The federal court ruled that the officers acted in good faith, but if the policy or custom of the city was shown to have encouraged such unwarranted arrests, the city could be held liable. In similar cases (see, for example, *Powe v. City of Chicago*, 664 F.2d 639, 1981), when a person's name was not properly removed from the police warrants list and so he or she later was arrested, the plaintiff has been successful in Section 1983 suits for false arrest.

Search and seizure, an especially complicated area of criminal procedure, is ripe for Section 1983 suits, primarily because of the ambiguous nature of the probable cause doctrine. In *Duncan v. Barnes*,[26] police officers obtained a warrant to search a suspect's home for heroin. The officers executed the warrant in early morning hours, using a sledge hammer to break down the rear door of the apartment. With guns drawn, officers entered two bedrooms, forcing the two females and one male inside to stand nude, spread-eagled against a wall, while their rooms were searched. Soon the officers realized that they had entered the wrong apartment, and they left. They also left the apartment in total disarray—several doors, a television, and a tape recorder were broken. The occupants, students at a court-reporting school, were so upset they were forced to miss classes for two weeks; as a result, their certification and employment as court reporters were delayed. The court had little difficulty finding that the officers had acted in an unreasonable manner.

Negligence by police officers is another cause of action under Section 1983. Negligence can be found, among other places, in the supervision and training of personnel. A few examples will illustrate this point. In *Sager v. City of Woodlawn Park*,[27] an officer accidentally killed a person when the shotgun he was pointing at the head of the prisoner discharged; the officer was attempting to handcuff the prisoner with his other hand. At trial, the officer stated he had seen the technique in a police training film. The training officer, however, testified that the film was intended to show how *not* to handcuff a prisoner; unfortunately, no member of the training staff made that important distinction to the training class. The court ruled that improper training resulted in the prisoner's death. In *Popow v. City of Margate*,[28] an innocent bystander was killed on his front porch at night by a police officer engaged in foot pursuit. The court held the city negligent because the officer had had no training on night firing, shooting at moving targets, or using firearms in a residential area. And in *Beverly v. Morris*,[29] a police chief was held liable for improper training and supervision of his officers following the blackjack beating of a citizen by a subordinate officer.

In June 1998, the Supreme Court rendered a significant decision concerning police liability in high-speed pursuits. Each year, police pursuits result in accidental deaths of hundreds of citizens.[30] In the past, courts have disagreed about when the police are liable for those deaths. This decision involved police in Sacramento, California, who chased a motorcyclist who abruptly braked in front of the officers;

a passenger who fell from the motorcycle was struck and killed by the pursuing police car. A lower court held in this case that the police could be sued for "reckless indifference to life," but the Supreme Court reversed, noting the need for split-second decisions by officers and that officers could be held liable only for activities that "shock the conscience."[31]

CRIMINAL PROSECUTIONS FOR POLICE MISCONDUCT

Whereas Section 1983 is a civil statute, Title 18, U.S. Code, **Section 242**, makes it a *criminal* offense for any person acting willfully "under color of law," statute, regulation, or custom to deprive any person of those rights and privileges guaranteed under the Constitution and laws of the United States. This law, like Section 1983, dates from the post–Civil War era and applies to all persons regardless of race, color, or national origin. Section 242 not only applies to police officers but also to the misconduct of public officials; prosecutions of judges, bail-bond agents, public defenders, and even prosecutors are possible under the statute.

An example of the use of Section 242 is the murder of a drug courier by two U.S. customs agents, while the agents were assigned to the San Juan International Airport. The courier flew to Puerto Rico to deposit approximately $700,000 in cash and checks into his employer's account. He was last seen being interviewed by the two customs agents in the airport; 10 days later, his body was discovered in a Puerto Rican rain forest. An investigation revealed that the agents had lured the victim away from the airport and murdered him for his money, later disposing of the body. They were convicted under Section 242 and related federal statutes, and both agents were sentenced to prison terms of 120 years.[32]

The two major federal entities involved in investigating civil rights cases are the Civil Rights Division of the Department of Justice and the FBI's Civil Rights Unit. The FBI investigates these cases and presents them to the Department of Justice to review and decide if any prosecution is warranted.

A great amount of weight is attached to the willfulness of the misconduct. The Supreme Court has ruled that in Section 242 prosecutions, the federal government must prove the defendant's *specific* intent to engage in misconduct. When the misconduct is deliberate and willful (for example, when a suspect is beaten to obtain a confession or an arrestee is beaten in retaliation for resisting arrest), the Justice Department will not hesitate to prosecute.[33]

A frequent complaint by police officers in the past was the lack of prosecutions against persons who falsely accused the police of misconduct. In 1984, however, the Supreme Court held that Title 18, U.S. Code, Section 1001, does cover false statements made to FBI agents, thus providing for the prosecution of people who make false statements. One such case occurred in Louisiana, where a jail inmate falsely reported to the FBI that he had been assaulted and kicked by a deputy sheriff; in reality, the inmate had received his injuries during a fight with another inmate. He was convicted under Section 1001 and sentenced to three additional years of imprisonment for his false statements.[34]

LIABILITY OF POLICE SUPERVISORS

As already shown, Section 1983 allows for a finding of personal liability on the part of police supervisory personnel if improper training is shown or it is proven that they knew, or should have known, of the misconduct of their officers yet failed to take corrective action and prevent future harm.

McClelland v. Facteau,[35] a Section 1983 suit against a state police agency chief as well as a local police chief was such a case. McClelland was stopped by Officer Facteau (a state employee) for speeding. He was taken to the city jail; there he was not allowed to make any phone calls, he was questioned but not advised of his rights, and he was beaten and injured by Facteau in the presence of two city police officers. McClelland sued, claiming the two police chiefs were directly responsible for his treatment and injuries due to their failure to properly train and supervise their subordinates. Evidence was produced of prior misbehavior by Facteau. The court ruled that the chiefs could be held liable if they knew of prior misbehavior yet did nothing about it.

Another related case was that of *Brandon v. Allen.*[36] In this case, two teenagers who were parked in a "lovers' lane" were approached by an off-duty police officer, Allen, who showed his police identification and demanded that the boy exit the car. Allen struck the boy with his fist and stabbed him with a knife, then attempted to break into the car where the girl was seated. The boy was able to reenter the car and manage an escape. As the two teenagers sped off, Allen fired a shot at them with his revolver; the shattered windshield glass severely injured the youths to the point that they required plastic surgery. Allen was convicted of criminal charges, and the police chief was also sued under Section 1983. The plaintiffs charged that the chief and others knew of Allen's reputation for being mentally unstable; none of the other police officers wished to ride in a patrol car with him. At least two formal charges of misconduct had been filed previously, yet the chief failed to take any remedial action or even to review the disciplinary records of officers. The court called this behavior "unjustified inaction," held the police department liable, and allowed the plaintiffs damages. The U.S. Supreme Court upheld this judgment.[37]

Police supervisors have also been found liable for injuries arising out of an official policy or custom of their department. Injuries resulting from a chief's verbal or written support of heavy-handed behavior resulting in excessive force by officers have resulted in such liability.[38]

Today's police supervisors are definitely in a "need to know" position where the law is concerned. They are caught in the middle; not only can they be sued for improper hiring, training, and supervision of their officers, but there are other civil rights laws that can be used by officers who believe they were improperly disciplined or terminated. Indeed, Section 1983 can also be used by unsuccessful job applicants if they can show that the administrator's tests were not job related, included inherent bias, and were not properly administered or graded. The same holds true if it can be shown that proper testing methods were not used in promotions or in discipline or firing of personnel. Police supervisors have lost in suits where they disciplined male and female officers who were having a private relationship,[39] where they disciplined black officers who removed the U.S. flag from their uniforms to protest perceived discriminatory acts by the city,[40] and where they disciplined officers for "improper" political party membership.[41]

NEW AREAS OF POTENTIAL LIABILITY

Recently, three areas of potential liability were established by the courts: pursuits by police vehicle, handling computer evidence, and providing information to the public via press releases.

VEHICLE PURSUITS

It's quiet, about 2:00 on Sunday morning. The Saturday-night party traffic is dwindling. While on routine patrol, a police officer, bored from inactivity and barely able to stay awake, sees a motorist run a red light. The offender refuses to yield to the officer's own red light and siren, and soon the chase is on. The incident ends only when the violator runs off of the roadway in a rollover at a high rate of speed 30 miles away in another state. On another night, at least five police vehicles are chasing a person who is driving with a suspended driver's license; the officers set up a roadblock in a remote area, and, as the offender tries to go around the road-block, an officer fires his handgun, trying to shoot the vehicle's tires. The offender stops immediately, because his 2-year-old son is standing beside him in the front seat of the car. In a third case, officers chase a traffic violator about 20 miles—across state lines; one of the police vehicles, with worn tires, has a tire blow out while traveling nearly 100 miles per hour.

Each of these incidents happened, and each might have resulted in a lawsuit. These kinds of police activities, or others similar to them, are not uncommon. This is another area of policing—that of so-called hot or vehicle pursuits—that is currently both a trend as well as an issue. A considerable amount of controversy has recently been generated because of the tremendous potential for injury, property damage, and liability that accompanies such pursuits. Police administrators are revamping their policies and procedures accordingly. As one sheriff stated, "For so long, administrators thought there was really one lethal weapon they gave their officers, and that was the gun, and yet there was another weapon . . . their car."[42]

As one police procedure manual describes it, "The decision by a police officer to pursue a citizen in a motor vehicle is among the most critical that can be made."[43] The decision to pursue often must be made in tense, uncertain, and rapidly changing circumstances. The officer has a duty to enforce the law and apprehend offenders, but that duty must be balanced with the right of the public and other police officers to be free from risks in which it is possible that serious harm will occur.

In sum, pursuit is justified only when the necessity of apprehension outweighs the degree of danger created by the pursuit. That is the reason why there is a nationwide trend by agencies to place more restrictions on pursuits and to have

PRACTITIONER'S PERSPECTIVE

Police and Civil Liability

Samuel G. Chapman
Samuel G. Chapman is presently a police consultant and continues to teach police officers at home and abroad. He was a Berkeley, California, police officer and a student at U.C. Berkeley, where he took degrees in criminology and studied under August Vollmer. He later taught at Michigan State University, served as chief of the Multnomah County (Portland), Oregon, Sheriff's Office, and was assistant director of the President's Commission on Law Enforcement and the Administration of Justice in Washington. In 1967, he began a 24-year career as professor of political science at the University of Oklahoma, where he authored a number of books and monographs on various aspects of policing and the use of police dogs.

"Sue? Me? For what? Why? What do I do now? Will I lose everything?" Such thoughts preoccupy police officers when they learn that they are named as defendants in a civil rights lawsuit. Officers need help and direction at this juncture, including assurance that this isn't the end of the world. But they need to know that being sued is not a parlor game, either.

Police departments and their personnel must take civil rights litigation seriously, for such actions have been on the increase over the past three decades. Actually, civil rights lawsuits are seen by many as an occupational hazard in policing. Just when this virtual floodtide of litigation will ebb, if it ever will, is not clear.

When a lawsuit has been filed, the allegations should be evaluated by the government's defense attorneys. Fact-finding may disclose that the allegations appear to have little merit. It could be that the plaintiff counsel has seemingly filed a case of dubious substance, really seeking what is called a "convenience settlement," which occurs when a defendant pays the plaintiff a dollar amount less than what the defendant's costs would be to prepare for trial. But if after fact-finding it appears that the department and its officers are culpable, the defense team should start settlement negotiations early. The defense should make a meaningful offer, keeping it in the range of settlements in cases of a similar sort elsewhere.

At the same time, the defense must get underway with discovery within the context of a game plan aimed at minimizing loss should the cases eventually go to trial. Settlements that occur just before trial are invariably costly. The defense team should also evaluate the courtroom record of the plaintiff law firm, since some firms are more competent than others. Moreover, knowledge about the track record of the opposing attorneys may bear on whether or not to settle and the amount of an offer.

In many instances, fact-finding will reveal that a case is realistically defensible. When this is so, the defense team may decide to reject a convenience settlement and prepare for trial. Such a stance will cause the plaintiff law firm to evaluate whether to expend resources and time in pursuing a case not likely to be won. Clearly, when the defense feels it can be successful and decides to stand up and fight, it establishes the jurisdiction as a "hard target" and sends a message that lawsuits with little merit are going to be stridently defended.

Of course, it is urgent that whomever is named to defend officers and police agencies be skilled in handling civil rights cases. It is a grave mistake for the government to take a bargain-basement approach to defending civil rights lawsuits by assigning staff attorneys who have little or no experience working these highly technical types of litigation.

A popular misconception is that only officers in the nation's largest police forces are sued. The truth is that although personnel in major forces are frequent targets of litigation, officers in very small forces are named as defendants too, far more often than one imagines. In fact, persons serving very small forces may be more vulnerable to litigation because they may not be as well trained and supervised as their counterparts in large forces, and the rules, regulations, and procedures manuals may not be as thorough and up to date.

With the upsurge in litigation against police, officers may become demoralized, dispirited, and defensive. They are easy prey for cynical locker-room jockeys who proclaim that the best way to avoid trouble is to slow down, cool it, and do no more police work than one has to. Such advice, which is of the worst sort, poses a special urgent challenge for police supervisors, who must assure that officers execute their duties faithfully and in accord with departmental guidelines.

Yes, the police can fight back by suing those persons who sue them. But this means hiring counsel, which is expensive. And even if the lawsuit is successful and brings a dollar judgment against a person, such persons are usually notorious for being poor and hence unable to meet any financial judgment levied against them. In short, suing back is not a realistic course of action.

Government's best defense against an adverse judgment in a civil rights lawsuit is to thoroughly train and regularly retrain its police personnel and to supervise them well. In addition, the police department's rules and regulations, as well as its procedures manuals, must be kept current and as complete and practical as possible. The content of these important documents must be made known to and understood by all officers. If all these things are in place and if officers perform as trained and execute their roles consistent with departmental guidelines and policies, a persuasive defense can be mounted against any allegations of misconduct.

field supervisors (sergeants) call off the chase. Chases are often discontinued by supervisors because too many officers are involved. It is not uncommon for agency policy to limit the pursuit to two pursuit vehicles: a supervisor and another vehicle authorized by a pursuit monitor.[44]

Many veteran police officers, who are not used to having a policy at all, express their disdain for new, more restrictive policies that appear to give in to offenders. As one veteran stated, "We are police officers. It's our job to catch bad guys, not let them go." However, the veterans can applaud the recent U.S. Supreme Court decision discussed earlier, making it far more difficult for officers to be sued by citizens who are injured during police pursuits.[45]

HANDLING COMPUTER EVIDENCE

Today it is almost impossible to investigate a fraud, embezzlement, or child pornography case without dealing with some sort of computer evidence. Even evidence in a homicide or narcotics case may be buried deep within a computer's hard disk drive. As a reaction to this situation, many police agencies have recruited self-taught "experts" to fill the role of computer evidence specialists. Usually such persons are highly motivated and have some knowledge of the rules of evidence and experience in testifying in court. Other police agencies have enlisted the support of personnel at local universities or computer repair shops to help in dealing with computer evidence issues.[46]

This increased exposure to computer evidence by people both inside and outside policing brings an increase in potential legal liabilities. For example, if a police agency seizes the computer books and records of an ongoing business, there may be a negative financial impact on the operation of the business involved. And, if it can be shown that the police accidentally destroyed business records through negligence, a criminal investigation might well become the civil suit of the decade. Furthermore, if a computer to be seized contains a newsletter, draft of a book, or any computer bulletin board system, there may be liability under the Privacy Protection Act.[47]

The risk of liability in these kinds of cases may be reduced substantially if police investigators follow generally accepted forensic computer science procedures. Guidelines approved by the Department of Justice's Computer Crime and Intellectual Property Section dictate how the police are to search, seize, or analyze computers.

It is crucial that the police be trained in the proper procedures for handling computers as well as rules of evidence, and the federal government has made computer evidence training of federal, state, and local law enforcement officers a priority.[48]

DISSEMINATING PUBLIC INFORMATION

The Louisiana Supreme Court recently held that police department public information officers (PIOs) can be held liable for unfounded statements they make in news releases. The case involved a defamation suit against the state police, where the PIO told a reporter that the defendant was running a large-scale illegal gambling operation and bilking customers. The court found that the PIO had no reasonable basis for saying the defendant had cheated customers and thus made injurious statements that he had no reason to believe were true. Although the ruling applies only to PIOs in Louisiana, it could grow to have national implications in the future.[49]

SUMMARY

This chapter examined the high legal cost of misconduct or negligence in policing. Included were discussions of police actions promoting liability, a number of police civil liability cases decided by federal courts, and new areas of potential police liability: hot pursuits, computer evidence, and disseminating public information.

Implicit in this chapter are the dire consequences of failing to properly hire, train, and supervise police personnel. Furthermore, students and criminal justice practitioners must remember that the law is fluid, constantly changing; therefore, officers and students must keep abreast of the laws concerning liability and conduct themselves accordingly.

ITEMS FOR REVIEW

1. Explain the basic means by which laws are enacted and the difference between civil and criminal law.
2. Define *negligence* and *Section 1983*.
3. Give reasons for the recent increase in Section 1983 lawsuits.
4. Delineate some types of police actions that are vulnerable to Section 1983 actions.
5. How may police supervisors be liable under Section 1983?
6. Explain how police officers may be held *criminally* liable for their misconduct.
7. Describe some of the liability issues that arise when handling computer evidence.

NOTES

1. See, for example, John L. Worrall and Otwin Marenin, "Emerging Liability Issues in the Implementation and Adoption of Community Oriented Policing," *Policing: An International Journal of Police Strategies & Management* 22 (1998): 121–136.
2. H. E. Barrineau III, *Civil Liability in Criminal Justice* (Cincinnati, OH: Pilgramage, 1987), p. 58.
3. *Ibid.*, p. 5.
4. Charles R. Swanson, Leonard Territo, and Robert W. Taylor, *Police Administration: Structures, Processes, and Behavior,* 5th ed. (Upper Saddle River, NJ: Prentice Hall, 2001), p. 438.
5. *Ibid.*, pp. 357–358.
6. 403 U.S. 388, 29 L.Ed.2d 619, 91 S.Ct. 1999 (1971).
7. Swanson, Territo, and Taylor, *Police Administration*, p. 438.
8. *Ibid.*, pp. 438–439.

9. Kenneth Culp Davis, *Administrative Law of the Seventies* (Rochester, NY: Lawyers Cooperative, 1976), p. 551.
10. Swanson, Territo, and Taylor, *Police Administration*, p. 449.
11. 365 U.S. 167, 81 S.Ct. 473 (1961).
12. Wayne W. Schmidt, "Section 1983 and the Changing Face of Police Management," in *Police Leadership in America*, ed. William A. Geller (Chicago: American Bar Foundation, 1985), pp. 226–236.
13. *Ibid.*, p. 228.
14. *Ibid.*, p. 227.
15. 436 U.S. 658 7 (1978).
16. Per *Alabama v. Pugh*, 438 U.S. 781 (1978).
17. Swanson, Territo, and Taylor, *Police Administration*, p. 449.
18. Wayne County Circuit Court, Michigan (August 1979).
19. U.S. District Court, Southern District of Ohio (April 1982).
20. Wayne County Circuit Court, Michigan, No.77-728013 (December 29, 1980).
21. 569 S.W. 365 (1978).
22. 522 F.2d 438 (6th Cir., 1975).
23. U.S. Dist. Ct., E.D. Mich. (October 1981).
24. L.A. Cty. Sup. Ct., California (October 1982).
25. 634 F.2d 365 (1980).
26. 592 F.2d 1336 (1979).
27. 543 F.Supp. 282 (D.Colo., 1982).
28. 476 F. Supp. 1237 (1979).
29. 470 F.2d 1356 (5th Cir., 1972).
30. Gordon Witkin, "Police Chases Get a Green Light," *U.S. News and World Report*, June 8, 1998, p. 25.
31. *Sacramento County, Calif. v. Lewis*, 66 LW 4407 (May 26, 1998).
32. On appeal, the Section 242 convictions were vacated, as the victim was not an inhabitant of Puerto Rico; therefore, he enjoyed no protection under the U.S. Constitution. On resentencing, in January 1991, the agents each received 50 years in prison for convictions of several other federal crimes under Title 18.
33. John Epke and Linda Davis, "Civil Rights Cases and Police Misconduct," *FBI Law Enforcement Bulletin* 60 (August 1991): 14–18.
34. *Ibid.*, p. 17.
35. 610 F.2d 693 (10th Cir., 1979).
36. 516 F.Supp. 1355 (W.D. Tenn., 1981).
37. *Brandon v. Holt*, 469 U.S. 464, 105 S.Ct. 873 (1985).
38. See, for example, *Black v. Stephens*, 662 F.2d 181 (1991).
39. *Swope v. Bratton*, 541 F.Supp. 99 (W.D. Ark., 1982).
40. *Leonard v. City of Columbus*, 705 F.2d 1299 (11th Cir., 1983).
41. *Elrod v. Burns*, 427 U.S. 347 (1975).
42. Oklahoma County sheriff John Whetsel, quoted in Nicole Marshall, "Hot Pursuit," *Tulsa World*, June 15, 1998, p. A11.
43. Tulsa, Oklahoma, Police Department (procedure manual), Ronald Palmer, Chief of Police, June 10, 1998, p. 1.
44. Marshall, "Hot Pursuit," p. A11.
45. *Ibid.*
46. Michael R. Anderson, "Reducing Computer Evidence Liability," *Government Technology* 10 (February 1997): 24, 36.
47. *Ibid.*
48. *Ibid.*
49. "Be Careful What You Say," *Law Enforcement News*, December 15, 1997, p. 1.

CHAPTER 12

Issues and Trends

Key Terms and Concepts

Civilianization
Collective bargaining
Double marginality
Fair Labor Standards Act
  (FLSA)
Grievances
Impasse resolution

Job actions
Labor relations
Peace Officer Bill of Rights
Private police
Stress
Unionization
Women and minority officers

Debate on public issues should be uninhibited, robust, and wide open.
—*Justice William Brennan*

History has many cunning passages, contrived corridors, and issues.
—*T. S. Eliot*

INTRODUCTION

This chapter addresses a number of important policing matters that could have been included in earlier chapters. But because of their common nature—all involve a substantial degree of change, controversy, or influence over their agencies—they are consolidated here for discussion.

The chapter is divided into two major segments: contemporary policing trends and contemporary policing issues. In the former, we examine the rights of police officers, labor relations (including unionization and collective bargaining), women and minorities in policing, the private police, and the civilianization and accreditation of police agencies.

The issues portion of the chapter includes two primary subjects: higher education for the police and police stress—not as an issue per se but because of its myriad causes, physical and emotional effects, and possible solutions. In this section, we also briefly discuss what officers can do to maintain an edge. Note also that Appendix III discusses a recently inaugurated program that is related to higher education: the Police Corps.

The line separating what is a "trend" and an "issue" is fuzzy; some of the topics covered in this chapter may be seen as *both* issues and trends. Therefore, the chapter is not always organized topically.

CONTEMPORARY POLICING TRENDS

LABOR RELATIONS: OFFICERS' RIGHTS, UNIONIZATION, COLLECTIVE BARGAINING

In an earlier time, police supervisors, middle managers, and chief executives were largely unrestricted and unchallenged in their treatment of rank-and-file officers. Disciplinary action was subjective, and the prevailing theme was often "do as I say, not as I do." Today the labor-management relationship has changed significantly. **Labor relations**—a broad term that includes officers' employment rights and the related concepts of unionization and collective bargaining—has become a major element of policing. This section explores those topics.

Rights of Police Officers

Chapter 10 delineated several areas (such as place of residence, religious practice, freedom of speech, search and seizure, and so on) where, because of the nature of the officer's position, the police may encounter treatment by employers—and federal courts—that is quite different from that received by regular citizens. An officer does give up certain constitutional privileges when he or she puts on a police uniform. In this section we look at some measures the police have taken to maintain their rights on the job to the extent that it is possible to do so.

In the 1980s and 1990s, police officers began to insist on greater procedural safeguards to protect themselves against what they perceived as arbitrary infringement on their rights. These demands have been reflected in a statute enacted in many states, generally known as the **Peace Officer Bill of Rights**. These statutes confer upon an employee a property interest in his or her position, and this legislation mandates due process rights for peace officers who are the subject of internal investigations that could lead to disciplinary action. These statutes identify the type of information that must be provided to the accused officer, the officer's responsibility to cooperate during the investigation, the officer's rights to representation during the process, and the rules and procedures concerning the collection of certain types of evidence. Following are some common provisions of state Peace Officer Bill of Rights legislation:

> *Written notice.* The department must provide the officer with written notice of the nature of the investigation, a summary of the alleged misconduct, and the name of the investigating officer.
>
> *Right to Representation.* The officer may have an attorney or a representative of his or her choosing present during any phase of questioning or hearing.
>
> *Polygraph Examination.* The officer may refuse to take a polygraph examination unless the complainant submits to an examination and is determined to be telling the truth. In this case, the officer may be ordered to take a polygraph examination or be subject to disciplinary action.

Officers expect to be treated fairly, honestly, and respectfully during the course of an internal investigation. In turn, the public expects the agency will develop sound disciplinary policies and conduct thorough inquiries into allegations of misconduct.

It is imperative that supervisors become thoroughly familiar with statutes, contract provisions, and existing rules between employer and employee, so that procedural due process requirements can be met, particularly in disciplinary cases when an employee's property interest might be affected.

Police officers today are also more likely to file a grievance when they believe their rights have been violated. Grievances may cover a broad range of issues, including salaries, overtime, leave, hours of work, allowances, retirement, opportunity for advancement, performance evaluations, workplace conditions, tenure, disciplinary actions, supervisory methods, and administrative practices. The preferred method for settling officers' grievances is through informal discussion during which the employee explains his or her grievance to the immediate supervisor; most complaints can be handled this way. Those complaints that cannot be dealt with informally are usually handled through a more formal grievance process, which may involve several different levels of action.

Unionization

Probably as a result of their often difficult working conditions and traditionally low salary and benefits packages, police have often elected to band together to fight for improvement. Another major force in the development and spread of **unionization** was the authoritarian, unilateral, and "do as I say, not as I do," management style that characterized many police administrators of the past.

The first campaign to organize the police started shortly after World War I, when the American Federation of Labor (AFL) reversed a long-standing policy and issued charters to police unions in Boston, Washington, D.C., and about 30 other cities. Many police officers were suffering from the rapid inflationary rate following the outbreak of the war and believed that if their chiefs could not get them long-overdue pay raises, then perhaps unions could. Capitalizing on their sentiments, the fledgling unions signed about 60 percent of all officers in Washington, 75 percent in Boston, and a similar proportion in other cities.[1]

The unions' success was short lived, however. The Boston police commissioner refused to recognize the union, forbade officers to join it, and filed charges against several union officials. Shortly thereafter, on September 9, 1919, the Boston police initiated a famous strike of three days' duration, causing major riots and a furor against the police all across the nation; nine rioters were killed and 23 were seriously injured. During the strike, Massachusetts governor Calvin Coolidge uttered his now-famous quote, "There is no right to strike against the public safety by anybody, anywhere, anytime."

During World War II, however, the unionization effort was reignited. Unions issued charters to a few dozen locals all over the country and sent in organizers to help enlist the rank and file. Most police chiefs continued speaking out against unionization, but their subordinates were moved by the thousands to join, sensing the advantage in having unions press for higher wages and benefits.[2] But in a series of rulings, the courts upheld the right of police authorities to ban police unions.

The unions were survived in the early 1950s by many benevolent and fraternal organizations of police. Some were patrolmen's benevolent associations (PBAs), like those formed in New York, Chicago, and Washington, while others were fraternal orders of police (FOPs). During the late 1950s and early 1960s, a new group of rank-and-file association leaders came into power. They were more vocal in articulating their demands. Soon a majority of the rank and file vocally supported higher salaries and pensions, free legal aid, low-cost insurance, and other services. For the first time, rank-and-file organizations were legally able to insist that their administrators sit down at the bargaining table.[3]

Since the 1970s, the unionization of the police has continued to flourish. Today the majority of U.S. police officers belong to unions.[4] The International Conference of Police Associations (ICPA) is the largest organization, with more than 100 local and state units representing more than 200,000 officers.[5] This dramatic rise in union membership was fomented by several factors: job dissatisfaction, the belief that the public is hostile to police needs, and an influx of younger officers who hold less traditional views on relations between officers and the department hierarchy.[6]

Collective Bargaining: Types, Relationships, Negotiation, Job Actions

Three Models. Each state is free to decide whether and which public-sector employees will have collective bargaining rights and under what terms; therefore, there is considerable variety in collective bargaining arrangements across the nation. In states with comprehensive public-sector bargaining laws, the administration of the statute is the responsibility of a state agency such as a public employee relations board (PERB) or a public employee relations commission (PERC). Three basic models are used in the states: *binding arbitration, meet and confer,* and *bargaining not required.*[7]

The binding arbitration model is used in 25 states. There, public employees are given the right to bargain with their employers. If the bargaining reaches an impasse, the matter is submitted to a neutral arbitrator, who decides what the terms and conditions of the new collective bargaining agreement will be.[8]

Only three states use the meet-and-confer model, which grants very few rights to public employees. As with the binding arbitration model, police employees in meet-and-confer states have the right to organize and to select their own bargaining representatives.[9] However, when an impasse is reached, employees are at a distinct disadvantage. Their only legal choices are to accept the employer's best offer, try to influence the offer through political tactics (such as appeals for public support), or take some permissible job action.[10] The 22 states that follow the bargaining-not-required model either do not statutorily require or do not allow collective bargaining by public employees.[11] In the majority of these states, laws permitting public employees to engage in collective bargaining have not been passed.

The Bargaining Relationship. In those states and agencies seeking to organize for **collective bargaining,** the process is as follows. First, a union will begin an organizing drive seeking to get a majority of the class(es) of employees it seeks to represent to sign authorization cards. At this point, agency administrators may attempt to convince employees that they are better off without the union. Questions may also arise, such as whether or not certain employees (for example, police or prison lieutenants) are part of management and therefore ineligible for union representation.

Once a majority (50 percent plus one of the eligible employees) have signed cards, the union notifies the criminal justice agency. If management believes that the union has obtained a majority legitimately, it will recognize the union as the bargaining agent of the employees it has sought to represent. Once recognized by the employer, the union will petition the PERB or other body responsible for administering the legislation for certification.

Negotiations. Figure 12–1 depicts a typical configuration of the union and management bargaining teams. Positions shown in dashed boxes typically serve in a support role and may or may not actually partake in the bargaining. The management's labor relations manager (lead negotiator) is often an attorney assigned to the human resources department who reports to the city manager or assistant city manager and represents the city in grievances and arbitration matters. The union's chief negotiator will normally not be a member of the bargaining unit; rather, he or she will be a specialist brought in to represent the union's position and provide greater experience, expertise, objectivity, and autonomy. The union's chief negotiator may be accompanied by individuals who have conducted surveys on wages and benefits, trends in the consumer price index, and so on.[12]

Management's chief negotiator may be the director of labor relations or human resource director for the unit of government involved or a professional labor relations specialist. The agency's chief executive should not appear at the table personally; it is extremely delicate for the chief to represent management one day and then return to work among the employees the next. Rather, management is represented by a key member of the command staff who has the executive's confidence.

In the Event of an Impasse . . . The purpose of bargaining is to produce a bilateral written agreement to which both parties will bind themselves during the lifetime of the agreement. Even parties bargaining in good faith may not be able to resolve their differences by themselves, and an impasse may result. In such cases, a neutral third party may be inserted to facilitate, suggest, or compel an agreement. Three major forms of **impasse resolution** are *mediation, fact-finding*, and *arbitration*.

- Mediation occurs when a third party, called the mediator, comes in to help the adversaries with the negotiations.[13] This person may be a professional mediator or someone in whom both parties have confidence. In most states, mediation may be requested by either labor or management. The mediator's task is to build agreement about the issues involved by reopening communications between the two sides. The mediator has no way to compel an agreement, so an advantage of the process is that it preserves the nature of collective bargaining by maintaining the decision-making power in the hands of the involved parties.[14]

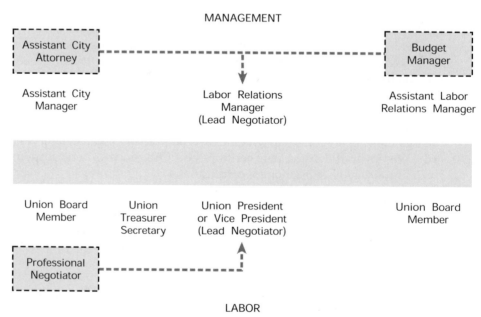

FIGURE 12–1 Union and Management Collective Bargaining Teams.
(*Source:* Jerry Hoover, Chief of Police, Reno, Nevada)

- Fact-finding primarily involves the interpretation of facts and the determination of what weight to attach to them. Appointed in the same way as mediators, fact-finders also do not have a way to impose a settlement of the dispute. Fact-finders may sit alone or as part of a panel, which normally consists of three people. The fact-finding hearing is quasi-judicial, although less strict rules of evidence are applied. Both labor and management may be represented by legal counsel, and verbatim transcripts are commonly made. In a majority of cases, the fact-finder's recommendations will be made public at some point.[15]
- Arbitration parallels fact-finding, but it differs in that the "end product of arbitration is a final and binding decision that sets the terms of the settlement and with which the parties are legally required to comply."[16] Arbitration may be voluntary or compulsory. It is compulsory when mandated by state law and is binding upon the parties even if one of them is unwilling to comply. It is voluntary when the parties undertake of their own volition to use the procedure. Even when entered into voluntarily, arbitration is compulsory and binding upon the parties who have agreed to it.

Grievances. The establishment of a working agreement between labor and management does not mean that the possibility for conflict no longer exists; the day-to-day administration of the agreement may also be the basis for strife. Questions can arise about the interpretation and application of the document and its various clauses, and **grievances**—complaints or expressions of dissatisfaction by an employee concerning some aspect of employment—may arise. The grievance procedure is a formal process that involves the seeking of redress of the complaints through progressively higher channels within the organization. The sequence of grievance steps will be spelled out in the collective bargaining agreement and would typically include the following five steps: (1) the employee presents the grievance to the immediate supervisor; and, if he or she does not receive satisfaction, (2) a written grievance is presented to the division commander, then (3) to the chief executive officer, then (4) to the city or county manager, and, finally, (5) to an arbiter, who is selected according to the rules of the American Arbitration Association.[17]

The burden of proof is on the grieving party except in disciplinary cases, for which it is always on the employer. The parties may be represented by counsel at the hearing, and the format will include opening statements by each side, examination and cross-examination of any witnesses, and closing arguments in the reverse order in which opening arguments were made.[18]

Job Actions. A **job action** is an activity in which employees engage to express their dissatisfaction with a particular person, event, or condition or to attempt to influence the outcome of some matter pending before decision makers.[19] Job actions are of four types: *the vote of confidence, work slowdowns, work speedups, and work stoppages.*

- The vote of confidence is used sparingly; a vote of no confidence signals employees' collective displeasure with the chief administrator of the agency. Although such votes have no legal standing, they may have high impact due to the resulting publicity.

- In work slowdowns, employees continue to work, but they do so at a leisurely pace, causing productivity to fall. As productivity declines, the unit of government is pressured to resume normal work production—for example, a police department may urge officers to issue more citations so revenues are not lost.[20]
- Work speedups involve accelerated activity in the levels of services. For example, a police department may conduct a "ticket blizzard" to protest a low pay increase, pressure government leaders to make more concessions at the bargaining table, or abandon some policy change that affects their working conditions.
- Work stoppages constitute the most severe job action. The ultimate work stoppage is the strike, or withholding of all employees' services. This tactic is most often used by labor to force management back to the bargaining table when negotiations have reached an impasse. However, criminal justice employee strikes are now rare. Briefer work stoppages, which do not involve all employees and are known in policing as "blue flu," last only a few days.

Fair Labor Standards Act

An area of policing that is at the heart of management-labor relations is the **Fair Labor Standards Act** (FLSA). For some police administrators, the Fair Labor Standards Act has become a budgetary and operational nightmare. Indeed, one observer referred to the FLSA as the criminal justice administrator's "worst nightmare come true."[21] The act provides minimum pay and overtime provisions covering both public- and private-sector employees and contains special provisions for firefighters and police officers. Although there is currently some discussion in Congress concerning the repeal or modification of the act, at the present time it is still a legislative force that administrators, mid-level managers, and supervisors must reckon with.

The act was passed in 1938 to protect the rights and working conditions of employees in the private sector. During that time, long hours, poor wages, and substandard work conditions plagued most businesses. The FLSA placed a number of restrictions on employers to improve these conditions. In 1985, the U.S. Supreme Court brought local police employees under the coverage of the FLSA. In this major (and very costly) decision, *Garcia v. San Antonio Transit Authority*,[22] the Court held, 5 to 4, that Congress imposed the requirements of the FLSA on state and local governments.

Police operations, which take place 24 hours per day, seven days per week, often require overtime and participation in off-duty activities such as court appearances and training sessions. The FLSA comes into play when overtime salaries must be paid. It provides that an employer must generally pay employees time and a half for all hours worked over 40 per week. Overtime must also be paid to personnel for all work in excess of 43 hours in a seven-day cycle or 171 hours in a 28-day period. Public safety employees may accrue a maximum of 480 hours of "comp" (compensation) time, which, if not utilized as leave, must be paid off upon separation from employment at the employee's final rate of pay or at the average pay over the last three years, whichever is greater.[23] Furthermore, employers usually cannot require employees to take compensatory time in lieu of cash. The primary issue with FLSA is the rigidity of application of what is compensable work. The act prohibits an agency from taking "volunteered time" from employees.

Today, an officer who works the night shift must receive pay for attending training or testifying in court during the day. Furthermore, officers who are ordered to remain at home in anticipation of emergency actions must be compensated. Notably, however, the FLSA's overtime provisions do not apply to persons employed in a bona fide executive, administrative, or professional capacity. In criminal justice, the act has generally been held to apply to detectives and sergeants but not to those of the rank of lieutenant and above.

Garcia prompted an onslaught of litigation by police and fire employees of state and local government entities. The issues are broad but may include paying overtime compensation to K-9 and equestrian officers who care for department animals while "off duty," overtime pay for officers who access their work computer and conduct business from home, pay for academy recruits who are given mandatory homework assignments, and standby and on-call pay for supervisors and officers who are assigned to units that require their unscheduled return to work. These are just a few of the many FLSA issues that are being litigated in courts across the nation.

WOMEN WHO WEAR THE BADGE

Over the past 30 years, the proportion of **women police officers** has grown steadily. During the 1970s, some formal barriers to hiring women were eliminated, such as height requirements; also, subjective physical agility tests and oral interviews were modified.[24] Some job discrimination suits further expanded women's opportunities. The 1980 lawsuit by Los Angeles police officer Fanchon Blake opened up the ranks above sergeant to women; the case of New York City detective Kathleen Burke, who in May 1991 won a settlement of $85,000 and a promotion to detective first grade, broadened their possibilities even more.[25]

The proportion of sworn female employees in local police agencies has risen to about 11 percent, up from just 1.4 percent in 1971, or an 800 percent increase.[26] Furthermore, the federal government leads in employing female officers; about 14 percent of all sworn federal officers are women.[27] (Table 12–1 shows the gender and race of full-time, sworn federal officers.) State patrol and police agencies lag behind; women represent only about 6.3 percent of all troopers.[28]

But serious obstacles remain. Women are still overwhelmingly employed in the lowest tier of sworn police positions. The proportion of women holding top command positions (captain and higher) is only 7.5 percent, and they fill only 1.6 percent of supervisory positions (sergeant and lieutenant).[29]

The female "baby-boomer officers" who were able to get hired in the 1970s have now either retired or will be retiring soon. As the second generation of women began to enter policing, Peter Horne identified several key issues that may have been problematic in the past and need to be addressed at this crucial time:[30]

TABLE 12-1 Gender and Race or Ethnicity of Federal Officers with Arrest and Firearm Authority, Agencies Employing 500 or More Full-Time Officers, June 1998

PERCENT OF FULL-TIME FEDERAL OFFICERS WITH ARREST AND FIREARM AUTHORITY

	NUMBER OF OFFICERS*	Total	GENDER		RACE/ETHNICITY				
			Male	Female	White	American Indian	Black or African American	Asian or Pacific Islander	Hispanic or Latino of any race
Immigration and Naturalization	16,888	100%	88.3%	11.7%	59.2%	0.5%	5.3%	2.3%	32.0%
Federal Bureau of Prisons	12,751	100	87.9	12.1	62.6	1.4	23.4	0.9	11.7
Federal Bureau of Investigation	11,451	100	84.1	15.9	83.9	0.5	6.3	2.5	6.9
U.S. Customs Service	10,863	100	81.4	18.6	66.3	0.7	7.2	3.5	22.3
U.S. Secret Service	3,594	100%	91.4%	8.6%	79.7%	0.8%	12.9%	1.3%	5.3%
U.S. Postal Inspection Service	3,537	100	85.5	14.5	67.1	0.4	22.5	3.0	7.0
Internal Revenue Service	3,370	100	74.8	25.0	80.2	0.9	9.4	3.3	6.1
Drug Enforcement Administration	3,396	100	92.1	7.9	80.5	0.6	8.2	2.0	8.8
U.S. Marshals Service	2,755	100%	88.6%	11.4%	83.8%	0.7%	7.1%	1.9%	6.5%
National Park Service	2,207	100	86.8	13.2	86.4	1.3	6.5	2.4	3.4
Ranger Activities Division	1,534	100	85.0	15.0	90.0	1.9	3.1	2.2	2.8
U.S. Park Police	673	100	90.8	9.2	78.2	0.0	14.3	3.0	4.6
Bureau of Alcohol, Tobacco and Firearms	1,732	100	87.8	12.2	78.6	1.3	10.8	1.7	7.6
U.S. Capitol Police	1,055	100%	82.1%	17.9%	67.4%	0.4%	29.8%	0.8%	1.7%
GSA — Federal Protective Service	904	100	91.2	8.8	57.7	0.2	30.4	2.1	9.5
U.S. Fish and Wildlife Service	836	100	90.0	10.0	91.7	1.9	1.4	0.6	4.3
U.S. Forest Service	604	100	83.9	16.1	82.5	7.3	3.1	1.0	6.1

Note: Gender and race/ethnicity data for Drug Enforcement Administration are estimates based on Department of Justice data. Data on gender and race or ethnicity of officers were not provided by the Administrative Office of the U.S. Courts. Detail may not add to total because of rounding.

*Includes employees in U.S. Territories.

Source: Brian A. Reaves and Timothy C. Hart, *Federal Law Enforcement Officers, 1998* (Washington, D.C.: U.S. Department of Justice, Bureau of Justice Statistics Bulletin, 2000), p. 6.

Women in Policing: Past, Present, and Future

Chief Penny Harrington

Chief Penny Harrington, author of Triumph of Spirit, *her auto-biography, is the director of the National Center for Women & Policing, a division of the Feminist Majority Foundation. She spent 23 years in the Portland, Oregon, Police Bureau, where she started in 1964 as a policewoman in the Women's Protective Division. Based on her record in the bureau and her support from the community, Penny was named chief of police in 1985, making her the first woman to become chief of a major U.S. city. After leaving the Portland Police Bureau, Penny became the assistant director of investigations for the State Bar of California. As director of the National Center for Women & Policing, Penny is working nationally to bring more women into policing and to help women reach the higher levels of command within their agencies. For more information on the National Center for Women & Policing, see www.feminist.org/police.*

When Lola Baldwin joined the Portland, Oregon, Police Bureau in 1905, she was the first woman in the field of law enforcement. In 1985, the Portland Police Bureau again made history by promoting me to be the first female police chief of a major U.S. city. In those 77 years, a lot of progress was obviously made. Yet women remain underrepresented in policing, and the current rate of progress is such that we will not achieve equality with men for about another 77 years. Here, I'll briefly outline the past and present status of women in policing. I'll also chart a course for the future that includes increasing women's representation and transforming the police force toward a more community-oriented style of policing.

A Brief History

Pioneers such as Lola Baldwin led the way for women in policing, but even when I began my career in the 1960s I was hired as a "policewoman." The position wasn't equal in pay or stature to that of "patrolman," which was explicitly described as being available only to men. In 1968, the Indianapolis Police Department was the first to assign women to patrol on an equal basis with the men. Other departments followed, including the FBI, which hired its first woman agent in 1972.

Throughout this period of change, women entered the field of law enforcement in larger numbers, but their service was nonetheless seen as an experiment. In fact, several research studies were conducted in the early 1970s to see whether women could actually perform the job successfully. The answer was a unanimous "yes." Not only were women officers equally competent as their male peers, but the research began to illuminate some of the unique advantages that women bring to the field. For example, women officers were often better at communicating with citizens, and they used these skills to de-escalate situations that might otherwise have turned violent. This is something that I have certainly found to be true, and women officers from around the country all say the same thing—they prefer to talk through a situation rather than resort to physical force because it is a smarter way to do business.

As women made gains in the field of policing, we were helped by ground-breaking legislation. Perhaps the most important was the 1964 Civil Rights Act, which was passed by Congress to extend legal protection to both women and minorities in the workplace. This was followed in 1972 by the guidelines published by the Equal Employment Opportunity Commission for law enforcement agencies about the employment of minorities and women.

However, much of the progress has been the result of brave women officers who used the court system to challenge the discrimination and harassment they experienced. Over the years, lawsuits have been filed around the country and improved the practices involved in hiring, assignment, and promotion. In Portland, we filed many civil rights complaints and negotiated agreements in each of these areas. It is perhaps unfortunate that women have so often had to file a lawsuit to improve their situation, but by doing so they have reduced discrimination and opened the field to other women who followed them.

CURRENT STATUS

Despite the many important gains women have made in policing, research consistently reveals that we are still vastly underrepresented. As director of the National Center for Women & Policing, I oversee our annual research into the current status of women in policing. Each year, we identify a random sample of police agencies with 100 or more sworn officers, and we ask them to provide detailed information regarding the women within their agency. In 2000, we found that women represented only 13 percent of sworn personnel in these large agencies.[a] This is a paltry four percentage points higher than 1990, when women represented 9 percent of sworn law enforcement. Considering that women accounted for 46.5 percent of the adult labor market in 2000,[b] it is clear that we have a long way to go before we will be equally represented. Worse, the pace of increase is so slow that women gain only half a percentage point every year. This means that it will be decades before women will achieve equality with men in policing.

The reasons for women's underrepresentation in policing are many. Research indicates that discrimination continues to exist in hiring and promotion practices and that many women experience a wide range of harassing behaviors

[a]*Equality Denied: The Status of Women in Policing 2000* (National Center for Women & Policing, 2000), available at www.feminist.org/police.
[b]Bureau of Labor Statistics, 2000. Available at www.bls.gov.

once on the job. Only a small number of women who experience harassment report the misconduct, however, out of fear of retaliation. Unfortunately, this fear is well founded. Women who report discrimination or harassment do often experience retaliation, and it is sometimes so severe that they leave the field as a result.

There are signs, however, that this is beginning to change. In recent years, many agencies have had a difficult time recruiting qualified personnel, and police executives are beginning to realize that they cannot overlook half of the pool of potential applicants. There is also a growing body of research documenting the many advantages that women officers bring. As I've previously described, research conducted both in the 1970s and more recently demonstrates that women often use a style of policing that relies more on communication and less on physical force. Women also tend to ascribe more to the ideals of community policing and are much less likely to use excessive force compared to their male colleagues. Finally, women officers respond more effectively to crimes of violence against women, and our very presence often brings about change in the culture of policing.[c] For all of these reasons, police executives across the country are redesigning their agencies to better recruit and retain women.

LOOKING TO THE FUTURE

There is still a long way to go. We all read the headlines and know that police agencies around the country are struggling with problems of excessive force, corruption, and mistrust from the community. Much of the problem is due to the structure of the agencies themselves, which often provide too little supervision and reinforce all the wrong behaviors. However, another significant part of the problem is the type of person that police agencies have historically sought. With a "he-man" stereotype of policing as a profession for only those with the thickest necks, police agencies have placed too much emphasis on factors such as upper body strength and military experience.

As agencies shift toward a more community-oriented style of policing, we must rethink these stereotypes and look for people who possess the kinds of skills this new kind of officer will need. Recruitment must begin to focus on interpersonal skills, including communication and an appreciation for diversity. Agencies must seek out individuals who can creatively solve neighborhood problems and work with citizens to build trust and partnerships. In many cases, this individual will be a woman.

I have a lot of hope for the future of policing. I believe that agencies will (perhaps slowly) improve their recruitment and retention of women. At the National Center for Women & Policing, we do everything we can to promote that change and transform the face of policing. Our communities deserve the best police services, and women and men must work together to see that this happens.

[c]*Hiring and Retaining More Women: The Advantages to Law Enforcement Agencies* (National Center for Women & Policing, 2000), available at www.feminist.org/police.

Recruitment: This is the first critical step in getting qualified women to enter policing. Unfocused, random recruiting is unlikely to attract diversity. Literature such as flyers, posters, and brochures should feature female officers working in all areas of policing. The variety of assignments (e.g., victim/witness assistance, victims of domestic violence, community policing) is often cited by prospective policewomen as attractive, and recruitment should stress these factors. Furthermore, agencies should go anywhere (local colleges, women's groups, female community leaders, gyms, martial arts schools) and use all types of media to attract qualified applicants.

Pre-employment Physical Testing: Historically, women have been screened out disproportionately in the pre-employment screening physical testing used by many agencies. Tests that include such components as scaling a six-foot wall, bench pressing one's own weight, and throwing medicine balls are likely to have an adverse impact on female applicants, so agencies should examine their physical tests to determine the reasons women are disproportionately screened out. Also, agencies should permit all candidates to practice for the pre-employment physical exam.

Academy Training: Recruits must be trained in sexually integrated academy classes to ensure full integration between female and male officers. Female instructors are especially important during academy training because they are positive role models and help female rookies develop skills and confidence. These instructors should teach more than the sterotypical "feminine" parts of the curriculum (such as child abuse or juvenile delinquency). Involving female instructors in firearms and physical/self-defense training will send a message that trained, veteran female officers can effectively handle the physical aspects of policing.

Field Training: Field training officers (FTOs, discussed in Chapter 3) play a crucial role in transforming the academy graduate to a competent field officer. FTOs should be both supportive of female rookies and effective evaluators of their competence. Females should also serve as FTOs.

Assignments: A rookie's assignments have long-term effects on his or her policing career; agencies should routinely review the daily assignments of all probationary officers to assure they have equal opportunity to become effective patrol officers. If women are removed from patrol early in their careers (for any reason), they will miss vital field patrol experience. The majority of supervisory positions exist in the patrol division, and departments generally believe that field supervisors must have adequate field experience in order to be effective and respected by subordinates. Special assignments are often quite attractive, with better working shifts and days off, greater independence, and increased salaries and benefits. These positions are highly sought after, and male officers may resent women who receive coveted assignments merely because of their gender. Furthermore, agencies should conduct audits of such assignments to ensure that women are not ghettoized into stereotypical jobs, as was the case historically.

Promotions: For a number of reasons, women are not getting promoted in proportion to their numbers. The so-called glass ceiling continues to restrict women's progress through the ranks. Performance evaluations and the overall promotional system utilized by agencies should be scrutinized for gender bias. For example, studies show that the more subjective the promotion process, the less likely women are to pass it. To provide more objectivity (in terms of ability to measure aptitude), the process may include more hands-on (practical, applied) tasks and the selection of board members who represent both sexes.

Today about 11 percent of the nation's sworn municipal and county officers are women, who occupy a variety of roles. (*Courtesy Reno, Nevada, Police Department*)

Harassment and Discrimination: Police agencies across the nation have paid millions of dollars in compensation for acts of harassment and discrimination against women officers. Aside from the obvious financial toll, sexual discrimination and harassment exact a human cost from the women involved—including a negative impact on their performance and careers (and, probably, a negative impact on the recruitment of other women into policing). A Police Foundation study found that "most women officers have experienced both sex discrimination and harassment on the job."[31] Departments need to have policies in place concerning sexual harassment and gender discrimination—and they need to enforce them. Furthermore, all supervisory candidates should be screened for their own treatment of women and attitudes about women in policing.

Mentoring: Retention issues are also significant in policing. Women officers have a higher turnover rate (about 6 percent) than men (about 4 percent). The employee's experience as he or she transitions into the organization can be a deciding factor in whether or not the employee remains with the organization. Formal mentoring programs—which can begin even before the rookie enters the academy—have helped some agencies raise their retention rates for women; such programs can include having a veteran employee provide new hires with information concerning what to expect at the academy and beyond during field training and the probationary period. Mentors can also act as confidants and provide access to informal organizational networks.

Career and Family: Police work can create a considerable amount of conflict between one's work life and one's family life. The facts that women have a slightly higher turnover rate and use somewhat more sick leave than their male counterparts are related to their family lives and responsibilities that go with their roles as wives and mothers. The treatment of pregnant officers has been one of the more complex (and heavily litigated) personnel issues in recent years. To reduce this conflict, police agencies should have a leave policy covering pregnancy and maternity leave. Light-

or limited-duty assignments (e.g., desk, communications, records) can be made available to female officers when reassignment is necessary. Other issues include the availability of quality child care and shift rotation (which can more heavily burden single parents) and uniforms, body armor, and firearms that fit women.

As the community oriented policing and problem solving (COPPS, examined in Chapter 6) concept continues to expand, female officers can likewise play an increasingly vital role. Former Austin, Texas, police chief Elizabeth Watson stated that "Women tend to rely more on intellectual than physical prowess. From that standpoint, policing is a natural match for them."[32] Experts maintain that the verbal skills many women possess often have a calming effect that defuses potentially explosive situations. Such a style is especially effective in handling rape and domestic violence calls.[33] As former Madison, Wisconsin, police chief David Couper put it, female officers have helped usher in a "kinder, gentler organization."[34]

Still, this clearly remains an area in which policing has been slow to change. For women to serve effectively as police officers, police managers must see the value of utilizing women and vigorously recruiting, hiring, and retaining them. As shown earlier, a basic task for the chief executive is to consider how departmental policies impact female officers with respect to selection, training, promotion, sexual harassment, and family leave. Most important, executives must set a tone of welcoming women into the department.

MINORITIES AS POLICE OFFICERS

Minority police officers, like females, are slowly but steadily increasing their representation in state and local police agencies. For example, in local (city and county) police agencies, African Americans account for 11.3 percent of sworn officers, compared to 10.5 percent in 1990 and 9.3 percent in 1987.[35] And, as with women, federal law enforcement agencies are major employers of members of racial or ethnic minorities—29.2 percent overall.[36] (See Table 12–1.)

Several developments spurred the growth of African Americans in policing in the early 1970s, including federal reports (such as that from the U.S. Commission on Civil Rights),[37] the Supreme Court's decision in *Griggs v. Duke Power Company* (1971) (banning the use of intelligence tests and other artificial barriers that were not job-related),[38] and congressional legislation such as the federal Equal Employment Opportunities Act of 1972. The mid-1970s witnessed the advent of the National Black Police Association and the National Organization of Black Law Enforcement Executives (NOBLE), which advocated increased hiring of minority officers and improvement of community relations. In the 1980s, several lawsuits successfully challenged such requirements as height, age, weight, sex, and a clean arrest record.[39]

Recruitment of minority officers remains a difficult task. Probably the single most difficult barrier has to do with the image that police officers have in the minority communities. Unfortunately, police officers have been seen as symbols of

Police departments must also strive to enhance
diversification in their ranks. (*Courtesy NYPD Photo Unit*)

oppression and have been charged with using excessive brutality; they are often
seen as an army of occupation. Many minorities view black police officers as people who have "sold out." For many of those blacks in uniforms, a so-called **double
marginality** has existed, whereby the black officers feel accepted neither by their
own minority group nor the white officers. However, many blacks view police
work as an opportunity to leave the ghetto and enter the middle class. As one
black officer put it, "There were two ways to get out of my neighborhood and not
end up dead or in prison. You either became a minister or a cop. I always fell
asleep in church so I decided to become a cop."[40]

Black police officers face problems similar to those of women who attempt
to enter and prosper in police work. Until more black officers are promoted and
can affect police policy and serve as role models, they are likely to be treated
unequally and have difficulty being promoted—a classic catch-22 situation. Blacks
considering a police career may be encouraged by a survey of black officers,
which found that most thought their jobs were satisfying and offered opportunities for advancement.[41]

Interestingly, a study of black high school students found that although negative experiences with the police increased negative attitudes toward the police,
these attitudes did not in themselves reduce interest in police work. The findings

suggested that police recruiters need not expend a lot of effort trying to neutralize perceptions that black people are treated unfairly by the police. A greater concern among black youths was the danger of the job itself. Therefore, it may be more important for police departments to educate the parents of potential black recruits about the benefits and objectives of police work rather than focus excessively on dealing with negative attitudes.[42]

ON GUARD: THE PRIVATE POLICE

A burgeoning industry in this country is one that is both a "trend" and an "issue," as we will see; this industry is that of the **private police**, which provide protective services for profit.

A Historical Overview

Since prehistoric times, people have sought security from enemies and from beasts of prey by developing weapons; building and fortifying caves; and using fire, stone, hemp, and water to punish and control their fellow humans. We have sought to protect ourselves from a hostile environment.[43]

Later, in the early 1800s in England, private police were hired as watchmen by shipping docks, industrial firms, and railroad companies. In the mid-1800s, America's western frontier was awash with train robbers, pickpockets, bootleggers, and common outlaws; as a result, railroad companies were allowed to establish their own in-house, proprietary security forces. This was also the beginning of the contract (fee-for-service) security system for private concerns.[44]

In 1851, a pioneer in the private security industry, William Allan Pinkerton, initiated the Pinkerton National Detective Agency, which specialized in railroad security. Pinkerton established the first private security contract operation in the United States. His motto was "We Never Sleep," and his logo, an open eye, was probably the genesis of the term "private eye." Pinkerton also established a code of ethics for his agents. Other pioneers in the security industry were Henry Wells and William Fargo, who in 1853 formed their contract security company, Wells Fargo, which provided private detective and shotgun riders on stagecoaches west of the Missouri River.

Later, William J. Burns, a former head of the forerunner agency of the FBI, founded a detective agency in 1909. (Pinkerton and Burns had the only national investigative agencies until the formation of the FBI in 1924.) In 1917, the Brink's armored car was unveiled. In 1954, George R. Wackenhut and three other former FBI agents formed the Wackenhut Corporation, another large contract guard and investigative firm in the country.[45]

Types and Extent of Crime in "A Nation of Thieves"

Saul Astor said, "We are a nation of thieves."[46] And, it might be added, a nation of rapists, robbers, and other types of offenders. There are nearly 25.9 million

William A. Pinkerton, Principal of the
Western Branch of Pinkerton's National
Detective Agency.

offenses committed in America each year, 6.6 million of which are violent in nature.[47] As a result, this nation has become security minded concerning its computers, lotteries, celebrities, college campuses, casinos, nuclear plants, airports, shopping centers, mass transit systems, hospitals, and railroads.

Given this crime picture and level of concern and the limited capabilities of the nation's 733,000 full-time sworn federal, state, and local (municipal and sheriff's) agents, officers, and deputies (see Chapter 2),[48] it is not surprising that Americans have increasingly turned to protection from the "other police"—those of the private sector.

More than 1.5 million persons are now employed by the 4,000 agencies that constitute the private police industry; private security is estimated to be a $10-billion-a-year enterprise.[49] Thus, the private policing field is larger in both personnel and resources than the federal, state, and local forces combined.[50] Indeed, the private police now outnumber public police by more than two to one, and both employment and expenditures for private security are expected to greatly increase well into the twenty-first century.[51]

In-house security services, directly hired and controlled by the company or organization, are called *proprietary services;* conversely, *contract services* are outside firms or individuals who provide security services for a fee. The most common security services provided include contract guards, alarm services, private investigators, locksmith services, armored-car services, and security consultants.[52]

Although some of the duties of the security officer are similar to those of the public police officer, their overall powers are entirely different. First, because security officers are not police officers, recent court decisions have stated that the security officer is not bound by the Miranda warnings of rights for suspects. Furthermore, security officers generally possess only the same authority to effect an arrest as does the common citizen (the exact extent of citizen's arrest power varies, however, depending upon the type of crime, jurisdiction, and the status of the citizen). In most states, warrantless arrests by private citizens are allowed when a felony has been committed and reasonable grounds exist for believing that the person arrested committed it. Most states also allow citizen's arrests for misdemeanors committed in the arrester's presence.[53]

Recruitment and Training

With the boom of the industry, both in terms of numbers of employees and in dollars spent, there have also been problems and concerns. As a federal task force noted, "Of those individuals involved in private security, some are uniformed, some are not; some carry guns, some are unarmed; some guard nuclear energy installations, some guard golf courses; some are trained, some are not; some have college degrees, some are virtually uneducated."[54]

Studies have shown that security personnel are generally recruited from among persons with minimal education and training; because the pay is often quite low, the work often attracts only people who cannot find other jobs or who seek only temporary work. Thus, most of the work is done by the young and the retired.[55] A RAND Corporation study also found that fewer than half of all private security guards have a high school education, and their average age was 52. Ninety-seven percent of the respondents failed to pass a "simple examination designed to test their knowledge of legal practice in typical job-related situations."[56] All too often, the new security officer is taught only the security "primer": the chain of command, how to use a radio, and uniform dress standards. Some organizations "train" their employees by merely having them accompany other security officers around the company's facility.[57] For these reasons, security training has been an issue in court cases involving claims of negligent security.

Carrying Weapons

A long-running debate is whether the private police should even be armed at all. When the critical decision is made by an employer to arm security personnel, then the type and level of firearms training that is provided is justifiably called into closer scrutiny. Much ado has been made about security officers who have received little or no prior training or have undergone no checks on their criminal history records but are carrying a weapon. This is probably a debate that is not going to be settled in the foreseeable future. However, both sides agree that there are some roles in which security officers must be armed, such as protecting high-risk or high-value assets, for example, nuclear power plants and certain shipments in transit, such as money in armored cars. Twenty-three hours of firearms instruction is recommended for all security personnel, as well as another 24-hour block on general matters and proper legal training.[58]

The Problem of Liability

Civil liability has also become an increasing concern in the private security industry. Civil actions may be brought against any private security personnel who commit an unlawful action against another person. Often the officer's employer is sued as well.[59] The typical scenario develops when a person is attacked on property that is either patrolled by a private security company, has a security officer stationed on the premises, or has no security present. Then, when the assailant is not caught or has no assets to compensate the injured victim, the victim sues the landowner and security company for recovery. Thus, reasonable precautions must be taken to try to prevent innocent people from being victimized.

CIVILIANIZATION

Police administrators are increasingly realizing that more and more tasks can be performed as efficiently—and at far less cost than using a sworn officer—through **civilianization**. Agencies are encouraging and obtaining greater citizen interaction through the use of civilian personnel to perform a variety of police-related tasks, such as investigations of crime scenes or traffic accidents. Civilianization is a cost-effective way to capitalize on citizens' capabilities. Functions involving police administration, report-writing, dispatching, animal control, and jail functions are among those that could be civilianized. In keeping with the community oriented policing and problem solving (COPPS) philosophy, increasing numbers of civilians will likely be hired in the future to assist police agencies with tasks that do not require sworn officers. This trend will be beneficial in helping to break down barriers between the police and the general public and will foster the belief that civilians have an important role in police functions.

Another use of nonsworn persons in police-related work involves volunteerism, which can build a sense of community, break down barriers between people, and raise our quality of life. But volunteers will lose interest if not given meaningful works; this is typically not hard to do in police work. An excellent example of the kinds of assistance that can be provided by volunteers is the Volunteer Services Division of the Redding, California, police department. Forty active volunteers donate nearly 10,000 hours each year, working with abandoned vehicle abatement (data entry, filing), court liaisons (case preparation, processing traffic citations), warrants (records, research), vehicles (inspecting for defects), records (compiling statistics, data entry, microfilming), crime analysis and prevention (computer- or pin-mapping crimes, preparing parolee tracking records, maintaining Neighborhood Watch files), and investigations (missing persons followup, maintaining gang files, assisting with D.A.R.E., maintaining mug shots).[60]

ACCREDITATION

In 1979, accreditation of police agencies began slowly with the creation of the Commission on Accreditation for Law Enforcement Agencies (CALEA). In mid-1994, CALEA reduced the number of standards to be met for accreditation from 897 to 498, which cover the role and responsibilities of the agency; its organization and administration; law enforcement, traffic, and operational support; prisoner- and court-related services; and auxiliary and technical services. This voluntary process is still quite expensive, both in actual dollars and human resources; it often takes a year to 18 months for the agency to prepare for the assessment.[61] Today more than 500 police and sheriff's departments have met the standards for accreditation.

Once the agency believes it is ready to be accredited, an application is filed and the agency receives a questionnaire that is used for a self-evaluation to determine its current status. If the self-evaluation indicates that the agency is ready to attempt accreditation, an on-site team appointed by CALEA conducts an assessment and writes a report on its findings. After becoming accredited, the agency must apply for reaccreditation after five years.[62]

Several unanticipated consequences of the accreditation process have emerged. A number of states have formed coalitions to assist police agencies in the process of accreditation. Some departments report decreased insurance costs as a result of accreditation.[63] And, finally, the accreditation process has raised questions among many practitioners in terms of how it squares with the movement toward COPPS. The question to be answered is whether or not a COPPS orientation and the accreditation process—with its emphasis on adherence to standards—will be compatible.

CONTEMPORARY POLICING ISSUES

HIGHER EDUCATION FOR POLICE

An Enduring Controversy

One of the most enduring and controversial issues in policing is whether or not police officers benefit from, and should pursue, higher education.

Efforts to involve college-educated personnel in police work were first made by August Vollmer in 1917, when he recruited University of California students as part-time police officers in Berkeley.[64] However, few departments elsewhere in the country took any immediate steps to follow Vollmer's example. Rank-and-file officers strongly resisted the concept of college-level studies for police, and officers with a college education remained very much an exception; they were often referred to derogatorily as "college cops."[65]

However, the movement toward higher education for police continued to spread; by 1975, there were 729 community college and 376 four-year programs.[66] The Law Enforcement Education Program (LEEP) provided tuition assistance for in-service police officers and pre-service students. In 1973, 95,000 college and university students were receiving LEEP assistance; these were unquestionably the "glory days" of higher education in criminal justice.[67] Many patrol officers who otherwise could not have afforded to do so received quality higher education.

Rationales For and Against Police Higher Education

In 1973, the National Advisory Commission on Criminal Justice Standards and Goals, recognizing the need for college-educated police officers, recommended that all police officers have a four-year college education by 1982.[68] That goal obviously went unmet. Nevertheless, from 1967 to 1986, every national commission that studied crime, violence, and police in America was of the opinion that a college education could help the police to do their jobs better.[69] Advocates of higher education for the police maintain that it will improve the quality of policing by making officers more tolerant of people who are different from themselves; in this view, educated officers are more professional, communicate better with citizens, are better decision makers, and have better written and verbal skills.

A ringing endorsement for higher education for the police came in 1985, when a lawsuit challenged the Dallas, Texas, Police Department's requirement that all applicants for police officer positions possess 45 credit hours and at least a 'C' average at an accredited college or university. The Fifth Circuit Court of Appeals and, eventually, the U.S. Supreme Court upheld the educational requirement, the circuit court saying:

> Even a rookie police officer must have the ability to handle tough situations. A significant part of a police officer's function involves his ability to function effectively as a crisis intervenor, in family fights, teen-age rumbles, bar brawls, street corner altercations, racial disturbances, riots and similar situations. Few professionals are so peculiarly charged with individual responsibilities as police officers. Mistakes of judgment could cause irreparable harm to citizens or even to the community. The educational requirement bears a manifest relationship to the position of police officer. We conclude that the district court's findings . . . are not erroneous.[70]

Conversely, critics believe that educated officers are more likely to become frustrated with their work and that their limited opportunities for advancement will cause them to leave the force early. Furthermore, they argue that police tasks that require mostly common sense and/or street sense are not performed better by officers with higher education.[71]

Police administrators fear that having a formal educational requirement will prompt discrimination suits from minorities who do not possess such an education, that the police officers' union would seek higher pay and benefits for such a requirement, and that academy classes would go unfilled; many administrators also lack conviction that higher education is beneficial for officers.

However, abundant empirical evidence indicates that college-educated police officers are better officers. College-educated officers have significantly fewer founded citizen complaints;[72] have better peer relationships;[73] are likelier to take a leadership role in the organization;[74] tend to be more flexible;[75] are less dogmatic and less authoritarian;[76] take fewer leave days, receive fewer injuries, have less injury time, have lower rates of absenteeism, use fewer sick days, and are involved in fewer traffic accidents than non–college-educated officers;[77] have greater ability to analyze situations and make judicious decisions; and have a more desirable system of personal values.[78] Furthermore, college graduates violated their department's internal regulations regarding insubordination, negligent use of a firearm, and absenteeism significantly less often than did officers who lacked a college degree.[79]

Some studies have found negative effects of higher education, however. These studies found that it had no positive effect on officers' public service orientation (those with a degree displayed less orientation toward public service than those without a degree),[80] and that college-educated officers attach less value on obedience to supervisors than officers without a college education.[81]

Because a number of views and arguments for both sides of the higher education issue remain, it is unlikely that this question will be resolved soon. Perhaps the entry of federal courts into the foray will help to expedite a resolution. Change is also indicated by such developments as the Tulsa, Oklahoma, Police Department requirement as of January 1998 that all new hires possess a bachelor's degree. The Tulsa police chief, Ron Palmer, has observed "a world of difference between a high school graduate and a college graduate in regard to the skill levels and the handling of people."[82] Palmer also found that the requirement has not hampered the department's efforts to attract more minority recruits. Another indicator of change is the existence of the American Police Association, based in Alexandria, Virginia, which is composed of college-educated officers.

It seems paradoxical that as we reach the new millennium our society—given the research, the increase in educational level of society in general, and the need for police to recruit the best people—still does not compel its police to raise their educational standards. No true profession requires less than a college degree. Police, in their quest for greater professionalism, should take notice.[83]

A RELATED PROGRAM: THE POLICE CORPS

The era of police higher education may be returning, through the Violent Crime Control and Law Enforcement Act of 1994—the so-called Crime Act—which provided $200 million for a Police Corps program and an in-service scholarship program. The Police Corps program is discussed more thoroughly in Appendix III.

STRESS: SOURCES, EFFECTS, MANAGEMENT

There is a side of policing that we would prefer to ignore: the stress that is induced by the job. Indeed, Sir W. S. Gilbert observed that "when constabulary duty's to be done, the policeman's lot is not a happy one."[84]

Stress may be defined as "a force that is external in nature that causes strain upon the body, both physical and emotional, from its normal process." The late Hans Selye, who is known as "the father of stress research," defined stress as a nonspecific response of a body to demands placed upon it.

William A. Westley observed long ago that "the policeman's world is spawned of degradation, corruption and insecurity. He walks alone, a pedestrian in Hell."[85] Today, police adversaries are more heavily armed and more arrogant than ever. Being a cop today has been described as a "stop-and-go nightmare."[86] As noted in the dangers of patrol section of Chapter 5, the job of policing has never been easy, but the danger, frustration, and family disruption of the past have been made worse by the drug war and violent criminals who are heavily armed and hold more contempt for the police than ever before.

Sources of Stress

There are several potential sources of stress for police officers:

I. Stressors Originating Within the Organization
 A. **Intra-agency Practices and Characteristics**
  1. *Poor Supervision:* Stress can be fostered by supervisors who go by the book and issue a constant flurry of memos, won't back up their subordinates or fail to attend to their basic needs. These behaviors can contribute substantially to the stress levels of their workers.
  2. *Absence of Career Development Opportunities:* The vast majority of police officers start and end their careers as patrol officers. Promotional opportunities are limited, and the promotion process can lack fairness and generate frustration.

Dr. Hans Selye, the "father of stress research." (*Courtesy Edward Donovan*)

3. *Inadequate Reward System:* Recognition and compensation are limited for work that is well done. A lot of negative reinforcement exists in police work; punishment—demotion, reprimand, suspension, termination, reassignment—is commonly meted out when things go wrong.

4. *Offensive Policy:* Police organizations possess numerous policies and guidelines that officers find offensive, such as locking and unlocking municipal parking lots, emptying parking meters, and running all kinds of errands.

5. *Offensive Paperwork:* The volume of paperwork that police officers must complete is incredible. Computerization has generally created more paperwork for police officers because additional kinds of information are now desired by management.

6. *Poor Equipment:* In policing, inadequate vehicles, communication equipment, and safety equipment (flares, for example) can become sources of frustration.

II. Stressors External to the Organization

Stressors that come from outside of the organization can cause a great deal of tension; with few exceptions, the individual officer or agency normally has little or no control over them.

A. **Interorganizational Practices and Characteristics**

1. *Absence of Career Development Opportunities:* Few chances for mobility exist in police work; rank and seniority are seldom tranferable to another agency, so opportunities for lateral moves are limited. Thus, the officer must remain with the same agency unless he or she retires or is willing to start anew.

2. *Jurisdictional Isolationism:* Police agencies are not always the tight, congenial fraternity that the public often sees or perceives. Jurisdictional boundaries can be guarded zealously. Some police agencies are notorious for withholding information or grabbing credit for good arrests.

B. **Criminal Justice System Practices and Characteristics**

1. *Seemingly Ineffective Corrections System:* The police often see corrections agencies (for example, probation and parole, prisons) as a revolving door for criminals; attempts to rehabilitate offenders at these agencies often fail.

2. *The Courts:* Unfavorable court decisions and misunderstood judicial procedure (for example, plea bargaining) lead to considerable frustration for the police.

C. **Distorted Press Accounts of Police Incidents**

Reporters often distort the news because they hastily try to gather the facts to meet deadlines, do not understand the facts, or simply do not care to report them; if too overzealous, they can interfere with investigations.

D. Unfavorable Civilian Attitudes

Minorities often accuse police of over- or under-policing their neighborhoods or being brutal and racist. Similarly, the majority often complain about police response time and argue that the police should be out trying to catch bank robbers instead of issuing traffic citations.

E. Derogatory Remarks by Neighbors and Others

Police officers in a sense are never completely off duty; they are always cops, accosted at home and while attending social functions by people who want to complain about the police or want legal advice.

F. Adverse Government Actions

Units of government often make decisions without consulting police officers or managers, including decisions about budgeting, recruitment, hiring, firing, and downsizing. Public officials may monitor and chase police calls, make and enforce policy, check on officers, and seek to "fix" arrests of friends or family members.

III. Stressors Connected with the Performance of Police Duties

    A. Police Work Itself

        1. *Role Conflict:* Police must determine when to maintain a crime-fighter image and when to portray an order-maintenance role. The public

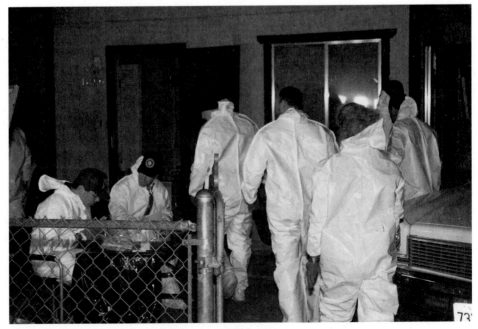

The omnipresent danger of policing is heightened even more during drug operations; here, officers in protective clothing raid a methamphetamine lab. (*Courtesy Washoe County, Nevada, Sheriff's Office*)

expects officers to be all things to all people—social worker, minister, physician, babysitter, psychologist, and so forth.

2. *Adverse Work Scheduling:* Shift work can adversely affect an officer's health as well as family and social life, especially the graveyard shift and a rotating shift schedule,[87] causing constant adjustment to new sleeping and eating patterns.

3. *Fear and Danger:* As noted in Chapter 5, fear and danger are constant companions of police officers, so officers need to keep their survival instincts honed. The annual *Law Enforcement Officers Killed and Assaulted* publication of the FBI regularly reveals an unsettling fact: Police officers often are overpowered, outmuscled, disarmed, and then murdered by offenders whom they try to arrest. The type of assignment can also bring greater stress; for example, working undercover[88] and homicide investigation[89] have been identified as particularly stressful. Also, many times officers are simply in poorer physical condition than the criminals they chase. Today more than ever, officers need to maintain a good physical state. In addition, they must also properly maintain their equipment; use de-escalation techniques (sometimes termed verbal judo); employ physical defense, approach, and positioning tactics; properly use backup personnel; deploy proper weapon and driving skills; use safe techniques when searching buildings; and handle prisoners with care.[90]

4. *Sense of Uselessness and Inefficiency of Referral Agencies:* Many officers entered the police service in order to help people. However, in many jurisdictions few agencies are available to which the police can refer people with problems. Police see firsthand the outcomes of the absence of such social services—suicides, murders, spousal or child abuse, and so on.

5. *Absence of Closure:* Police can seldom close their cases through disposition; furthermore, plea bargaining affects much of what they did in the first place (what is believed to be a "righteous" arrest is often reduced down by the prosecutor).

6. *People Pain:* The police see firsthand the gamut of violence humans employ against one another and witness both crime victims and offenders at their worst. Police deal constantly with people in pain and are expected to remain calm and collected during the entire affair.

IV. Stressors Particular to the Individual Officer
   A. **The Police Officer Himself/Herself**
      1. *The Fear-Ridden:* Some, although not many, police officers are overcome by the fear and danger of the job; they are in constant fear for their physical well-being. These officers often, but not always, leave the force well before retirement.
      2. *The Nonconformist:* Police work demands a certain amount of conformity and adherence to a chain of command, allowing little room for

An Orlando, Florida, police officer cries after telling a mother that her child died in a house fire. (*Courtesy Bobby Coker*/The Orlando Sentinel)

individualism or freedom of expression. Court decisions even allow police administrators to have residency requirements, a dress code, and so on. Pressures to conform can be severe, and some nonconforming officers cannot endure such pressure.

3. *The Minority Officer:* Both African American and women police officers suffer discrimination by virtue of entering a predominately white, male-dominated occupation.

V. Effects of Critical Incidents

Several stressful repercussions can follow a police shooting. The officer will probably be secluded from work and peers, and his or her gun, badge, and other trappings of the job will probably be taken away; the officer is suspended from work pending a full investigation.[91] Relationships with family may become strained, and the officer may experience time distortion, emotional numbing, a feeling of isolation, denial, flashbacks, sleep disturbances, legal problems, and guilt.[92] A high percentage of police officers who kill another person leave the force within five years.[93]

Effects and Management

It has been estimated that at any one time, 15 percent of a department's officers will be in a burnout phase. These officers account for 70 to 80 percent of all the complaints against their department, including physical abuse, verbal abuse, and misuse of firearms.[94] If they do not relieve the pressure they may eventually suffer heart attacks, nervous breakdowns, back problems, headaches, psychosomatic illnesses, and alcoholism.[95] They may also manifest excessive weight gain or loss; combativeness or irritability; excessive perspiration; excessive use of sick leave; excessive use of alcohol, tobacco, or drugs; marital or family disorders; inability to complete an assignment; loss of interest in work, hobbies, and people in general; more than the usual number of "accidents," including vehicular and other types; and more shooting incidents.[96] An extreme reaction to stress is suicide. Police are at a higher risk for committing suicide because of their access to firearms, continuous exposure to human misery, shift work, social strain and marital difficulties, physical illness, and alcohol problems. Police are eight times more likely to commit suicide than to be killed in a homicide, three times more likely to kill themselves than to die in job-related accidents.[97]

No human being can exist in a continuous state of alarm. The body, striving to maintain its normal state, or homeostasis, and adapt to the alarm, can actually develop a disease in the process. Often, however, the problem of stress is exacerbated by the reluctance of police to admit they have problems. They often avoid seeking professional psychological help; officers fear being placed on limited duty or being labeled "psychos" by their peers.[98] As they start to acknowledge the problem, many police departments are attempting to improve their recruit screening and provide better counseling programs.

It is imperative that officers learn to manage their stress before it causes deep physical and/or emotional harm. One means is to view the mind as a "mental bucket" and strive to keep their mental buckets full through hobbies or activities that provide relaxation. Exercise, proper nutrition, and positive lifestyle choices (such as not smoking and using alcohol with moderation) are also essential for good health.

The federal government has taken official notice of the problem of police stress. In 1997, the National Institute of Justice awarded grants to eight police agencies and organizations for devising effective stress-reduction programs.[99]

SUMMARY

This chapter has examined contemporary trends and issues in policing. Clearly, it was demonstrated that policing, after existing for more than a century and a half, still has not resolved several long-standing problems and areas of controversy. As examples, women and minorities have yet to be accepted in this occupation, even though studies have proven their tremendous value. Furthermore, higher education remains a questionable element of policing for many people, irrespective of the fact that major national commissions have been advocating this requirement for nearly three decades.

One must wonder what the early years of the new millennium will bring in these and other regards.

ITEMS FOR REVIEW

1. Review employment rights that are possessed by today's police officers, including the common provisions of state Peace Officer Bill of Rights legislation.
2. Describe some advantages and problems within private policing.
3. List some advantages of hiring officers who have completed higher education, using civilians for some police tasks, and having an accredited police agency.
4. Explain the major reasons for the development and expansion of police unions and their impact today.
5. Describe the three models of collective bargaining and what happens when there is a bargaining impasse.
6. Describe the four kinds of job actions that can occur.
7. Define the Fair Labor Standards Act and how it operates in policing.
8. Discuss some problems that are involved with recruiting and retaining women and minorities into policing and the primary issues that concern women who are already working in policing.
9. Describe some of the major causes of stress and how officers can manage it.

NOTES

1. W. Clinton Terry III, *Policing Society: An Occupational View* (New York: Wiley, 1985), p. 168.
2. *Ibid.*
3. *Ibid.*, pp. 170–171.
4. Samuel Walker, *The Police in America: An Introduction*, 3d ed. (Boston: McGraw-Hill, 1999), p. 368.
5. George F. Cole and Christopher E. Smith, *The American System of Criminal Justice*, 8th ed. (Belmont, CA: West/Wadsworth, 1998), p. 240.
6. *Ibid.*, p. 318.
7. Will Aitchison, *The Rights of Police Officers*, 3d ed. (Portland, OR: Labor Relations Information System, 1996), p. 7.
8. *Ibid.*
9. *Ibid.*
10. *Ibid.*, p. 8.
11. *Ibid.*, p. 9.
12. Charles R. Swanson, Leonard Territo, and Robert W. Taylor, *Police Administration: Structures, Processes, and Behavior*, 5th ed. (Upper Saddle River, NJ: Prentice Hall, 2001), p. 410.
13. Arnold Zack, *Understanding Fact-Finding and Arbitration in the Public Sector* (Washington, D.C.: Government Printing Office, 1974), p. 1.
14. Thomas P. Gilroy and Anthony V. Sinicropi, "Impasse Resolution in Public Employment," *Industrial and Labor Relations Review* 25 (July 1971): 499.
15. Robert G. Howlett, "Fact Finding: Its Values and Limitations—Comment," *Arbitration and the Expanded Role of Neutrals, Proceedings of the twenty-third annual meeting of the National Academy of Arbitrators* (Washington, D.C.: Bureau of National Affairs, 1970), p. 156.
16. Zack, *Understanding Fact-Finding*, p. 1.
17. Charles W. Maddox, *Collective Bargaining in Law Enforcement* (Springfield, IL: Charles C. Thomas, 1975), p. 54.
18. Swanson, Territo, and Taylor, *Police Administration*, p. 411.
19. *Ibid.*, p. 423.
20. *Ibid.*
21. L. Lund, "The 'Ten Commandments' of Risk Management for Jail Administrators," *Detention Reporter* 4 (June 1991): 4.
22. 469 U.S. 528 (1985).
23. Swanson, Territo, and Taylor, *Police Administration*, p. 310.

24. Barbara Raffel Price, "Sexual Integration in American Law Enforcement," in *Police Ethics: Hard Choices in Law Enforcement*, eds. William C. Heffernan and Timothy Stroup (New York: John Jay Press, 1985), pp. 205–214.
25. Jeanne McDowell, "Are Women Better Cops?" *Time*, February 17, 1992, p. 71.
26. Peter Horne, "Policewomen: 2000 A.D. Redux," *Law and Order*, November 1999, p. 53.
27. Brian A. Reaves and Timothy C. Hart, *Federal Law Enforcement Officers, 1998* (Washington, D.C.: U.S. Department of Justice, Bureau of Justice Statistics Bulletin, 2000), p. 6.
28. Horne, "Policewomen," p. 53.
29. *Ibid.*
30. Adapted from *ibid.*, pp. 54–60.
31. Quoted in Horne, "Policewomen," p. 59.
32. McDowell, "Are Women Better Cops?" p. 70.
33. *Ibid.*, p. 71.
34. *Ibid.*, p. 72.
35. U.S. Department of Justice, Bureau of Justice Statistics, "Local Police Departments, 1993" (April 1996), http://www.ojp.usdoj.gov/bjs/.
36. Reaves and Hart, *Federal Law Enforcement Officers, 1998*, p. 6.
37. Richard Margolis, *Who Will Wear the Badge? A Study of Minority Recruitment Efforts in Protective Services: A Report of the United States Commission on Civil Rights* (Washington, D.C.: U.S. Government Printing Office, n.d.).
38. 401 U.S. 424 (1971).
39. Robert Pursley, *Introduction to Criminal Justice* (Encino, CA: Glencoe Press, 1977).
40. Quoted in Peggy S. Sullivan, "Minority Officers: Current Issues," in *Critical Issues in Policing: Contemporary Readings*, ed. Roger G. Dunham and Geoffrey P. Alpert (Prospect Heights, IL: Waveland Press, 1989), p. 338.
41. Lena Williams, "Police Officers Tell of Strains of Living as a 'Black in Blue,'" *The New York Times*, February 14, 1988, pp. 1, 26.
42. Robert J. Kaminski, "Police Minority Recruitment: Predicting Who Will Say Yes to an Offer for a Job as a Cop," *Journal of Criminal Justice* 21 (1993): 395–409.
43. Ken Peak, "The Quest for Alternatives to Lethal Force: A Heuristic View," *Journal of Contemporary Criminal Justice* 6 (1990): 8–22.
44. *Ibid.*
45. *Ibid.*
46. Saul D. Astor, "A Nation of Thieves," *Security World* 15 (September 1978).
47. U.S. Department of Justice, Bureau of Justice Statistics, *National Crime Victimization 2000* (Washington, D.C.: Author, June 2001), p. 3.
48. Reaves and Hart, *Federal Law Enforcement Officers, 1998*; U.S. Department of Justice, Bureau of Justice Statistics Executive Summary, *Local Police Departments, 1997*, October 1999; U.S. Department of Justice, Bureau of Justice Statistics Executive Summary, *Sheriffs' Departments, 1997*, October 1999; U.S. Department of Justice, Bureau of Justice Statistics Bulletin, "Census of State and Local Law Enforcement Agencies," June 1998.
49. Cole and Smith, *The American System of Criminal Justice*, p. 241.
50. *Ibid.*
51. K. Peak and K. Braunstein, "On Guard: The Private Security Industry in America," in *Introduction to Criminal Justice: Theory and Practice*, ed. Dae Chang and Michael Palmiotto (Wichita, KS: midContinent Academic Press, 1997), pp. 281–300.
52. *Ibid.*
53. Lawrence J. Fennelly, ed., *Handbook of Loss Prevention and Crime Prevention*, 2d ed. (Boston: Butterworths, 1989).
54. *Ibid.*, in foreword.
55. Cole and Smith, *The American System of Criminal Justice*, p. 245.
56. James Kakalik and Sorrell Wildhorn, *Private Police in the United States: Findings and Recommendations* (Washington, D.C.: U.S. Government Printing Office, 1972), p. 155.
57. Thomas T. Chuda, "Taking Training Beyond the Basics," *Security Management* 39 (February 1995): 57–59.
58. National Advisory Commission on Criminal Justice Standards and Goals, *Private Security* (Washington, D.C.: U.S. Government Printing Office, 1976), p. 99.
59. Karen M. Hess and Henry M. Wrobleski, *Introduction to Private Security*, 3d ed. (St. Paul, MN: West, 1992).

60. Kenneth J. Peak and Ronald W. Glensor, *Community Policing and Problem Solving: Strategies and Practices*, 2d ed. Upper Saddle River, NJ: Prentice Hall, 1999), p. 53.
61. Steven M. Cox, *Police: Practices, Perspectives, Problems* (Boston: Allyn and Bacon, 1996), p. 90.
62. *Ibid.*
63. *Ibid.*
64. Albert Deutsch, *The Trouble with Cops* (New York: Crown Publishers, 1955), p. 122.
65. Herman Goldstein, *Policing a Free Society* (Cambridge, MA: Ballinger Publishing Company, 1977), pp. 283–284.
66. Deutsch, *The Trouble with Cops*, p. 213; *Law Enforcement and Criminal Justice Education: Directory 1975–76* (Gaithersburg, MD: International Association of Chiefs of Police, 1975), p. 3.
67. Law Enforcement Assistance Administration, *5th Annual Report, Fiscal Year 1973* (Washington, D.C.: GPO, 1973), p. 119.
68. National Advisory Commission on Criminal Justice Standards and Goals, *Police* (Washington, D.C.: GPO, 1973), p. 369.
69. Gerald W. Lynch, "Why Officers Need a College Education," *Higher Education and National Affairs* (September 20, 1986): 11.
70. *Davis v. City of Dallas*, 777 F.2d 205 (5th Cir. 1985).
71. Robert E. Worden, "A Badge and a Baccalaureate: Policies, Hypotheses, and Further Evidence," *Justice Quarterly* 7 (September 1990): 565–592.
72. Victor E. Kappeler, Allen D. Sapp, and David L. Carter, "Police Officer Higher Education, Citizen Complaints and Departmental Rule Violations," *American Journal of Police* 11 (1992): 37–54.
73. Charles L. Weirman, "Variances of Ability Measurement Scores Obtained by College and Non-College Educated Troopers," *The Police Chief* 45 (August 1978): 34–36.
74. *Ibid.*
75. Robert Trojanowicz and T. Nicholson, "A Comparison of Behavioral Styles of College Graduate Police Officers v. Non-College Going Police Officers," *The Police Chief* 43 (August 1976): 56–59.
76. A. F. Dalley, "University and Non-University Graduated Policemen: A Study of Police Attitudes," *Journal of Police Science and Administration* 3 (1975): 458–468.
77. Wayne F. Cascio, "Formal Education and Police Officer Performance," *Journal of Police Science and Administration* 5 (1977): 89–96; Bernard Cohen and Jan M. Chaiken, *Police Background Characteristics and Performance* (New York: RAND, 1972); B. E. Sanderson, "Police Officers: The Relationship of College Education to Job Performance," *The Police Chief* (August 1977): 62–63.
78. James W. Sterling, "The College Level Entry Requirement: A Real or Imagined Cure-all?" *The Police Chief* 41 (April 1974): 28–31.
79. Gerald W. Lynch, "Cops and College," *America* (April 4, 1987): 274–275.
80. Jon Miller and Lincoln Fry, "Reexamining Assumptions about Education and Professionalism in Law Enforcement," *Journal of Police Science and Administration* 4 (1976): 187–198.
81. John K. Hudzik, "College Education for Police: Problems in Measuring Component and Extraneous Variables," *Journal of Criminal Justice* 6 (1978): 69–81.
82. Quoted in "Men and Women of Letters," *Law Enforcement News*, November 30, 1997, p. 1.
83. National Advisory Commission on Criminal Justice Standards and Goals, *Police*, p. 367.
84. John Bartlett, ed., *Familiar Quotations*, 16th ed. (Boston: Little, Brown, 1992), p. 529.
85. William A. Westley, *Violence and the Police* (Cambridge, MA: The MIT Press, 1970), p. 3.
86. Gordon Witkin, Ted Gest, and Dorian Friedman, "Cops Under Fire," *U.S. News and World Report*, December 3, 1990, pp. 32–44.
87. Michelle Ingrassia and Karen Springen, "Living on Dracula Time," *Newsweek*, July 12, 1993, pp. 68–69.
88. See, for example, Stephen R. Band and Donald C. Sheehan, "Managing Undercover Stress: The Supervisor's Role," *FBI Law Enforcement Bulletin*, February 1999, pp. 1–6.
89. See, for example, J. Sewell, "The Stress of Homicide Investigations," *Death Studies* 18 (1994): 565–582.
90. Adapted from Gerald W. Garner, "Prepare to Survive," *Police*, January 1994, pp. 18–19, 90.
91. Anne Cohen, "I've Killed That Man Ten Thousand Times," in *Annual Editions: Criminal Justice, 88/89* eds. John J. Sullivan and Joseph L. Victor (Guilford, CT: The Dushkin Publishing Group, Inc., 1988), pp. 86–90.
92. John G. Stratton, *Police Passages* (Manhattan Beach, CA: Glennon Publishing Co., 1984), pp. 235–236.

93. James M. Horn and Roger M. Solomon, "Peer Support: A Key Element for Coping with Trauma," *Police Stress* 9 (Winter 1989): 25–27.
94. Ben Daviss, "Burnout," *Police Magazine*, May 1982, p. 10.
95. *Ibid.*, p. 50.
96. Gerald L. Fishkin, *Police Burnout: Signs, Symptoms and Solutions* (Gardena, CA: Harcourt Brace Jovanovich, 1988), pp. 233–239.
97. "What's Killing America's Cops?" *Law Enforcement News*, November 15, 1996, p. 1.
98. *Ibid.*
99. "Stressed Out? Help May Be on the Way," *Law Enforcement News*, January 31, 1997, p. 5.

Comparative Perspectives

Key Terms and Concepts

Holy Koran	Royal Ulster Constabulary
Interpol	Taazir, Quesas, Diyya offenses
IRA	Ulster
The law of Sharia	Writ of *amparo*

Courts and camps are the only places to learn the world in.
—*Earl of Chesterfield*

INTRODUCTION

Crime is an international problem. Life in general and crime in particular have become globalized. Crime easily transcends national borders, and the fruits of those crimes are also dispatched with ease across geographical boundaries, to be laundered and used in other criminal enterprises. In an era of rapidly advancing technology and mobility, the world order has become interdependent, and crime and justice issues are clearly transcontinental.[1]

It has therefore become increasingly important to understand the structure and function of international criminal justice systems. It has been said that the comparative approach provides the opportunity to search for order.[2] A comparative view of foreign police systems also provides the opportunity to better assess the methods, role, and functions of our democratic American policing system. We can also look for the common properties of all systems and determine where each political system and, consequently, each policing system stands in its regard for human rights and the extent it intrudes into the private lives of its citizens.[3]

We will look at policing in four countries: China, Saudi Arabia, Northern Ireland, and Mexico. One might ask, "Why were those countries chosen? What do they have to do with policing in the United States?" The answer is that they actually have little in common with the system of policing in our country, and that is precisely why they were chosen. These countries all employ police systems that are uniquely different from ours, making them ideal for comparison; also, it is hoped that an understanding of these systems will give the reader a deeper appreciation for policing methods in a democracy.

China is included in this overview because of its cultural emphasis on self-policing and its harsh treatment of offenders; Saudi Arabia was selected for similar reasons as well as its unique political and social structures and its reliance on religion for providing the basis for law and punishment. Northern Ireland is discussed because of its unique religious, political, and terrorist-group influences over policing. Police officers there, caught in a civil war, are prime targets of terrorist bombs; thus, officers conduct themselves both professionally and personally in a very different way than do police in more stable locations. Finally, Mexico is included because of its obvious proximity to the United States, the joint Mexican–U.S. drug interdiction efforts, and the frequency of encounters U.S. tourists have long had with the Mexican authorities. (A related problem—the daily attempts of Mexican citizens to cross into American territory illegally—is discussed in Chapter 7.)

Finally, we analyze the International Criminal Police Organization, or Interpol, the most effective international effort at crime fighting in the world. Included are the history, organization, and methods of this unique organization.

POLICING IN CHINA

POLICING A VAST LAND

China is a vast country, encompassing 3.7 million square miles (sharing borders with 14 other nations) and about 1.2 billion people—the world's largest population.[4] The Chinese police bear the responsibility for maintaining law and order in the world's most populous nation.

Although there is no official estimate as to the numbers of police or police-to-population ratio in China, unofficial estimates put the ratio at 1:745 to 1:1,400. By

Chinese police officers.
(*Courtesy OICJ*)

whatever estimate, the ratio is much lower than that in almost all Western nations. Even so, since the founding of the People's Republic in 1949, the Chinese police have maintained one of the lowest crime rates in the world.[5]

REFORM UNDER POLICE LAW 1995

Since the early 1980s, China has implemented a series of police reforms. But the most significant is the passage of Police Law 1995, which contains specific provisions regarding police organization, duties, recruitment and training, powers, accountability, and styles. Next we briefly discuss each of these provisions.

With respect to *police organization*, according to Police Law 1995, the Chinese police consist of five components: public security police, state security police, prison police, judicial police in people's courts, and judicial police in people's procuratorates (prosecutors' offices). Public security police are the largest and oldest police force, performing a wide range of ordinary police duties as well as administering the household registration system. When people talk about police in China, they usually think of the public security police.[6] At the helm of this police force is the Ministry of Public Security, which represents the central government and directs and operates police operation throughout the country. Figure 13–1 shows the organization of public security police in China.

State security police are responsible solely to protect state security—to prevent foreign espionage, sabotage, and conspiracies. Prison police supervise convicted offenders, make general regulations with regard to prison management,

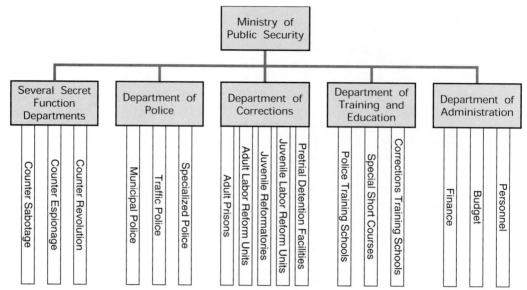

FIGURE 13–1 Organization of the Chinese Police. (Source: Richard A. Myren, "The Developing Legal System of China," *CJ International 2* (March–April 1986), p. 13. Reproduced with permission.)

and make policies concerning supervision and rehabilitation of inmates. Judicial police are officers who work in both people's courts and people's procuratorates; they provide physical security, serve subpoenas, conduct searches, and execute court orders (including death sentences).

Because of the relatively low police-to-population ratio and China's increasing problems with law and order, police face an urgent need with respect to *recruitment and training*. Recently the quality of new recruits has been deemed less than satisfactory, resulting in incidences of police corruption, abuse of powers, and other misconduct.[7] Police Law 1995 set out basic qualifications of police officers: One must be over 18 years old, support the Constitution, have good character, be in a state of excellent physical condition, and be at least a high school graduate. Qualified applicants must take a competitive entrance examination and complete a one-year probationary period, during which time training is received in an academy. The Police Law sets out higher qualifications for those who seek to hold leadership positions: to do so, one must have a college degree, legal knowledge, practical police experience, administrative talent, management skills, and requisite in-service training.[8]

Powers and functions held by the Chinese police are much broader than those of their counterparts in many other nations; indeed, the list of functions of China's police appears to represent a combination of the kinds of work found in an array of federal and state law enforcement agencies in the United States. In addition to routine police duties of investigation and traffic control, the Chinese police also engage in fire prevention; controlling firearms, ammunition, knives, and explosive and radioactive materials; supervising the operation of certain types of professions and industries; guarding high-ranking government officials, dignitaries, buildings, and facilities; administering the household registration system as well as handling entry into or exit from the country; maintaining borders; supervising offenders on parole; supervising the security and protection of computer information; and guiding and supervising the security of government offices and social organizations.[9]

The Chinese police's wide powers have been a subject of controversy and criticism, stemming from the fact that these police officers are given wide detention power, which can result in an individual's loss of freedom for months or even years. Although the Criminal Procedural Law (CPL) regulates the power of police to arrest and detain without judicial or prosecutorial approval, these regulations are nevertheless compromised by the wide detention power of the police. Administrative detention can be used by the police for up to 15 days for investigation or interrogation; this is a very powerful weapon in the police arsenal. The use of physical torture to force confessions is widely reported. But the most disturbing problem is extended detention, which in many cases runs for more than three months. After taking suspects into custody, police often do not engage in a diligent investigation. The police may also send offenders to farms or camps for reeducation through labor for up to three years; this measure is the most severe administrative sanction police may take against an offender without court approval.[10]

Police Law 1995 also granted the police a new power: If police have reasonable suspicion to believe that a person has committed a crime, they may stop and question that person after identifying themselves. If their suspicions are not allayed, they may then take the individual to a police station for further questioning for up to 24 hours—and up to 48 hours under "exceptional circumstances."

Police accountability is obviously a matter of concern in China. The Chinese police in general enjoy a positive image in the eyes of the public; one national survey showed that 70 percent of the public expressed confidence that if they reported personal victimizations to the police, their complaints would be addressed.[11] The Chinese police, however, are not immune from misconduct; there have been increasing numbers of reported cases of police corruption, abuse of power, and infringement of citizens' rights. The Public Law also established an internal police supervisory system, however, providing that higher-level police agencies must oversee and inspect the legality of law enforcement activities of lower-level police agencies. The law also established an internal supervisory system within each police bureau; a committee has the authority to receive and investigate complaints against officers and to impose disciplinary sanctions against them as necessary.

The *policing styles* and strategies have also been transformed in order to deal with the country's changing social and economic environment. The key to the success of the strategies remains with the police's ability to keep a tight control over the population through the household registration system and extensive surveillances. Every citizen must register his or her residence in a locality with the police; neighborhood committees are established in all neighborhoods. This system has the advantage of enabling officers to keep close contacts with community residents and to keep a tight surveillance on every neighborhood. No strangers can enter a neighborhood without being noticed immediately and reported promptly to the police.[12] Although this might seem quite intrusive to Americans, the system does allow the police to monitor large city populations and criminals, especially gangs.

Community policing (examined in Chapter 6) has also worked well in China; to combat the current rise in crime, the government advocates an overall strategy known as "comprehensive management"—the mobilization of all possible social forces to strengthen public security and prevent crime.[13]

POLICING IN SAUDI ARABIA

SOCIAL BEHAVIOR IN A PATRIARCHAL LAND

The kingdom of Saudi Arabia is the largest country in the Middle East and occupies four-fifths of the Arabian Peninsula—873,000 square miles. It covers an area roughly the size of the United States east of the Mississippi, about one-third of America's size.

EXHIBIT 13–1

Changing Bad Cops to Good

Faculty at John Jay College of Criminal Justice in New York City have developed a course meant to reverse the effects of years of bad police practices in countries with authoritarian rule. Since 1994, more than 3,000 police officers in 50 troubled countries have taken the course "Human Dignity and Policing," funded by the U.S. Department of Justice. Realizing that straight lectures cannot change these international students, the course emphasizes group therapy and role playing to help them confront themselves and their past. In Bosnia, police noted that their superiors strip-searched them at the end of each shift to take whatever bribes they had collected. In Latin America, officers were regularly forced to do menial work for their bosses, such as building houses. An El Salvador officer admitted that his superiors told him a prisoner he was escorting should be killed—which he did, receiving a hero's medal for murder. These officers need to return to their homelands wanting to do good for others; as a college dean says, "All cops want to have their jobs mean something."

Source: Adapted from "Teaching Cops Right From Wrong," *Time*, March 19, 2001, p. 59.

The government is patriarchal with attributes of democracy. At a given hour of the day, the king makes himself accessible to his subjects. Sitting as a supreme court of appeal, he hears the complaints brought before him by even the humblest members of society. The king's decisions cannot be appealed.

Arabic is the language of the country, but many of the younger educated businessmen were educated in English-speaking countries; thus, English is widely spoken in business and the military. The population in most of the cities of Saudi Arabia is a cosmopolitan mixture of Arabs from all over the Muslim world. The family structure is undergoing marked changes. Monogamy is becoming the norm, even though Saudi Arabian men still have the right to take four wives and divorce is comparatively simple. The social life is also very much male oriented—women still wear the veil and must be covered from ankles to wrists in public to avoid calling attention to themselves. Women are not allowed to drive or work, and it is still considered a primary function for women to bear a son.[14]

Westerners will find some matters of Arabian social etiquette to be curious, particularly those forms of etiquette relating to Arab courtesy and hospitality. For example, to admire an Arab's possessions is to put him under obligation to present the item to the visitor as a gift; to refuse the offer would be to offend the host. A famous Arab hospitality is the serving of coffee, black and sweetened in tiny cups and poured from a brass pot; it is customary to accept two cups. Handshaking is more frequent than in the West, and before discussing business, Arabs will normally indulge in sincere small talk about each other's health, work, and interests; undue haste in a business transaction is considered bad form. There are innumerable anecdotes about Arabs who gave their last morsel to feed a guest. One popular story involves a poor Bedouin who had endured the loss of his flocks until only one horse was left. A man, who had long coveted the horse and offered to buy him, visited one day to attempt the purchase and was invited by the Bedouin to dine. After eating, the man offered again to buy the horse from the poor Bedouin, who listened attentively and then replied, "I thank you for your generosity, but we have just consumed my horse!"[15]

Westerners are considered a temporary commercial colony in Saudi Arabia today and are generally concentrated in Jidda, Riyadh, Taif, and Khamis Mushayt. Many Americans work in oil-related and other business and commercial enterprises, and thus they must have schools and other amenities for their children and families.

RELIGIOUS UNDERPINNINGS

To describe Saudis is to speak of their religion, culture, and customs, all of which are bound closely together. The national flag of Saudi Arabia, bearing the Islamic creed and unsheathed sword, symbolizes strength rooted in faith. Saudi Arabia is the center of the Islamic faith and contains its two holiest cities, Mecca (Mohammed's birthplace) and Medina (his burial place). Mecca is the focal point of religious devotion; five times daily, Moslems throughout the world turn toward the holy city to pray.

Saudi Arabian police mannequin, displayed at the International Police Museum, Central Police University, Taoyuan, Taiwan.

Few acts are performed that do not have their basis in the **Holy Koran**. *Islam* means complete submission to the will of God. Five primary duties, known as the Five Pillars of Faith, are required of every Muslim: the profession of faith (a statement of loyalty to God and Mohammed, printed on the country's flag); prayer; almsgiving (helping poor Muslims); fasting (during Ramadan, the ninth month of the Muslim year); and pilgrimage (for those who are able to make the *hajj*, or pilgrimage, to Mecca or Medina, the second holiest city).[16]

LAWS AND PROHIBITIONS

To provide a well-ordered society, the interpretation of Islamic law is stricter in Saudi Arabia than in any other Islamic country. Saudi Arabian law, like other facets of Saudi life, is governed by the Koran. If an American or any other foreigner breaks the law in Saudi Arabia, there is little anyone can do to assist; he or she must expect to receive the full penalty for any offense. In fact, in March 1993, the U.S. Supreme Court held that a U.S. worker could not sue the Saudi government in an American court for torture and other abuses received at the hands of Saudi authorities.[17] There, the appellee, Scott Nelson, charged that he had been detained and tortured by Saudi police for 39 days for complaints about safety he issued while working in a Saudi-run hospital. Writing for the majority, Justice David Souter observed that the police power of a nation was "particularly sovereign in nature" and was sheltered from lawsuits in U.S. federal courts according to the Foreign Sovereign Immunities Act, a 1976 law governing suits against foreign states.

Alcohol is forbidden in Saudi Arabia; those persons who brew, sell, smuggle, or drink alcohol are punished. Such infractions will result in the police conducting an investigation to determine the source of the alcohol. There are three kinds of narcotics offenses: possession, trafficking, and smuggling. Possession includes small amounts of hashish or marijuana, for which the maximum punishment is two years' imprisonment. An individual arrested for trafficking faces a maximum sentence of five years in prison. Possession of large amounts of narcotics carries a possible death sentence by beheading in public. Many drug offenders are deported.[18]

The law of Islam (the **Sharia**) is recognized as the fundamental code in Saudi Arabia. Islamic criminal legislation is divided into three categories: crimes against the divine or God's rights—Hudud crimes, **Quesas** and **Diyya** crimes—those against the individual, and **Taazir** crimes—those that are left undetermined by religious law. Hudud offenses harm the property or security interests of society. The seven types of Hudud crimes are theft, slander (falsely accusing someone of adultery or defamation of a married woman), adultery, highway robbery, consuming alcohol, transgression (revolting against a legitimate religious leader, known as an imam), and apostasy (renouncing Islam).

For the crime of theft, even for the first offense, the penalty may be amputation of the left hand at the wrist. The penalty for slander is flogging, usually with 80 lashes, and the same penalty may be applied for consuming alcohol. The penalty for adultery is flogging 100 times. A woman who engages in adultery is liable to flogging or burial to the waist in a pit; stoning may follow. Highway robbery is punishable by execution or crucifixion, the amputation of opposite hands and feet, or exile from the land. Transgression will be confronted by the Saudi armed forces until the foes of the imam are defeated, and apostasy carries the death penalty.[19]

Public executions are commonplace in Saudi Arabia. The country's public executioner has carried out more than 600 executions since 1989. Hundreds of worshippers, including children and women, often gather to observe these and other punishments. The crowd usually applauds after the execution and some bystanders spit on the blood of the dead persons and curse them.[20]

Quesas and Diyya crimes involve compensation (Diyya) from the offender to the victim and normally involve acts of murder or intentional or unintentional bodily injury or maiming. Even murder, with the consent of the victim's family, may be punished by compensation.[21] Taazir offenses—a broad category meant to include those crimes that are harmful to the public order—carry penalties that are left to the discretion of the judge or other public authority.

GUARDIANS OF RELIGIOUS PURITY

Due in large measure to the harsh criminal code and its penalties, there is comparatively little for the Saudi police to do in terms of crime prevention or investigation, nor are they actively engaged in random patrolling or traffic enforcement. Thus, there is ample time for other endeavors. The religious police—Mutawin—wear short, coarse, white cotton robes and sandals; as one writer observed, they look like "desert nomads who have stumbled unexpectedly into the 20th century."[22] These police patrol in jeeps and stroll through the streets and malls looking for people who are improperly dressed or women who have a loose strand of hair falling across their face or who need to adjust their *tarhas* (head coverings).

For the religious police—officially known as the Committee for the Promotion of Virtue and the Prevention of Vice—such encounters are all part of a day's work. While these minions of the law view their corrective actions as a small victory for Islam, many citizens view the police actions as nothing more than harassment. The police are currently at the heart of a fierce debate about the country's future, one that pits a mostly Western-oriented elite against those who want a stricter Islamic rule. In an effort to draw the Muslim fundamentalist opposition into the system and bolster its own religious credentials, the government has steadily expanded the scope and power of the police in the last few years.[23] This expansion of powers does not bode well for those persons wanting to see liberal democratic reforms.

Around-the-clock patrols ensure that shops are closed in time for daily prayers and that only married couples are sitting in family sections of restaurants. The squads often follow persons suspected of being involved in what is deemed immoral behavior, such as drug use, homosexuality, gambling, and begging. Teams of religious police destroy home satellite dishes, which bring uncensored Western television broadcasts into Saudi homes. For young men and women, social life often becomes a cat-and-mouse game with the roving bands of religious police. The religious police, who are often accompanied by uniformed police officers, are viewed with open hostility by many Saudi youth; the police are seen as the enemy. University students in Jidda hang out at a shopping mall next to campus, a place that is frequently raided by special squads. But as evidence of their effectiveness, the police point to their ordered society, where violent crime and theft are rare.[24]

The religious police have a broad scope of authority in a country dominated by Wahabism, a puritanical sect of Islam embraced by the ruling Saud family and most of the population. Under this sect's influence, in addition to alcohol, theaters and many Western publications are banned. Women and men are segregated in workplaces, restaurants, and schools. Wahabism views daily life as a war against vice, including an "infusion" of imported pornographic videos. Police officials acknowledge that their officers commit some abuses—such as beating some guest workers from foreign countries—but they deny any widespread mistreatment.[25]

The police also assist with the so-called chop-chops, which are conducted in many cities and with greater frequency on the Justice Square in Riyadh on Fridays. Believing that these public amputations are a deterrent to other would-be offenders, persons—including Americans—who are in the vicinity of the Square are encouraged by the police to witness these affairs of state; thus, one may find Riyadh's streets to be largely empty on Fridays. When a thief's right hand is cut off in public, a string is tied to the middle finger and the hand is hung from a hook high on a streetlight pole on Justice Square for all to see; this announces the thief's transgression and casts shame on the family.[26]

Notwithstanding the harsh criminal code, forms of punishment, and police presence, many people residing in Saudi Arabia report feeling safer than they do living in the United States, largely owing to the absence in Saudi Arabia of gangs, drive-by shootings, purse snatchings, and forms of contraband (such as drugs, alcohol, and weapons) on the streets. Many Americans report that their biggest fear while living in Saudi Arabia is being caught by the police for improper dress. The police busily patrol the streets in Chevrolets, BMWs, and Volvos, looking for such infractions. The police also check vehicles for bombs near the U.S. Embassy building. Although there has been no such problem for many years, a concern with terrorists keeps the police alert.[27]

POLICING IN NORTHERN IRELAND

RECENT DEVELOPMENTS IN A LONG CIVIL WAR

About 3,200 people have been killed in Northern Ireland since 1970; fathers have been shot in front of their children and children have been blown up while playing in their own backyards.[28] More than 40,000 people were injured and at least 15,000 families driven from their homes in Belfast in the 1970s alone (more figures concerning the carnage—including that which has been directed toward the police—are presented later). As a result of the violence, Ireland is one of the most segregated countries in the world: two groups of people—speaking the same language, having the same skin color, and sharing a Christian heritage—regard each other as separate races.[29]

On April 10, 1998—Good Friday—voters in Ireland approved referendums that, it was hoped, would end the civil conflicts, creating a new regional legislature, a cross-border council, and an eventual repeal of articles in the Irish constitution that committed the Dublin government to seek a united Ireland.[30] The pact also allowed the early release of the country's 500 convicted terrorists.[31]

But both sides have extremists, and dissidents vowed to continue their violence.[32] Indeed, the carnage continued: In July 1998, two months after the pact, protests against the ban of a planned march through a Catholic neighborhood near Belfast led to the burning of 10 Roman Catholic churches by pro-British loyalists. In related rioting, Protestant mobs injured 42 police officers of the Royal Ulster Constabulary, 314 vehicles were set on fire or damaged, 110 homes and buildings were vandalized, and 63 rioters were arrested.[33] Then, in August 1998, the single deadliest blast in Northern Ireland's decades of conflict occurred via a car bomb in the center of a bustling market town, Omagh, killing more than 30 people and injuring more than 200.[34]

The next notable development occurred in November 1999, when, pursuant to the 1998 Good Friday peace pact, a new 12-member cabinet was formed with equal numbers of Protestants and Roman Catholics. The formation of the Cabinet, composed of six pro-British Protestants and six republican Catholics who want to break from the British, would have allowed—for the first time—Northern

Ireland's Protestants and Catholics to share power over local affairs.[35] However, just three days later, Britain stripped the new Cabinet of its power.

Then, in July 2000, the peace process seemed to tumble further into chaos. In a week of rioting, thousands of hard-line Protestants lit towering bonfires and fired volleys of shots in connection with marches and parades across Northern Ireland. The police arrested 146 people and seized 1,000 gasoline bombs. Nearly 100 vehicles were hijacked and burned, and 57 officers were wounded.[36] It is clear that the political process that created a power-sharing government in Northern Ireland remains on course, although a shaky one.

A DIVIDED LAND

Since 1921, Ireland has been divided into two political units; Northern Ireland, with 1.5 million people and 5,241 square miles, and the Republic of Ireland, with 2 million people and 21,895 square miles. Northern Ireland is composed of the six northeastern counties of Ireland; this area, often referred to as **Ulster**, enjoys a measure of self-government within the United Kingdom of Great Britain. As with the other international venues discussed in this chapter, one must first

Northern Ireland's Royal Ulster Constabulary. (*Courtesy OICJ*)

understand some of the political history of Ireland before attempting to grasp the policing system that has evolved.

In the twelfth century, the pope gave all of Ireland to the English crown. In 1171, Henry II of England was acknowledged Lord of Ireland, and the English monarchy's control over the entire island became absolute in the seventeenth century with a Protestant military victory over the Catholics in 1690. In 1801, England and Ireland became the United Kingdom of Great Britain and Ireland. This entire period was in fact characterized by several centuries of anti-British agitation and demands for Irish home rule. The advent of World War I delayed the institution of home rule and resulted in the Easter Rebellion in April 1916, in which Irish nationalists unsuccessfully attempted to end British rule. The Government of Ireland Act of 1920 attempted to give self-government to Ireland and prohibit control of Ulster by the Roman Catholic majority of southern Ireland. This plan was accepted in Northern Ireland but was rejected by the south. Nevertheless, the act continued to serve as the basis of a constitution for Northern Ireland. The essential feature of the constitution is that Northern Ireland is a province of the United Kingdom. The Irish Republican Army (IRA) has argued since 1920 that its right to kill people comes from this partition of Ireland into two illegitimate states.[37]

POLITICAL FACTIONS AND VIOLENCE

The Irish Free State (Eire) was established by King George V in June 1921. From that point on, there was civil war in Ireland that included acts of bombing, shooting, and incendiarism. In 1922, 232 persons were killed and 1,000 were injured. In 1949, the Republic of Ireland withdrew from the British Commonwealth. This gave Northern Ireland greater status within the United Kingdom but further soured its relationship with the Irish Republic.

During and prior to this time, most Irish Protestants regarded themselves as British and valued their membership in the United Kingdom. They were and are opposed to a united Ireland. Thus, inhabitants of Northern Ireland, who are mostly Protestant, of British ancestry, and in the majority, have historically maintained the strongest opposition to the return of home rule and Irish self-government. Most Catholics regard themselves as Irish and cherish the prospect of a united Ireland. Roman Catholics in all of Ireland still hold a fervent desire to become an independent nation.

Terrorist groups seek to bring Northern Ireland into the Republic through violent methods. Many Protestant Loyalists (who are pro-British) consider the police to be traitors to their cause and have employed terrorist attacks against them. Some of the predominant terrorist groups in this camp are the Ulster Defense Association, the Ulster Freedom Fighters, and the Ulster Volunteer Force. The majority of the terrorist murders and other acts of violence have been carried out by the **IRA**, the secret terrorist group comprised mainly of Catholics. Other groups in this camp include Sinn Fein, the political mouthpiece of the IRA, and the Irish National Liberation Army.

Beginning in 1963, the prime minister of Northern Ireland made several attempts to establish cordial relationships with the Republic of Ireland; the Republic, however, refused to recognize the Northern Ireland government. A civil rights campaign was inaugurated in Ulster in 1968, and demonstrators collided with the police and Protestant extremists. Catholics, who comprise one-third of the population of Northern Ireland, complained of housing and voting discrimination, as they had done for many decades.

In April 1970, the controversial Ulster Special Constabulary was replaced by a part-time Ulster Defense Regiment. The Constabulary, usually referred to as the B-Specials, were commonly regarded by Catholic terrorists as a militant Protestant organization with the primary goal of keeping the British in power. Even with their replacement, IRA street riots and killings continued to increase.

ATTACKING THE ROYAL ULSTER CONSTABULARY

Since 1922, it has been the task of the **Royal Ulster Constabulary (RUC)**—the Northern Ireland police force—to combat acts of terrorism and other forms of criminal activity. (Contrary to what may seem to be the case, acts of terrorism have never accounted for more than one-third of all crime in Northern Ireland.) These acts of terrorism have brought a bunker attitude to the RUC, who view themselves as fighting a terrorist war. They must guard against acts of terrorism directed against themselves as well. Most, if not all, officers refuse to allow their children to answer the door at their homes, and before leaving for work in the morning, the officers routinely check their vehicles for bombs.[38]

It would probably be impossible to find a more dangerous jurisdiction in which to be a police officer. From 1969 to 1990, 267 RUC regular and reserve police officers were killed and 6,598 were injured by terrorist acts. Furthermore, 422 British soldiers and 1,910 civilians died at terrorists' hands, and nearly 26,000 soldiers and civilians were injured.[39] These figures would undoubtedly be much higher were it not for the vast amounts of munitions and people seized by the RUC during that period: 1,337,506 bullets, 10,050 firearms, 95.5 tons of explosives, and 14,116 people charged with terrorist activities.[40]

With all of this carnage, it is difficult to believe that the RUC was once an unarmed police organization. On the recommendation of a special committee, the RUC experimented with disarmament in 1971. After several terrorist attacks, however, it was hastily allowed to rearm itself. Paradoxically, the RUC still basically considers itself an unarmed force that is temporarily armed. In reality, the RUC is one of the most heavily armed police organizations in the world.[41]

Today there are approximately 8,300 RUC regular officers and 4,700 reserve officers (of whom 3,000 are full-time). It quickly becomes clear that the RUC is in a precarious position, locked between two groups of paramilitary extremists. As one RUC officer stated, they are truly "the piggies in the middle."[42]

Many officers have been attacked by the pro-British Protestant Loyalists, and some officers have had their cars bombed or homes burned. If a plainclothes officer happens to be in the "wrong" neighborhood (Catholic areas known as "hard green" neighborhoods that are hotbeds of terrorist activity) and happens to be recognized, his or her life is immediately in danger. The type and method of police patrol varies, but most patrol officers wear flak jackets and carry carbines; they often patrol with an armored Land Rover. The danger of attack always exists.[43]

Terrorists understand that the RUC will always respond to emergency calls from citizens. Therefore, a typical ambush involves a terrorist takeover of two houses directly in line of fire from each other. A resident in one home is then forced to call an RUC station and report some crime. When the police arrive, they are fired upon from the other residence.[44] Civilians have often felt the savagery of terrorism as well. British prime minister Margaret Thatcher was almost the victim of an IRA bomb at a Brighton hotel in 1984.[45]

Terrorists employ other forms of savagery besides bombing. One man in Northern Ireland had his tongue stapled to a table, and another had his kneecaps pulverized by an electric drill. Both Protestant paramilitants, who are battling to keep Northern Ireland British, and Irish nationalists, who are seeking unification with the Irish Republic, use such attacks for discipline and control. The favorite punishment for informers is "knee-capping," where the victim is shot in the knees. So frequent has this form of punishment been used that Belfast surgeons have developed sophisticated surgical techniques to save victims, and in

An RUC roadblock in West Belfast, Northern Ireland.
(*Courtesy Ron Glensor*)

Limerick, a factory produces plastic kneecaps.[46] RUC data show that from 1972 to 1990, there were 1,497 cases of knee-capping in Northern Ireland; another 151 people were tarred and feathered for their "crimes."[47]

The RUC's banning and rerouting of traditional parades have also sparked numerous attacks on officers and their homes; during the 1985 marching season in Northern Ireland (from May to August) when up to 1,500 parades were scheduled, 260 officers were injured.[48]

POLICING THE TERRORIST WAR

Indeed, there are some unique twists in the terrorism war. First, the RUC is much more successful in capturing or killing Loyalist (Protestant) terrorists than it is with Republican (IRA) terrorists, but the RUC is still viewed as being in the Loyalist camp. It is also much easier for the RUC to get intelligence information from the Protestant side. Second, statistics reveal that most Catholic terrorist acts kill other *Catholics*. About 17 percent of the RUC is comprised of Roman Catholic officers. Many of them are viewed as traitors and can never return to their childhood neighborhoods.[49]

The RUC has been granted considerable power in investigating terrorism. Suspects of any crime may be held up to 24 hours and longer, if a court order is issued. Suspected terrorists, however, may be held for up to seven days, primarily to isolate them from other free terrorists. Most terrorism suspects have been found difficult to interrogate; they simply do not acknowledge the interrogator's presence and remain silent during the questioning. Those charged with terrorist-type offenses are tried by a judge only; this is necessary because in the past jurors have been intimidated and witnesses threatened and murdered. If a terrorist is convicted, he or she has an automatic appeal to a higher court, where three judges will reexamine the case and the decision and sentence of the lower court and consider any new evidence. Convicted terrorists may be sentenced to specific terms of detention, which is served in modern, well-equipped prisons. Sentences under five years can be cut in half, while those of five years or more can qualify for a one-third reduction. Prisoners serving life sentences—usually for murder, as the death penalty has been abolished in the United Kingdom—are given a thorough review every year.[50]

ORGANIZATION OF THE RUC

Notwithstanding its uniquely dangerous situation, the RUC is recognized by police experts and terrorists alike as one of the world's most experienced and best police forces. With 13,000 personnel (4,500 of whom are full- or part-time sworn officers), the department adheres to high standards of professionalism and behavior. To be hired, one must be 18 to 30 years of age, in good health (tattoos may disqualify one), and a British subject or a native of the Republic of Ireland. Entrance exams include arithmetic, spelling, verbal reasoning, and intelligence examinations. All police officers serve a two-year probationary period as police

A fortified police station in West Belfast, Northern Ireland.
(*Courtesy Ron Glensor*)

constables—two years of intensive training in the classroom and on the street before confirmation as full-fledged officers. One may achieve a variety of police ranks, beginning at sergeant and rising to chief constable. Officers eventually may specialize in traffic, investigation, community relations, communications, or canine work. After 25 years of service (if one can successfully elude the terrorists), a pension equal to one-half of the final year's salary is payable at age 50. Full-time officers are employed on three-year contracts.[51]

By international standards, the ratio of police to population is relatively high—5 per 1,000. It is even higher (7 per 1,000) if full-time members of the police reserves are included in the calculation. The RUC is organized into eight departments, which are designated by alphabetical symbols. *A* designates Administration; *B*, Personnel; *C*, Crimes; *D*, Operation; *E*, Intelligence; *F*, Force Control and Information; *G*, Complaints and Discipline; and *H*, Community Relations and Traffic. In addition to these areas of specialization, there is a Headquarters Mobile Support Unit and a Divisional Mobile Support Unit (DMSU); both provide mobile reserve forces used in counterinsurgency and riot situations. The use of DMSUs by police to reroute parades away from Catholic areas has placed these units at the center of controversy.[52]

Although it appears to be very militaristic, and at times actually is so, the RUC has enjoyed separation from and primacy over the Army of Northern Ireland since 1976. The recent surge of popularity of the police among the people of Northern Ireland is due in large measure to this policy. Police primacy over the military has transformed the RUC into a force recognized internationally for its role in combatting terrorism, though at times its methods, including the use of and killings with plastic bullets in the early 1980s and a shoot-to-kill policy against terrorists, have provoked controversy.[53]

The RUC and Human Rights

But the RUC has not been without scandal in the area of human rights. In 1979, the U.S. Department of State suspended sales of handguns to the RUC in response to congressional pressure to end sales of arms to countries violating human rights. Overall, however, many observers believe that the RUC has stood up well against tremendous pressure and scrutiny over the years—a testimonial to the model of policing that evolved in Northern Ireland during the 1970s and through its emphasis on professionalism and autonomy.[54]

Epilogue

Today the violence continues in Northern Ireland. And for what? It has long been clear that the IRA could not force the British government to abandon the Protestants of Northern Ireland; it has also been clear for decades that the British could never stamp out entirely the IRA campaign of bombs and shootings in Ulster and Britain.[55]

Obviously, much remains to be seen as this drama continues to unfold. Meanwhile, the "piggies in the middle"—the Royal Ulster Constabulary—must remain vigilant and prepared for bloody violence at any time.

POLICING IN MEXICO

The official name of this country is the United Mexican States. It covers an area of 756,000 square miles and has a population of almost 100 million people. The country consists of 31 states and a federal district; a limited democracy form of government exists. The country's capital is Mexico City.

A New President: A "Revolution of Hope"

If one word were to be used by most Americans to historically and stereotypically describe Mexico's government and criminal justice system, it would probably be corruption. A new president, Vicente Fox, hopes to reverse that image—and reality—by creating a new government commission to investigate past scandals, overhaul police forces top to bottom, and bring about much greater effort in combatting drug trafficking.[56]

The election of Fox, a charismatic Harvard MBA who assumed office on December 1, 2000, concluded the world's longest-running political dynasty by defeating former Mexican President Ernesto Zedillo and the 71-year reign of the National Action Party. He promises major social and economic reform in a country where more than 40 million of its 100 million people live in poverty, 39 percent are malnourished, and 10 percent are illiterate.[57]

In addition, Fox promises to closely examine one of Mexico's unassailable institutions, the military, to root out corruption, examine its human rights record, and review its struggle against the nation's drug lords. This attempt by Fox to create greater accountability within the military will be closely watched by the United States because of the corruption that has existed within Mexico's military and the fact that the army still represents the primary bulwark against the northward flow of narcotics. Furthermore, Mexico's military receives more U.S. military aid than any Latin American country except Colombia.[58]

POLICE ORGANIZATION

Mexican police forces exist at the federal, state, and municipal levels through many overlapping levels of authority. The primary federal police, the General Directorate of Police and Traffic, are part of the Ministry of Government. The General Directorate is divided into four divisions: preventive, riot, auxiliary, and traffic and investigation (formerly called the Secret Service). The total strength of the General Directorate is about 30,000, including about 1,000 female officers. The Federal Judicial Police are also on the federal level, under the command of the Public Ministry, which performs criminal investigations and represents Interpol in Mexico.[59] The Federal Highway Police, who number about 2,000, patrol the federally designated highways and investigate auto accidents. Small police forces are also maintained in the health, railway, and petroleum industries.[60]

The federal police are organized along military lines and comprise 33 battalions, 31 of which are numbered and include the preventive police, auxiliary police, auxiliary private police, administrative police, and women's police. The unnumbered battalions are the Grenadiers (a riot-control force that includes a motorized brigade) and the Transport Battalion.

The federal district and each of the states has its own police force. The state police enforce state laws within their jurisdiction and assist the federal police in enforcing federal laws. Large cities have special units, such as the Park Police and the Foreign Language Police. Large urban areas have many precincts, called police delegations. A typical delegation has between 200 and 250 police assigned to it, under the command of a *comandante*, who is usually an officer with the rank of first captain. Lesser officers, usually lieutenants, are in charge of each eight-hour shift. They are assisted by first sergeants, second sergeants, and corporals. Most of the officers operate out of a command headquarters called a *comandancia*, but two-officer kiosks are scattered about as well. Auxiliary police also patrol on the night shift.[61]

RECRUITMENT AND TRAINING

Generally, the Mexican police are underfunded and understaffed at all levels and jurisdictions. As a result, the recruitment and training of police officers are severely affected. Corruption among Mexican police is a long-standing problem, and the media on both sides of the border report regularly on police corruption, immorality, and incompetence. Bribes are commonplace and even required for the performance of some tasks. A close relationship exists between the police and the criminal elements of Mexico; indeed, a former police chief of Mexico City is in prison for organized graft on a massive scale.[62]

The General Directorate of Police and Traffic operates a police academy, where four- to six-month-long courses are taught. A few states have academies for state police; the best are in the states of Nuevo Leon, Jalisco, and Mexico. Courses at these academies usually last for four months. In cities where no formal academies exist, police officers are appointed by political bosses.

LEVEL AND TYPES OF CRIMES

Although major crimes in Mexico are reported regularly, complete statistical data on the incidence of crime are not available; data on minor infractions are not compiled. However, it is known that the leading crimes in Mexico are assault and battery, which accounts for about 32 percent of the total; robbery, 25 percent; homicide, 15 percent; sex-related crimes, 6 percent; property damage, 3 percent; fraud, 2 percent; and embezzlement, under 2 percent. Kidnappings reportedly are on the increase, partly by dissident political groups and partly by common criminals. Drug abuse and narcotics traffic emerged in the 1970s as one of the country's major social problems, involving both domestic use and smuggling into the United States. Mexican police and army elements reportedly destroy thousands of acres of marijuana and opium poppies annually, but farmers still manage to grow these crops in isolated mountain areas.[63]

There are many youth gangs in Mexico City and other urban areas; professional thieves specialize in burglarizing houses and in picking pockets, usually during fiestas. In rural areas, the most prevalent crimes are assaults and homicides, usually as a result of intoxication. Machetes, knives, and daggers are frequently used. About 92 percent of all criminals are male. Prostitution is not illegal in Mexico, but soliciting clients is. Prostitutes must be registered and examined periodically, and they are sent to hospitals if they are found to have diseases. Most prostitutes, however, do not register and choose to ply their trade illegally.[64]

CRIMINAL CODES AND THE LEGAL SYSTEM

Mexico has separate federal and state codes of criminal procedure. Crimes are broadly categorized as those against persons, property, the state, public morals, and public health. Capital punishment was prohibited by the Federal Penal Code of 1931; although the Constitution provides for the death penalty for parricide, abduction, and highway robbery, it does not require that the penalty be imposed. The next highest punishment is 30 years in prison, which may be imposed for kidnapping.[65]

The most distinctive feature of the Mexican legal system is the **writ of *amparo***, similar to the American writ of habeas corpus used to challenge a prisoner's confinement and to Section 1983, the instrument used against police officers (see Chapter 10). Considered to be Mexico's contribution to jurisprudence, the *amparo* has been adopted in part by several Latin American countries. An *amparo* may be sought by any citizen for the redress of an infringement of his or her civil rights or against the act of any official, tribunal, police officer, legislature, or bureaucrat; it is issued only by a federal judge. No class action suits are permitted in Mexico, so each person aggrieved by a law, a body, or individuals must file a separate *amparo* for redress.[66]

Another distinctive feature of the Mexican criminal process is the Public Ministry, which dates back to the nineteenth century and exists at the federal, state, and municipal levels. Agents of the Public Ministry, called *fiscales*, are charged with suppressing crime and initiating proceedings in criminal cases. Usually they are assigned to police stations to determine the facts of a given case. They turn over their findings to a central office, which determines whether to initiate prosecution in a criminal case. Public prosecutors are independent of the courts and may not be censured by the judiciary. Since the Public Ministry represents the public, the office may not be sued by defendants if they are found innocent. The Public Ministry also commands the Federal Judicial Police.[67]

Except when an offender has been apprehended in the act of committing a crime (in flagrante delicto), arrest and detention must be preceded by a warrant issued by a competent judicial authority. The accused is entitled to counsel from the moment of arrest, and detention cannot exceed three days without a formal order of commitment. Except in cases of offenses that carry a prison term of more than five years as a penalty, bail is permitted.

Horrors in Mexican Jails

Human rights activists have long implored the Mexican government to clamp down on the country's brutal police and crooked judges. The Mexican/U.S. "War on Drugs" initiative that began in the early 1990s resulted in the capture and imprisonment of greater numbers of U.S. citizens.[68] Thousands of cases have been logged in which the police have allegedly taken innocent citizens off the streets and held them incommunicado in jail without allowing them to see a judge. Many have also reported paying exorbitant "legal fees" while seeking their release from Mexican jails; only after signing a confession are they allowed to contact the U.S. Embassy. Vigorous protests by the U.S. Embassy against these practices appear to have had little effect.[69] It has also been alleged that the Mexican police use torture to obtain confessions, including beating, electrical shocks, and the *tehuacanazo*—blasts of chili powder and seltzer water up the nose. Amnesty International reports that another technique is to put a suspect's head inside a plastic bag filled with ammonia fumes.[70]

Many U.S. citizens spend at least 20 months in prison before conviction, adding to rights violations and prison overcrowding. The prisoner exchange treaty between the United States and Mexico applies only to U.S. citizens who have been convicted.[71] Thus, it may be several years before U.S. offenders can transfer to the United States to serve their sentences.

INTERPOL

Tracking International Criminals

Donald Martin, alias Arman Esterhazy, alias Rene Tronter, alias Estaban Villarejo, left Kennedy Airport six months ago, posing as a businessman. He went from New York to London to Paris to Wiesbaden to Tel Aviv to Beirut in a crime spree that netted him $1 million, by conning dozens of importers into ordering commodities ranging from crude rubber to machine tools. Fleeced merchants began contacting police, who eventually contacted the **International Criminal Police Organization (Interpol)**. This organization checked its master file and concluded that Martin, Tronter, and Villarejo were all the same man. His photograph, his fingerprints, and a national alert were distributed to member countries. An airport stakeout spotted Martin in Bombay; he was Mr. Tronter on the passenger list. Extradition proceedings commenced immediately.[72]

This account is fictional, but the general scenario is not. Many crimes are conducted on an international scale, such as drug trafficking, bank fraud, money laundering, and counterfeiting, for which the crooks may escape detection for years while they enjoy their profits. Interpol is the oldest, best-known, and probably only truly international crime-fighting organization charged with the responsibility of capturing the "Martins" of the globe who seek to keep one country or continent ahead of the long arm of the law. To most people, the term Interpol probably conjures up a *Man from U.N.C.L.E.* image of worldwide secret agents; indeed, a *Man from Interpol* television show popularized the organization. In reality, however, Interpol agents do not patrol the globe, nor do they make arrests or engage in shoot-outs. They are basically intelligence gatherers who have helped many nations work together in attacking international crime since 1923.

Today, Lyon, France, serves as the headquarters for Interpol's crime-fighting tasks. One cardinal rule is observed—Interpol deals only with common criminals; it does not become involved with political, racial, or religious matters. Today 176 member countries constitute Interpol. U.S. participation in Interpol began in 1938 when Congress, under 22 U.S.C. 263a, authorized the attorney general to accept membership in the organization on behalf of the federal government. With the onset of World War II, however, the United States delayed membership until 1947, under the jurisdiction of the Federal Bureau of Investigation. In 1958 the Treasury Department was designated the official representative to Interpol, and in 1969 Treasury created the Interpol–U.S. National Central Bureau (USNCB). In 1977, dual authority was established between Treasury and Justice Department officials in administering the USNCB, and in 1981, the USNCB was again placed under the purview of the Justice Department.[73]

Police departments frequently request assistance in locating a fugitive or obtaining information about a criminal. In those cases, Interpol headquarters may issue an international circulation of information, known as a "diffusion"— an electronic dissemination of information about a wanted person to agencies in a particular country or area; the information is immediately broadcast to the police officers in that country. Interpol may also issue a "notice," which is similar to a diffusion and communicates various types of information that is color-coded into 10 categories (see Figure 13–2). Using the color code, officers receiving the notice immediately know the nature of the alert. For example, Interpol issues a red notice alerting officers at any location, especially border and immigration checkpoints, that their subject has outstanding arrest warrants.[74]

Red	Seeks arrest of subjects for whom arrest warrants have been issued and where extradition will be requested (e.g., fugitives)
Blue	Seeks information (e.g., identity, criminal records) for subjects who have committed criminal offenses and is used to trace and locate a subject where extradition may be sought (e.g., unidentified offenders, witnesses)
Green	Provides information on career criminals who have committed, or are likely to commit, offenses in several countries (e.g., habitual offenders, child molesters, pornographers)
Yellow	Seeks missing or lost persons (includes missing and abducted children)
Black	Provides details of unidentified dead bodies or deceased people who may have used false identities
White	Circulates details and descriptions of all types of stolen or seized property, including art and cultural objects
Purple	Provides details of unusual modus operandi, including new methods of concealment
Gray	Provides information on various organized crime groups and their activities
Orange	Provides information on criminal activity with international ramifications but not involving a specific person or group
FOPAC	Provides money-laundering information for use in countering international money laundering

FIGURE 13–2 Types of Interpol Notices. (*Source:* John J. Imhoff and Stephen P. Cutler, "INTERPOL: Extending Law Enforcement's Reach Around the World," *FBI Law Enforcement Bulletin,* December 1998, p. 13.)

Drug trafficking, terrorism, arms trafficking, and money laundering are unquestionably the most pressing international crime problems at present. Interpol has those crimes at the top of its list of concerns. The issue of terrorism raised some problems in Interpol because of its constitutional provisions against its becoming involved in political matters. The member countries decided it was imperative nonetheless to strengthen cooperation in the area of terrorism. Lately, Interpol has been fighting international crime "on a shoestring," as one author stated.[75] Its level of technology has been wanting throughout most of its history, especially in comparison to the resources employed by criminals.[76]

Drug trafficking now represents about 60 percent of all Interpol message traffic and presents a formidable law enforcement task. In a hypothetical case, cocaine may be grown in Bolivia and Peru, refined in Colombia, shipped through Mexico and the Bahamas, sold and consumed in the United States, and the profits deposited in a Panamanian bank, then invested in Britain, while the mastermind of the whole scheme lives in Honduras. The United States has become a leader in advocating the hard-line response to international crime, but the United States also has a history of attempting to resolve problems of terrorism, drugs, and money laundering by throwing money into foreign aid, by diplomatic relations, and by spying activities: the "big bucks, striped pants, and spooks" approach.

A FORMULA FOR SUCCESS

Interpol's multinational police cooperation process has a basic three-step formula for offenses that all nations must follow for success: Pass laws specifying the offense is a crime; prosecute offenders and cooperate in other countries' prosecutions; and furnish Interpol with and exchange information about crime and its

perpetrators. This formula could reverse the trend that is forecast for the world at present: an increasing capability by criminals for violence and destruction.

The rule of thumb governing extradition treaties seems to be that the offenses specified must involve double criminality. They must be recognized as crimes by *both* parties—the country where the offender is caught and the country in which the crime was committed. The following crimes, because they are recognized as crimes by other countries, are covered by almost all U.S. treaties of extradition: murder, rape, bigamy, arson, robbery, burglary, forgery, counterfeiting, embezzlement, larceny, fraud, perjury, and kidnapping.

SUMMARY

The four national police systems discussed in this chapter—in China, Saudi Arabia, Northern Ireland, and Mexico—were all similar. Each police force, except for the RUC of Northern Ireland, is in effect a pawn of its national government, which seems to have little regard for the human rights of its citizens. And each, except for the RUC, receives low wages and benefits and is held in low esteem by the citizens. All of the four police organizations train their officers well for the heavy-handed duties they must perform. Most are granted considerable powers of arrest for investigation and interrogation; several, especially China, Mexico, and Saudi Arabia, are known to be brutal and unrelenting to their prisoners; thus their citizens fear them. All four countries have suffered high levels of poverty in harsh lands where murders and executions have been relatively commonplace. Most (again, with the exception of Northern Ireland) have evolved relatively complex national police systems for governing their people.

While some of these elements may seem to exist in relationships of U.S. citizens with their police officers, there are many differences between U.S. police methods and those in the four countries discussed here. While U.S. police are well trained on the whole (and probably also consider themselves to be underpaid and accorded low esteem by the public), U.S. citizens do not have to fear an omnipotent national police force as citizens of other countries obviously must.

In the United States, there are few known government-sanctioned violations of human rights. And when they occur, the public is outraged and demands that those officers responsible be held accountable. History has shown that people fear the police, especially when police powers are unchecked and used for political advantage. The police must be held accountable to a higher authority—an authority that is local, that is decentralized, and that exists outside the police agency itself.

U.S. police, we have seen in previous chapters, perform defined, limited functions under a constitutionally grounded rule of law that governs arrests, searches, and seizures and prohibits excessive methods of interrogation and detention by the police. This comparative perspective has provided the opportunity to compare and contrast various police systems. It has demonstrated that while U.S. police have their own problems, a democratic system appears to be unequivocally better than those systems discussed in this chapter.

Finally, the discussion of Interpol shows that, even with political problems and the isolation of countries such as the four discussed here, foreign countries must, can, and do succeed in tracking and prosecuting international offenders.

ITEMS FOR REVIEW

1. Why it is helpful and important to examine criminal justice systems in other countries?
2. Analyzing the four police systems discussed in this chapter (China, Saudi Arabia, Northern Ireland, and Mexico), what would be the main characteristics of each with regard to the following:
    a. Political and legal foundation
    b. Organization and jurisdiction
    c. Methods of operation
    d. Respect for human rights
3. How do each of these four countries differ significantly from the U.S. system of policing in the major areas listed in Item 2 above?
4. What organization exists to combat international crime and catch transient fugitives? What are the primary means at this organization's disposal to accomplish its mission?

NOTES

1. Ken Peak, "The Comparative Systems Course in Criminal Justice: Findings from a National Survey," *Journal of Criminal Justice Education* 2 (1992): 267–272.
2. Mark Kesselman, "Order or Movement? The Literature of Political Development as Ideology," *World Politics,* 26 (1973): 139–154.
3. Gabriel S. Almond and John S. Coleman, *The Politics of the Developing Areas* (Princeton, NJ: Princeton University Press, 1960), p. 2.
4. Richard J. Terrill, *World Criminal Justice Systems,* 4th ed. (Cincinnati, OH: Anderson, 1999), pp. 513–514.
5. Yue Ma, "The Police Law 1995: Organization, Functions, Powers and Accountability of the Chinese Police," *Policing: An International Journal of Police Strategy and Management* 20 (1997): 113–135.
6. *Ibid.,* p. 115.
7. Z. Ma and M. Tian, *The People's Police Law of the People's Republic of China: Explanations and Seminars* (Beijing: China University of People's Security University Press, 1995).
8. Yue Ma, "The Police Law 1995," p. 119.
9. *Ibid.,* pp. 120–121.
10. *Ibid.,* pp. 123–125.
11. X. Wang, "Preliminary Analysis of Current Situation of Chinese Young People's Feeling of Safety," *Public Security Studies* 4 (1991).
12. Yue Ma, "The Police Law 1995," pp. 130–131.
13. *Ibid.,* p. 132.
14. Saudi Arabian International Schools, "An Introduction to the Kingdom of Saudi Arabia," Spring 1992, p. 11.
15. *Ibid.,* p. 12.
16. *Ibid.,* p. 4.
17. *Saudi Arabia v. Nelson,* 113 S.Ct. 1471, 123 L.Ed.2d 47 (1993).

18. Saudi Arabian International Schools, "An Introduction to the Kingdom of Saudi Arabia," p. 17.
19. Adel Mohammed el Fikey, "Crimes and Penalties in Islamic Criminal Legislation," *CJ International* 2 (July–August 1986), p. 13.
20. *Gulf News*, May 19, 1989, p. 1.
21. el Fikey, "Crimes and Penalties in Islamic Criminal Legislation," p. 14.
22. Chris Hedges, "Everywhere in Saudi Arabia, Islam Is Watching," *New York Times*, January 6, 1993, p. A4(N), col. 3.
23. *Ibid.*
24. *Ibid.*
25. *Ibid.*
26. Confidential personal communication, July 27, 1994.
27. *Ibid.*
28. Fintan O'Toole, "A Novel Question for Ireland," *U.S. News and World Report*, May 25, 1998, p. 22.
29. *Ibid.*, p. 22.
30. Stryker McGuire, "The Easter Peace," *Newsweek*, April 20, 1998, p. 34.
31. Thomas K. Grose, "Are 'The Troubles' Over?" *U.S. News and World Report*, April 20, 1998, p. 38.
32. McGuire, "The Easter Peace," p. 34.
33. Associated Press, "Britain Bolsters Troops: Northern Ireland's Mayhem Unchecked," *Tulsa World*, July 8, 1998, p. A4.
34. Associated Press, "Phone Call Causes Dozens of Deaths in Terrorist Bomb Blast in Ireland," August 16, 1998.
35. T. R. Reid, "Ulster Gets New Government," *Washington Post*, November 29, 1999, p. 1.
36. Shawn Pogatchnik, "Bonfires in Belfast," *Associated Press*, July 12, 2000; see also Thomas K. Grose, "It's Back to the Streets," *U.S. News and World Report*, July 17, 2000, p. 34.
37. O'Toole, "A Novel Question for Ireland," p. 23.
38. Tim Lewis, personal interview, May 10, 1990.
39. "RUC: Statistical Information" (in-house report), February 1990.
40. "RUC: The Grim Statistics," RUC report, July 1989, p. 12.
41. Paul K. Clare, "The Royal Ulster Constabulary: Northern Ireland's Beleaguered Police Force," *Criminal Justice International* 3 (May–June 1987): 3–6.
42. *Ibid.*, p. 6.
43. *Ibid.*, pp. 4–5.
44. *Ibid.*, p. 5.
45. Jeffrey Bartholet, "New Targets for Terror," *Newsweek*, August 13, 1990, pp. 42–43.
46. Richard Ward, "Gangs Deal in Fear and Savagery," *Criminal Justice International* 4 (November–December 1988): 3.
47. "RUC: The Grim Statistics," p. 27.
48. *Ibid.*, p. 30.
49. Lewis, personal interview.
50. *Ibid.*
51. "A Career with the Royal Ulster Constabulary," RUC report, April 1990.
52. John D. Brewer, Adrian Guelke, Ian Hume, Edward Moxon-Browne, and Nick Wilford, *The Police, Public Order and the State* (New York: St. Martin's Press, 1988), p. 59.
53. *Ibid.*, p. 73.
54. *Ibid.*, p. 81.
55. Mortimer B. Zuckerman, "Pack Up Your Trouble," *U.S. News and World Report*, May 18, 1998, p. 78.
56. Elliot Blair Smith, "Fox Ready for 'Revolution of Hope,'" *USA Today*, December 1, 2000, p. 1.
57. *Ibid.*
58. Andrea Mandel-Campbell and Thomas Omestad, "A Time of Change for Mexico's Military," *U.S. News and World Report*, December 11, 2000, p. 48.
59. George Thomas Kurian, *World Encyclopedia of Police Forces and Penal Systems* (New York: Facts on File, 1989), p. 258.
60. *Ibid.*, pp. 258–259.
61. *Ibid.*, p. 258.
62. *Ibid.*, p. 259.
63. *Ibid.*, pp. 260–261.
64. *Ibid.*, p. 261.
65. *Ibid.*, p. 259.
66. *Ibid.*, pp. 259–260.

67. *Ibid.*, p. 260.
68. Americas Watch, *Human Rights in Mexico: A Policy of Impunity* (Los Angeles, CA: An Americas Watch Report, 1990).
69. J. Michael Olivero, "The 'War on Drugs' and Mexican Prisons," *Criminal Justice and the Americas* 4 (April–May 1991): 3–4.
70. Tim Padgett, "Tijuana's Midnight Express," *Newsweek*, November 23, 1992, p. 41.
71. U.S. Congress, House Committee on the Judiciary, *Providing Implementation of Treaties for Transfer of Offenders to or from Foreign Countries: Report of the Judiciary on S1682, October 19, 1977* (Washington, D.C.: U.S. Government Printing Office, 1977).
72. Michael Fooner, *Interpol* (Chicago: Henry Regnery Company, 1973), pp. 2–3.
73. INTERPOL, "An Overview of INTERPOL and the U.S. National Central Bureau," October 1994, pp. 6–7.
74. John J. Imhoff and Stephen P. Cutler, "INTERPOL: Extending Law Enforcement's Reach Around the World," *FBI Law Enforcement Bulletin*, December 1998, pp. 10–17.
75. Richard Ward, "Interpol: Fighting Crime on a Shoestring," *Criminal Justice International* (Winter 1985): 1.
76. Michael Fooner, *Interpol: Issues in World Crime and International Criminal Justice* (New York: Plenum Press, 1989), p. 179.

Technology Review

Key Terms and Concepts

Automated fingerprint
   identification systems (AFIS)
Computer-aided dispatching
Crime mapping
Geographic profiling

Homicide investigation and
   tracking system
Less-than-lethal weapons
Mobile data systems

Knowledge is power.—*Francis Bacon*

Intelligence . . . is the faculty of making artificial objects, especially tools
to make tools.—*Henri Bergson*

Give us the tools, and we will finish the job.—*Winston Churchill*

INTRODUCTION

Several photographs and some discussions in Chapter 7 demonstrated the kinds of technologies now available for forensic analyses. This chapter reviews more broadly the kinds of technologies that are either available to the police or on the horizon.

The new millennium brings exciting, ongoing opportunities for policing with regard to technology; since the advent of computers, we are limited only by our collective imagination and willingness to provide the funding necessary for technology to evolve. While our crime laboratories have arguably provided the greatest advancements in technologies in recent years, tremendous opportunities exist in other areas as well. It is not surprising that most technological developments in criminal justice were developed for and involve the police, given the nature of their work, tools, and problems.

We begin by considering some fundamental problems that exist in the area of policing and technology, particularly the need for better trained personnel who understand computer hardware and software. Next we discuss technologies that are being developed to combat terrorism; then we look at the development and status of less-than-lethal police weapons, followed by a look at wireless technology for use in databases, in crime mapping, and in dealing with serial offenders and gunshots. Then we examine how electronic capabilities are being applied to several traffic functions. Following is a brief look at technological advances with DNA, and then we review developments with fingerprints and mug shots. New uses of technologies in the area of crime scene investigations are presented, and we then consider how computers are assisting with regard to firearms, particularly in training officers and in solving cases involving guns. The chapter concludes with a short commentary on intelligence systems of gangs.

It should also be mentioned that several of the technologies discussed in this chapter may be—and are being—applied to community oriented policing and problem solving (COPPS, examined in Chapter 6); furthermore, some crime analysis methods and tools (such as computer mapping) briefly mentioned in Chapter 6 are examined in more detail in this chapter.

POLICE AND TECHNOLOGY OF THE FUTURE: PROBLEMS AND PROSPECTS

First, the good news. It is anticipated that police officers of the future will function in very different ways and on very different terms than officers of the past. Given existing technologies and what they bode for the future, every officer will function with few time and space constraints because all officers will be equipped with a pager, a cellular phone, and a laptop computer with software that includes encryption programs and sophisticated databases and search engines. These officers will have software that allows them to have real-time chats with officers from other agencies or in other states or even other countries. Before going on the streets, every rookie will be thoroughly computer literate and able to use crime analysis software.[1] Exhibit 14–1 describes how one form of hand-held technology—the hand-held minicomputer—already allows officers to be removed from their patrol car laptops.

With such tools, it is easy to envision that an officer's home or car or even a convenience store may become his or her workplace. Identification of suspects in the field will take a quantum leap with electronic telecommunication of fingerprints, scanning of retinal patterns, and facial ratio and heat patterns that say positively, "This is the criminal." Officers will also access maps and data; be able to bring up any call, crime type, or problem by geographic area; and sort out this information and compare similar incidents. They will be able to touch their computer keys and ask for the top 10 crimes in their beat area while receiving instant crime analysis for use in deployment and other operational decisions. All civilians will likewise be trained in computer usage.[2]

As the saying goes, however, "the fleas come with the dog." Aspects of the alliance between police and technology presage problems for the foreseeable future. From 1995 to 1998 (the last year for which such applications were received), the federal Office of Community Oriented Policing Services poured hundreds of million of dollars into police agencies to provide grants for equipment.[3] But this federal largesse brought problems with the arrival of new equipment into police station houses. First, many departments lacked, and still lack, the in-house computer expertise to install or run the software and equipment. Many police executives believe that they hire people to be police officers—not to be computer programmers or database experts. But the nature of the policing business is changing; officers with such skills are not only desirable and valuable but are also increasingly and rapidly becoming a *necessity* as a problem-solving aid.

In many police agencies, information technology staff are often civilians and are generally kept away from the operational side of the organization. They understand what computers do but not necessarily how that capability supports the operational needs of the police officer on the street.

These problems are in addition to other technology-related shortcomings. For example, the police are also playing catch-up to counter passwords, digital compression, remote storage, audit disabling, anonymous remailers, digital cash, computer penetration and looping, cloning of cellular phones and phone cards, and a host of other evasive criminal schemes.[4]

Nonetheless, we examine the kinds of advances that are rapidly being made in the field. It should be emphasized that several of the described systems are extremely expensive and still cost prohibitive. Furthermore, these systems are not a panacea and cannot replace the traditional forms of police work; rather, they are simply a way of managing information and focusing an investigation or search in a small area.

Exhibit 14–1

The Power of Information, in a Palm-Sized Package

What the newest hand-held minicomputers lack in heft, they more than make up for in the wealth of knowledge they can supply to police in those crucial moments before they approach a suspect. The devices, which tip the scales at a mere four ounces or so, are finding their way on to the equipment belts of law enforcement officers in a steadily growing number of jurisdictions, including New York City, Charleston, S.C., and Franklin County, Ohio. . . .

New York is the first city in the nation to have officers use the devices during routine street patrols. Worn on the officers' gun belts, they come with small keyboards that can be used to enter license numbers, names and other data. What sets them apart from existing NYPD computers that can provide the same information, however, is their speed and stealth.

Said Ari Wax, the NYPD's deputy commissioner for technology and development: "It can be just like the cop on the street is checking his e-mail."

In April, two housing officers demonstrated how effective the hand-helds could be when police are faced with quality-of-life crimes. Confronting a man drinking beer on a Harlem stoop, the officer entered his name into a device they were testing. It turned out that the subject, Adrian Bowman, was wanted for a triple homicide in St. Louis. . . .

Det. Chris Floyd noted that in the past, when suspects are detained, deputies might use a walkie-talkie to call the dispatcher for arrest information. Those results, transmitted over the radio, could then be heard by the suspect. "Now this guy knows I know who he is," said Floyd. "It's giving this guy a warning to run or fight."

While it is the portability of hand-held computers that has garnered the greatest appreciation thus far from police, they also cost much less than laptops, noted Sgt. Robert Flynn, director of computer services for the Charleston Police Department. The agency handed out 25 minicomputers last month to its traffic officers as part of a 60-day trial. It is the only department in the state to try out the devices so far, according to The (Charleston) Post and Courier. . . .

"It's tremendous in that respect," he said, "but the cost can't be ignored." To install a laptop in a vehicle costs approximately $7,100; the hand-held computers, with software and network data time through a provider, cost approximately $1,900. "It's substantial. We can do everything a laptop was doing for us in the palm of the officer's hand for less money."

Source: Reprinted with permission from *Law Enforcement News*, May 31, 2001, p. 11. John Jay College of Criminal Justice (CUNY) 555 West 57th St., New York, NY 10019.

TECHNOLOGY VERSUS TERRORISTS

The terrorist attacks on September 11, 2001, not only changed the way federal, state, and local law enforcement agencies approach their mission (in essence, they now also protect citizens against attack from the *outside*, rather than only from within) but also set in motion research and development of new technologies for protecting against terrorist acts. Polls conducted immediately following the New York attacks revealed that an overwhelming majority of Americans—about 75 percent—thought it necessary to give up some personal freedoms for the sake of security.[5] Following are some technologies that are being or have been developed to supplement luggage scanners and metal detectors to assist in detecting and foiling terrorists.

A low-dose X-ray imager sees through garments to detect stashed weapons, drugs, or other contraband (the September 11th terrorists slipped through checkpoints with box cutters). Surveillance cameras can scan faces and feed the images to a computer that scours a database of digital mug shots for a match (these devices were actually used during the Super Bowl in early 2001 in Tampa, Florida, and identified several petty criminals in the crowd, but soon after that many Floridians and the state's civil liberties union protested against their use).[6] Better bomb-detecting technology is also being developed, including machines that perform a cross-sectional X-ray scan of checked baggage. Ion-sniffing swabs can find bomb residue on hand luggage. Also being considered for limited use are "smart cards"—IDs equipped with memory chips that store personal data and can track movements and transactions.[7]

THE DEVELOPMENT
OF LESS-THAN-LETHAL WEAPONS

A HISTORICAL OVERVIEW

Since the dawn of time, people have sought to control the behavior of their fellows. Our recorded history is full of accounts of the need to protect ourselves from a hostile environment.[8] Certainly international police forces have attempted to employ many means to protect themselves and the public at large.

Although the term "nonlethal weapon" is often used in such discussions, this term is inappropriate because *any* tool or weapon can be lethal if it is used in an improper or unintended manner. For those reasons the term **less-than-lethal** is more properly used to designate a weapon when it produces only a temporary effect and minimal medical consequences for normally healthy persons.[9]

The Unveiling: 1829

When London's new bobbies began patrolling the streets in September 1829, they were armed with a baton ("truncheon") that still is a standard-issue weapon; it was traditionally made of hardwood, two to four centimeters in diameter and 30 to 65 centimeters in length.[10]

Social turmoil in America between 1840 and 1870 brought increased use of force by the police, who themselves were frequent victims of assault. The New York City police and other departments were armed with 33-inch clubs, which officers were not reluctant to use. Charges of police brutality were common in the mid-1800s, when officers allegedly clubbed "respectable" citizens with frequency.[11]

Even until the end of the 1800s, the baton remained a staple tool among American police. As late as 1900, when the Chicago Police Department numbered 3,225 officers, the only tools given the new patrolmen were "a brief speech from a high-ranking officer, a hickory stick, whistle, and a key to the call box."[12] Charges of police brutality continued.

Chemical Weapons: 1860s–1950s

International policing from 1860 to 1959 witnessed only one major addition to the small array of less-than-lethal police tools: chemical weapons. CN gas was synthesized by German chemists in 1869 and was the first tear gas; it produced a burning sensation in the throat, eyes, and nose. CN became available for use in aerosol cans in 1965: Chemical Mace was the most well-known variety. By 1912, chemical weapons were increasingly used in riot situations and when subduing criminals.[13]

CS gas was first synthesized in 1928 as a white powder that stored well. It was adopted by many police forces as it had a greater effect than CN, causing a very strong burning sensation in the eyes that often caused them to close involuntarily. Severe pain in the nose, throat, and chest; vomiting; and nausea are also associated with its use.[14]

A Technological Explosion: The 1960s

The 1960s witnessed major technological advances in less-than-lethal weaponry. This was the age of rioting, on both domestic and foreign soil. Aerosol chemical agents, developed to provide alternatives to police batons and firearms, became the most widely used less-than-lethal weapons to emerge for police use. Mace seemed like "manna from heaven."[15]

CR gas appeared in 1962; six times more potent than CS and 20 times more potent than CN, it caused extreme eye pressure and occasional hysteria. CS gas cartridges, which were fired by shotguns at a range of 125 meters, were used in the United States and Britain as early as 1968.

In 1967, alternatives to lethal lead bullets were first used in Hong Kong. Wooden rounds, fired from a signal pistol with a range of 20 to 30 meters, were designed to be ricocheted off the ground, striking the victim in the legs. The wooden rounds proved to be fatal, however, and direct fire broke legs at 46 meters. Rubber bullets were developed and issued to British troops and police officers in 1967, only a few months after the wooden rounds appeared. Intended to deliver a force equivalent to a hard punch at 25 meters, the rubber bullets caused severe bruising and shock. Also intended to be ricocheted off the ground, the rubber bullet was designed so that riot police could outrange stone throwers.[16] (Rubber bullets were employed by Seattle, Washington, police officers during the December 1999 protests in connection with the World Trade Organization meeting.)[17]

During the 1960s, British riot police also began employing water cannons, which are designed to fire large jets of water at demonstrators. Resembling armored fire engines, water cannons were also used in Germany, France, Belgium and the United States. Nontoxic blue dye was added to the water for marking the offenders; for a time the firing of a CR solution was contemplated.[18]

In 1968, another unique riot-oriented weapon was unveiled in America: the Sound Curdler, consisting of amplified speakers that produced loud shrieking noises at irregular intervals. Attached to vehicles or helicopters, the device was first used at campus disturbances.

The 1970s: Beanbag Guns, Strobe Lights, and "New Age" Batons

Another unique tool was literally unfurled in 1970: a gun that shot beanbags rather than bullets. The apparatus fired a pellet-loaded bag that unfurled into a spinning pancake, capable of knocking down a 200-pound person at a range of 300 feet. Although its manufacturer warned that it had potentially lethal capabilities, it quickly became popular in several countries, including South Africa and Saudi Arabia.[19]

Several other inventions were added to the less-than-lethal arsenal of the police in the 1970s. The Photic Driver, first used by police in South Africa, produced a strobe effect: Its light caused giddiness, fainting, and nausea. A British firm developed a strobe gun that operated at five flickers per second. Some of these devices also had a high-pitched screamer device attached. Another apparatus, known as the Squawk Box, had two high-energy ultrasound generators operating at slightly different frequencies that produced sounds that also caused nausea and giddiness.[20]

Another invention of the early 1970s included an electrified water jet; a baton that carried a 6,000 volt shock; shotgun shells filled with plastic pellets; plastic bubbles that immobilized rioters; a chemical that created slippery street surfaces for combatting rioters; and an instant "cocoon," an adhesive substance that, when sprayed over crowds, made people stick together.[21]

Two new types of projectiles were developed in 1974: the plastic bullet and the TASER. The "softer" plastic bullet could be fired from a variety of riot weapons at a speed of 160 miles per hour and a range of 30 to 70 meters, making it attractive to riot police. Like wooden bullets, however, it could be fatal. Seven Northern Ireland persons died during the early 1980s after being struck in the head with these bullets.[22]

The TASER resembled a flashlight and shot two tiny darts into its victim. Attached to the darts were fine wires through which a transformer delivered a 50,000 volt electrical shock, which would knock down a person at a distance of 15 feet. Police officers in every state except Alaska had used the device by 1985, but its use was not widespread because of limitations of range and on persons under the influence of drugs. Heavy clothing could also render it ineffective.[23] The current status of the TASER is discussed later.

The stun gun, also introduced in the mid-1970s, initially competed with the TASER as the police tool of choice. Slightly larger than an electric razor, it also delivered a 50,000-volt shock when its two electrodes were pressed directly against the body. Like the TASER, its amperage was so small that it did not provide a lethal electrical jolt. Its target, who was rendered rubbery-legged and fell to the ground, was unable to control physical movement for several minutes.[24]

TODAY: THE QUEST CONTINUES

The quest continues today for effective alternatives to lethal force. It is a paradox in this high-technology age that reliable substitutes for lethal police weapons have not yet been developed. The chronology presented shows that past attempts to do so were sometimes deadly or ineffective. So the search for the "perfect" less-than-lethal weapon continues.

The TASER underwent considerable revision in the late 1990s, evolving into a much better tool, one that is now widely adopted by police agencies. The newer models are handgun shaped, use 26 watts to deliver 50,000 volts (as opposed to the five- or six-watt earlier versions) for five seconds, and can penetrate up to three inches of clothing. Their range is up to 21 feet. Even if only one probe sticks to the person, the newer versions will still work most of the time.[25]

Also recently introduced are two types of foam that may hold promise. One is supersticky; offenders are drenched in a substance that, when exposed to the air, turns into taffy-like glue. The other creates an avalanche of very dense soap

The police use a variety of tools for addressing crime and disorder.
(*Source: Washoe County, Nevada, Sheriff's Office*)

bubbles that leave offenders unable to see or move but still able to breathe. Other chemical compounds, known as slick'ems and stick'ems, make pavements either too sticky or too slippery for vehicles to move.[26]

Police departments nationwide—many of them beleaguered by protests after fatal shootings—continue to look to new weapons that offer a less-than-lethal alternative, an "intermediate" weapon between the voice and the gun. Many less-than-lethal alternatives are relatively inexpensive for agencies to adopt and can significantly reduce a department's liability.[27]

Interest has been renewed in beanbag rounds (sometimes called kinetic batons) and capture nets. Beanbag rounds have become extremely popular; they have been adopted by police agencies across the country and have been used successfully to help end a number of standoffs.[28] A Spiderman-style net gun known as the WebShot is designed to wrap and immobilize a suspect. The WebShot gun has been on the market since 1999 and is in trial use with the Los Angeles and San Diego police. WebShot is ideally used against an unarmed person who is being combative.[29]

The Less Than Lethal program for the National Institute of Justice coordinates research on such weapons, ensuring that they are both technically feasible and practical for police officers. In addition to the WebShot gun, the program recently provided funding for research into the following technologies:

- The Laser Dazzler, a flashlight device designed to disorient or distract suspects with a green laser light
- The Sticky Shocker, a wireless projectile fired from a gas gun that sticks to the subject with glue or barbs that attach to clothing and delivers an electric shock
- The Ring Airfoil Projectile, a doughnut-shaped, hard rubber weapon that is under redesign to also disperse a cloud of pepper powder on impact[30]

Pepper spray, or oleoresin capsicum, is now used by an estimated 90 percent of police agencies, according to the National Institute of Justice.[31] This spray inflames the mucus membranes of the eyes, nose, and mouth, causing a severe burning sensation for 20 minutes or less.[32] The spray is highly effective in subduing suspects without causing undue harm or long-term aftereffects.[33]

Also newly invented (by a member of the Oxnard, California, Police Department's SWAT team) and having promise is "The Option," which offers lethal and less-than-lethal capabilities in a single unit; the weapon has a cylinder of pepper spray that is mounted to the barrel of a pistol or shoulder weapon.[34]

USE OF WIRELESS TECHNOLOGY

INSTANT ACCESS TO INFORMATION

Mobile data systems have been available since the 1970s. But the first-generation systems were based on large, proprietary computers that were very costly and were often beyond the reach of many small and medium-sized police agencies.

An officer uses a hand-held pocket computer.
(*Courtesy Aether Systems, Inc.*)

The first digital data was not transmitted from police headquarters to a cruiser until the mid-1980s. Today, armed with a notebook computer and a radio modem, police officers can have almost instant access to information in numerous federal, state, and local databases. Even small agencies can now afford a network and mobile data terminals (MDTs).[35]

A growing number of American police departments, including small agencies, are using laptop computers with wireless connections to crime and motor vehicle databases. These systems are believed to pay for themselves in increased fines and officer safety. Officers can access court documents, in-house police department records, and a **computer-aided dispatching (CAD)** system, as well as enter license numbers into their laptop computers. And, through a national network of motor vehicle and criminal history databases, they can locate drivers with outstanding warrants, expired or suspended licenses, and so on. Furthermore, rather than using open radio communications, police officers use their computers to communicate with each other via e-mail.[36]

INTEGRATED DATABASES

In 1999, a "railroad killer" was engaged in a rampage across the southwest. Texas Border Patrol agents unknowingly picked up the killer near the Mexican border and dutifully checked the agency's database for any outstanding warrants. Finding none, they released him, and within a few days he struck again, killing two women. Several federal and local law enforcement agencies wanted the suspect for questioning, but the Border Patrol had no way of searching those agencies' databases; had it done so, the killing spree might have ended sooner.[37]

Agencies will soon be able to cross-search databases when the U.S. Department of Justice builds its proposed Global Justice Information Network to link crime networks—including police, courts, and corrections agencies—at the local, state, national, and international levels. This is being termed a new course for the criminal justice community, which is at a historic turning point with these integration efforts. Such systems will improve the quality of information and decisions made by criminal justice officials.[38]

We will discuss integrating DNA databases later.

CRIME MAPPING

Conclusive evidence from clay tablets found in Iraq proves that maps have been around for several thousand years—perhaps tens of millennia.[39] **Crime mapping** has long been an integral part of the crime analysis process. The New York City Police Department, for example, has traced the use of maps back to at least 1900.

The traditional crime map was a jumbo representation of a jurisdiction with pins stuck in it. The old pin maps were useful for showing where crimes occurred, but they had several limitations as well. As they were updated, the prior crime patterns were lost; the maps were static, unable to be manipulated or queried. Also, pin maps could be quite difficult to read when several types of crime were mixed together. They also occupied considerable wall space (as an example, to make a single wall map of the 610 square miles of Baltimore County, 12 maps had to be joined, using 70 square feet).[40] Consequently, during the 1990s, pin maps largely gave way to desktop computer mapping, which has now become commonplace and fast, aided by the availability of inexpensive color printers.[41]

Computerized crime mapping has been termed "policing's latest hot topic."[42] For officers on the street, mapping puts street crime into an entirely new perspective; for administrators, it provides a way to involve the community in addressing its own problems by observing trends in neighborhood criminal activity. Crime mapping also offers crime analysts graphic representations of crime-related issues. Furthermore, detectives can use maps to better understand the hunting patterns of serial offenders and to hypothesize where these offenders might live.[43] Exhibit 14–2 briefly discusses how the federal government is assisting in the development of computer mapping technologies, and Exhibit 14–3 elaborates on this concept, explaining process mapping.

Computerized crime mapping combines geographic information from global positioning satellites with crime statistics gathered by the department's computer-aided dispatching system and demographic data provided by private companies or the U.S. Census Bureau (some agencies acquire information from the Census Bureau's Internet home page). The result is a picture that combines disparate sets of data for a whole new perspective on crime. For example, a map of crimes can be overlaid with maps or layers of causative data: unemployment

EXHIBIT 14–2

The Crime Mapping Research Center

In 1997, the National Institute of Justice established the Crime Mapping Research Center (CMRC) to promote research, evaluation, development, and dissemination of geographic information systems technology for criminal justice research and practice. The CMRC holds annual conferences on crime mapping to provide researchers and practitioners an opportunity to obtain practical and state-of-the-art information on the use and utility of computerized crime mapping. The CMRC provides fellowships, NIJ-funded grant awards; evaluation of best practices and current criminal justice applications and needs, training programs, a national geocoded data archive, and information through conferences, workshops, a Web site, and a listserv. The CMRC Web site address is: http://www.ojp.usdoj.gov/cmrc

Source: Adapted from U.S. Department of Justice, National Institute of Justice, *Crime Mapping Research Center* (Washington, D.C.: Author, 2000).

rates in the areas of high crime, locations of abandoned houses, density of population, reports of drug activity, or geographic features (such as alleys, canals, or open fields) that might be contributing factors.[44] Furthermore, the hardware and software are now available to nearly all police agencies and cost only a few thousand dollars.

In 1995, the Chicago Police Department implemented a system called Information Collection for Automated Mapping, or ICAM—a flexible, user-friendly system that enables all police officers to quickly generate maps of their beats, sectors, or districts and to search for and analyze crime patterns. ICAM—now officially ICAM2, having already been upgraded with 20 enhancements—is now in use in all of the city's 25 policing districts and can query up to two years and map from a selection of 300 crimes within specific time ranges; it can also reveal important neighborhood establishments, such as schools, abandoned buildings, and liquor stores. It provides mug shot images within a minute. ICAM can also be an informative tool for the community; many COPPS officers are providing maps detailing crime on a beat during community meetings.[45]

Another example of mapping success is the New York Police Department's highly touted CompStat program, which provides up-to-the-minute statistics, map patterns, and establishes causal relationships among crime categories. CompStat also puts supervisors in constant communication with the department's administration, provides updates to headquarters every week, and makes supervisors responsible for responding to crime in their assigned areas.[46]

EXHIBIT 14–3

21st Century Police Department

Naperville, Illinois

Over the last decade, companies in the manufacturing, entertainment, and defense industries have used a tool called "process mapping" to help them describe, analyze, and ultimately improve how their organizations operate. Recently members of the City of Naperville, Illinois, Police Department and 23 other police agencies were invited to attend training in process mapping, which involves the development of three different flowcharts that visually depict the series of activities involved in carrying out one of the organization's major functions:

> The *as-is map* describes the organization as it currently exists. This map is based on interviews and observations of people and is used to diagnose waste, duplication of effort, coordination of problems, or breakdowns in the flow of information.
>
> The *should-be map* makes short-term changes to reduce waste, remove duplication, and improve coordination and flow of information. This map is based on management analysis of the as-is map and suggestions gathered from field personnel during interviews.
>
> The *could-be map* describes the ideal process for the future. This map is based on the organization's vision and highlights the long-term changes that are needed to get there.

As an example, Naperville is focusing on "crime solving" as the major function to be mapped and is focusing on one crime type: burglary. Process mapping allows the agency to *increase the clearance rate for crimes* by identifying areas where new work methods or organizational changes might improve police ability to investigate crimes and arrest offenders; and it *makes more widespread and effective use of automation and technology* by identifying areas where work processes can be improved, such as automated case reporting.

Source: Adapted from City of Naperville, Illinois, Web page, "21st Century Police Department," 1997, pp. 1–2.

FIGURE 14–1 Street Gang-Motivated Homicide, Other Violence, and Drug Crime. (*Source:* U.S. Department of Justice, National Institute of Justice, *The Use of Computerized Mapping in Crime Control and Prevention Programs* [Washington, D.C.: Author, 1995], p. 4.)

Crime mapping is a major tool in other jurisdictions as well. Following are instances of successful outcomes:

- When an armored car was robbed in Toronto, Canada, dispatchers helped officers chase the suspects through a sprawling golf course using the mapping feature of the CAD system.
- The Illinois State Police map fatality accidents throughout the state and show which districts have specific problems. It can correlate that data with citations written, seat-belt usage, and other types of enforcement data.
- The Salinas, California, Police Department maps gang territories and correlates socioeconomic factors with crime-related incidents. Murders are down 61 percent, drive-by shootings by 31 percent, and gang-related assaults by 23 percent.[47]

The NIJ's Crime Mapping Research Center, established in 1997, provides information about research, evaluation, and training programs through its Web site address: http://www.ojp.usdoj.gov/cmrc.

LOCATING SERIAL OFFENDERS

Most offenders operate close to home. Offenders tend to operate in target-rich environments to "hunt" for their prey. **Geographic profiling**—a relatively new development in the field of environmental criminology—analyzes the geography of such locations and the sites of the victim encounter, the attack, the murder, and body dump and maps the most probable location of the suspect's home.[48]

Geographic profiling is most effective when used in conjunction with linkage analysis. For example, the Washington State Attorney General's office uses a **homicide investigation and tracking system (HITS)** that includes crime-related databases and links to vice and gang files, sex offender registries, corrections and parole records, and department of motor vehicle databases. HITS can simultaneously scan these databases. When an agency in the state has a major crime in its jurisdiction, the case is loaded into a central system, which scans every database and linking file for connections by comparing eyewitness descriptions of a suspect and vehicle. It then builds a dataset containing profiles of the offender, the victims, and the incidents. The dataset then goes into a geographic information system (GIS), where the program selects and maps the names and addresses of those suspects whose method of operation fits the crimes being investigated.[49]

GUNSHOT LOCATOR SYSTEM

A primary difficulty for the police is to determine the location of gunshots. Technology is now being tested that is similar to that used to determine the strength and epicenter of earthquakes. Known as a gunshot locator system, it uses microphone-like sensors placed on rooftops and telephone poles to record and transmit the sound of gunshots by radio waves or telephone lines. Software is then used to alert a dispatcher and to pinpoint the origin of gunshots via a flashing icon on a computerized map. Ideally, the system, which triangulates based on how long it takes the sound to reach the sensors, would greatly reduce police response time to crime scenes, meaning quicker aid for victims and a greater likelihood of arrests.[50]

Initial results of available technologies for gunshot detection have not been promising, however. An examination of about 750 gunshots in three areas of Dallas, Texas, found the following: Not a single offender was apprehended in response to reports of random gunfire; youths setting off fireworks resulted in officers being dispatched; there was a 16 percent reduction in the amount of time required to dispatch a random gunfire call and a one-minute decrease in the time it took police to respond to the scene. In fact, call takers, dispatchers, and patrol officers ended up spending less time processing citizen calls about random gunfire and more time processing such calls using the technology, suggesting that the technology tends to extend rather than reduce the time spent on random gunfire calls.[51]

Dogs provide many vital functions for the police, including finding lost persons. (*Courtesy Washoe County, Nevada, Sheriff's Department*)

DOGS AND SEARCHES FOR LOST PERSONS

Another related use of geographic technology—used extensively with military tanks during Operation Desert Storm and in the private sector (such as in taxicab firms and car rental express delivery companies)—is the global positioning system (GPS), which plots locations of vehicles and even people with remarkable accuracy. The police, for example, have adapted GPS to the problem of knowing exactly where search-and-rescue dogs are and how effective they have been in their search efforts. A small, specially designed backpack containing software is attached to the dog; after a search for lost persons, data from the backpack is downloaded to show where the dog has searched as well as areas where more searching may be needed.[52]

ELECTRONICS IN TRAFFIC FUNCTIONS

ACCIDENT INVESTIGATION

A multicar accident can turn a street or highway into a parking lot for many hours, sometimes days. It is necessary for the police to collect evidence relating to the accident, including measurement and sketches of the scene and vehicle and body positions, skid marks, street or highway elevations, intersections, and curves. These tasks typically involve a measuring wheel, steel tape, pad, and pencil. The cost of traffic delays—especially for commercial truck operators—for every hour traffic is stalled is substantial.

Some police agencies have begun using a version of a surveyor's "total station" to electronically measure and record distances, angles, elevations, and the names and features of objects. Data from the system can be downloaded into a computer for display or printed out on a plotter. In vehicular homicide cases requiring reconstruction of the crime scene, some courts now prefer the precision, scope, and professional appearance of court exhibits collected and generated by the electronic system. The system consists of four components: a base station, a data collector, a tripod, and a prism (which reflects an infrared laser beam back to the tripod-mounted base station).

With this device, officers can get measurements in an hour or so at major traffic accidents, push a button, and have lines drawn for them to scale; this process enables officers to get 40 percent more measurements in about 40 percent of the time, allowing the traffic flow to resume much more quickly. This system is also being used at major crime scenes, such as murders. It has only one drawback: The system cannot be used in heavy rain or snow.[53]

In a related vein, traffic enforcement has recently been raised to a higher level with the advent of traffic cameras, already nicknamed "photocops" and "electro-optical traffic enforcement devices" because they include both radar and cameras. Traffic cameras, either mounted on a mobile tripod or permanently fixed on a pole, emit a narrow beam of radar that triggers a flash camera when the targeted vehicle is exceeding the speed limit by a certain amount, usually 10 miles per hour. The ticket is then mailed to the vehicle's registered owner. If the owner has a question or defense to the citation (such as having loaned the vehicle to another person on the day of the violation), he or she may contact the police department or take the matter to court.

Several companies now manufacture and provide traffic cameras to the police. The devices cost nothing in taxes; the manufacturer receives approximately $25 per paid ticket, so violators actually finance the program. Some schemes are already in progress to fool "photocop." For example, a license plate cover advertised in auto magazines allows the license plate numbers to be read from straight ahead; however, from an angle like that of the traffic camera, the license numbers are obliterated.

ARRESTING IMPAIRED DRIVERS

During a vehicle stop for driving under the influence, officers might spend a lengthy period of time questioning a driver and conducting a barrage of screening tests. Then, if an arrest is made, the officer necessarily devotes a lot of time transporting and processing the arrestee at the jail, including formal testing of urine or blood. This delay in formal testing can skew test results because the alcohol has had time to metabolize.

New instruments are available for helping to automate the drunk-driving process. One tool used routinely during a drunk-driving stop is a breath-screening device—a small, portable machine that resembles a videogame cartridge. The DUI suspect blows into the device and the officer gets a reading of the amount of alcohol in the suspect's system. It is hoped that this device can be adapted in such a way that the test can be used in court. Fewer hours would then be spent transporting DUI suspects, testing them, and finding they were below the legal limit and thus unable to be prosecuted for DUI.

The revamped instrument would be attached to a notebook or laptop computer that officers would use to help speed them through the process; an officer would simply run a mag-stripe driver license through a reader on the computer to bring up all the driver's information. The computer would then prompt the officer to start the test and supply a readout of the results on the screen. The officer would then transfer the test results over telecommunications lines to a central location for recording.[54]

PREVENTING HIGH-SPEED CHASES

Chapter 11 included a discussion of the current controversy over the tremendous potential for injury, property damage, and liability that accompanies high-speed chases by police; indeed, this is such a concern that some agency policies completely prohibit such pursuits by officers. Such techniques as bumping, crowding, the three-cruiser rolling roadblock, and tire spikes can all result in significant damage to vehicles and/or personal injury to persons involved.

Tire spikes can be deployed when a fleeing vehicle is approaching and then retracted so that other vehicles and police cars can pass safely. However, tire spikes often do not work or result in the suspect losing control of his or her vehicle (although some spike devices are designed to prevent loss of control by breaking off in the tires and thus deflating them slowly);[55] furthermore, use of the spikes is limited to times and places where other traffic can be diverted.

One device has been developed using a short pulse of electrical current to disrupt electronic devices that are critical to the continued operation of the vehicle ignition system. Once pulsed, the vehicle rolls to a controlled stop similar to running out of fuel and will not restart until the affected parts are replaced. Since this device requires near-direct contact with the suspect vehicle, other nearby vehicles and persons are not affected. Demonstrations of this device are being performed for federal, state, and local law enforcement agencies across the country, as well as for transportation officials. Three versions of the device are available: the electronic roadblock, where a pad is rolled across the highway to be driven over; a car-chase version, where a wire is connected to the suspect vehicle; and a stationary car version, where a remote-controlled package is deployed under the suspect vehicle, allowing officers to determine the time and location when they shut down the vehicle's electronics.[56]

DNA

It would seem that new success stories involving the use of DNA in criminal justice are being reported each day, such as the following:

> A case file in the rape of a 12-year-old Englewood, Colorado, girl was starting to yellow. A sketch of the suspect had not produced anything concrete. Then the Colorado Bureau of Investigation linked a DNA "fingerprint" from the evidence file to a man recently imprisoned for another sexual assault.[57]

Today, some of the nation's most effective crime fighters wear white lab coats instead of blue uniforms and study DNA fingerprinting. Unfortunately, successes such as the Colorado case are severely limited by the scarcity of DNA fingerprints on file.

Integrated databases were discussed earlier; it was shown that today's police are exchanging information quickly and with good results. Certainly there have been successful outcomes in the area of DNA databases as well. For example, in the late 1980s, states began passing laws mandating that DNA samples be collected from persons convicted of sexual assault or other violent crimes. This data could then be used to solve existing cases and to identify the felons if they committed further crimes after they were released. All 50 states now have such laws and by the end of 1999 had gathered nearly 750,000 DNA samples and made the data accessible to police agencies. The downside, however, is that this figure represents less than 5 percent of the total number of DNA samples that need to be analyzed for 13 genetic markers.[58]

The U.S. Department of Justice has stepped in to help the states address their backlog of more than 700,000 DNA samples. The government has funded a private firm to develop and use a process called laser desorption mass spectrometry in conjunction with robotics; this process enables a sample to be analyzed in a matter of seconds. Once this process becomes available, states will be able to handle their backlogs in a matter of months and DNA will be able to be used much more broadly in identifying, apprehending, and prosecuting criminals.[59]

FINGERPRINTS AND MUG SHOTS

Though perhaps not as exotic as DNA identification, fingerprints are still a reliable means of positively identifying someone. Throughout the country, filing cabinets are filled with ink-smeared cards that hold the keys to countless unsolved crimes if only the data could be located. **Automated fingerprint identification systems (AFIS)** allow this legacy of data to be rapidly shared throughout the nation. One such system is the Western Identification Network (see www.winid.org), estab-

National automated fingerprint identification systems are used in the United States and abroad to connect hundreds of police agencies. (*Courtesy TRW Inc.*)

lished by nine states as a way to share their 17 million fingerprint records. These states were later joined by local agencies and the FBI, the Internal Revenue Service, the Secret Service, and the Drug Enforcement Administration. The system can generally provide a match within a few hours and has helped solve more than 5,000 crimes. A digital photo exchange facility known as WINPHO is now being added to supplement fingerprint data.[60]

The Boston Police Department, like most others in the country, was devoting tremendous resources to identify prisoners with mug shots and fingerprints. Then the department, with what is the first system of its kind in North America, replaced all filmed mug shots and inked fingerprints with a citywide, integrated electronic imaging identification system. The Boston Police Department is also the first city to receive the FBI's certification for electronic fingerprint submission.

Instead of transporting prisoners to a central booking facility in downtown Boston—a task that took 40,000 hours of officers' time per year—officers at the 11 district police stations can electronically scan a prisoner's fingerprints, take digital photographs, and then route the images to a central server for easy storage and access. This network gives investigators timely access to information and mug shot lineups and is saving the police department $1 million per year in labor and transportation costs while freeing officers from prisoner transportation duties.[61]

Technology involving mug shots and imaging systems can be useful in a variety of situations. For example, as a police officer responds to a domestic violence call, the CAD searches the address and finds a restraining order against the ex-husband or boyfriend; moments later, a street map and digital mug shot of the suspect appear on the officer's laptop computer monitor. The officer is thus aided with an image of the suspect and his criminal history prior to arriving at the scene.[62]

CRIME SCENES: COMPUTERS EXPLORE AND DRAFT EVIDENCE

Several technologies exist or are being developed that are relevant to crime scene investigations. Three-dimensional computer-aided drafting (3-D CAD) software is available and can be purchased for a few hundred dollars. By working in 3-D, CAD users can create scenes that can be viewed from any angle. Suddenly, very technical evidence can be visualized by nontechnical people. Juries can "view" crime scenes and see the location of evidence; they can view just what the witness says he or she saw. Police tell the CAD the exact dimensions and get a scaled drawing. A five-day, 40-hour course is available for police investigators, traffic accident reconstructionists, and evidence technicians.[63]

Often, crime scene evidence is too sketchy to yield an obvious explanation of what happened. New software called Maya has been developed for this field, called forensic animation; this software helps experts to determine what probably occurred. Maya is so packed with scientific calculations that it can create a virtual house from old police photos and replicate the effects of several types of forces, including gravity: how a fall could or could not have produced massive injuries; how flames would spread in a house fire; and the way smoke cuts a pilot's visibility in a plane crash.[64]

The federal Department of Energy is also testing a prototype laptop computer equipped with digital video and still cameras, laser range finders, and a global positioning system. A detective using it can beam information from a crime scene back to a laboratory to get input from experts. Researchers are also working on a so-called lab-on-a-chip that would give police the ability to process more evidence—including DNA samples—at the crime scene, eliminating the risk of contamination en route to the lab.[65]

DEVELOPMENTS RELEVANT TO FIREARMS

Handguns are, and nearly always have been, the weapon of choice for violent criminals, especially those who murder others. Handguns are also used with greatest frequency—in about seven of every 10 incidents—when police officers are murdered.[66] It is therefore important that everything that is technologically possible be done to train the police to use lethal force against citizens, identify persons who would use lethal force against others, and keep out of harm's way.

COMPUTER-ASSISTED TRAINING

A device known as FATS, for firearms training system, is said to be "as close to real life as you can get."[67] Recruits and in-service officers alike use the system. The students are given a high-tech lesson in firearms and can be shown a wide variety of computer-generated scenarios on a movie screen. An instructor at a console can control the scene. Using laser-firing replicas of their actual weapons, they learn not only sharpshooting but also judgment—when to shoot and when not to shoot. The system—consisting of a container about the size of a large baby buggy, with a computer, a laser disk player, a projector, and a hit-detect camera—can be transported to sites throughout the state. Some sites combine FATS with a driving simulator and have recruits drive to the scene of a bank robbery and then bail out of their car into a FATS scenario.[68]

Recent developments have made firearms simulators even more realistic than the basic FATS system just described. One version has been developed that has a synchronized "shootback cannon." If an individual on a projection screen pulls a gun and fires toward the trainee, a cannon, which hangs above the screen, pelts the trainee with painful .68-caliber nylon balls. The newest generation of firearms training systems—which are getting more and more realistic and better and better at breaking down how the trainee reacted under stress—now cost between $25,000 for a small system to $200,000 for a trailer that police departments can haul from station to station.[69]

Using "Fingerprints" of Guns to Solve Cases

Every gun leaves a unique pattern of minute markings on ammunition. If gun makers test-fire their weapons before the guns leave their factories, then the police can use spent ammunition recovered from crime scenes to trace guns, even when the gun itself is not recovered. This technology could even be taken to the next level: creating an automated database of the fingerprints of new guns. A $45 million multiyear contract was recently awarded to a Montreal, Canada, firm by the U.S. Bureau of Alcohol, Tobacco and Firearms to help fund the development of such technology.[70]

Several problems would need to be resolved, however: Criminals can easily use a nail file to scratch a new gun's firing pin; even if the system worked and turned up a serial number, the number could be used to trace the gun to its initial owner and not necessarily to the criminal; the database would not include the estimated 200 million weapons already in circulation; and gun lobbies would undoubtedly oppose the system as being "suspiciously like a national gun registry."[71]

GANG INTELLIGENCE SYSTEMS

Often a witness to gang violence has only a brief view of the incident—a glimpse of the offenders and their distinguishing characteristics and their vehicles and license plate number. Police are now armed with laptop computers and cellular phones to assist in solving gang-related crimes. Recently, for example, the California Department of Justice began installing CALGANG (known as GangNet outside California), an intranet-linked software package linked to nine other sites throughout the state. It is essentially a clearinghouse for information about individual gang members, the places they frequent or live, and the cars they drive. Within a few minutes, a police officer in the field, using a laptop and cellular phone, can be linked to CALGANG, type in information, and wait for matches. Other officers can be moving to make the arrest even before the crime laboratory technicians have dusted the scene for fingerprints.[72]

SUMMARY

This chapter examined the exciting high-technology developments in policing, including those in the areas of less-than-lethal weapons, wireless technology, electronics, imaging systems, firearms, and communications.

This is an exciting time for the police: More new technologies than ever are being made available to aid them in their efforts to analyze and address crimes. This chapter has shown the breadth of research and development that is under way.

The rapid expansion in computer technology, while certainly a strong advantage for society overall, bodes ill as well. The first problem with this technology lies in adapting the computer technologies to the needs of policing. While this nation's air force can drop a smart bomb down a smokestack and our army is rapidly moving toward the electronic battlefield, modern-day crooks and hackers engage in a variety of cybercrimes, and (even though the technology exists) the police are still unable to halt a high-speed chase that threatens the lives of officers and citizens.[73]

Obviously, many challenges remain. For example, we must continue to seek a weapon that can effectively and safely be employed that possesses less-than-lethal stopping power. We must also strive to enhance police effectiveness and efficiency through electronic means. As has been noted, "It is not enough to shovel faster. Criminal justice must enter the Information Age by incorporating technology as a tool to make the system run efficiently and effectively."[74]

ITEMS FOR REVIEW

1. Briefly, what were some of the major discoveries in less-than-lethal weapons for use by the police?
2. What did the development of wireless technology provide for the police?
3. In which areas can mapping and profiling systems aid the police, and what do these systems specifically provide?
4. How is the police traffic function assisted by technology? How is gang intelligence affected?
5. What developments are in place in the area of firearms in terms of how they are used by the police and the public?
6. How are the police training and communications functions enhanced by technology, and what role does FATS play in this function?

NOTES

1. Dave Pettinari, "Are We There Yet? The Future of Policing/Sheriffing in Pueblo—Or in Anywhere, America," http://www.policefuturists.org/files/yet.html (13 February 2001).
2. *Ibid.*
3. Jennifer Nislow, "Big Benefits, Huge Headaches," *Law Enforcement News*, May 15/31, 2000, p. 1.
4. "Cybergame for the Millennium: Cops 'n Robbers Playin' Hide 'n Seek on the Net," *Police Futurist* 8 (1): 4.
5. "Polls: Trade Some Freedom for Security," *Law Enforcement News*, September 15, 2001, p. 1.
6. "Surveillance Cameras Stir Up Static," *Law Enforcement News*, July/August 2001, p. 13.
7. Dana Hawkins and David LaGesse, "Tech vs. Terrorists," *U.S. News and World Report*, October 8, 2001, pp. 16–17.
8. Kenneth J. Peak, "The Quest for Alternatives to Lethal Force: A Heuristic View," *Journal of Contemporary Criminal Justice* 6, no. 1 (1990): 8–22.
9. S. Sweetman, *Report on the Attorney General's Conference on Less Than Lethal Weapons* (Washington, D.C.: Department of Justice, National Institute of Justice, 1987).
10. Peak, "The Quest for Alternatives to Lethal Force," p. 10.
11. *Ibid.*

12. *Ibid.*
13. T. S. Crockett, "Riot Control Agents," *The Police Chief* (November 1968): 8–18.
14. M. H. Haller, "Historical Roots of Police Behavior: Chicago, 1890–1925," *Law and Society Review* 13, no.2 (1976): 303–323.
15. T. F. Coon, "A Maze of Confusion over Amazing Mace," *Police* (November-December 1968): 46.
16. Sarah Manwaring-White, *The Policing Revolution: Police Technology, Democracy, and Liberty in Britain* (Brighton, Sussex: The Harvester Press, 1983).
17. Luis Cabrera, "Police Explore 'Less-Than-Lethal' Weapons," Associated Press, June 19, 2000.
18. *Ibid.*, pp. 141–142.
19. Peak, "The Quest for Alternatives to Lethal Force," pp. 15–16.
20. *Ibid.*, p. 16.
21. Manwaring-White, *The Policing Revolution*, pp. 145–146.
22. *Ibid.*, pp. 142–143.
23. Sweetman, *Report on the Attorney General's Conference on Less Than Lethal Weapons*, pp. 4–5.
24. M. S. Serrill, "ZAP! Stun Guns: Hot but Getting Heat," *Time*, May 1985, p. 59.
25. Darren Laur, "More Powerful, but Still Less Lethal," *Law Enforcement Technology*, October 1999, p. 10.
26. John Barry and Tom Morganthau, "Soon, 'Phasers on Stun,'" *Newsweek*, February 7, 1994, pp. 24–25.
27. "Rethinking Stopping Power," *Law Enforcement News*, November 15, 1999, pp. 1, 9.
28. *Ibid.*
29. Cabrera, "Police Explore 'Less-Than-Lethal' Weapons."
30. *Ibid.*
31. "Rethinking Stopping Power," pp. 1, 9.
32. "Effectiveness Times Three," *Law Enforcement News*, May 15, 1993, p. 1.
33. *Ibid.*, p. 6.
34. "Rethinking Stopping Power," pp. 1, 9.
35. Blake Harris, "Goin' Mobile," *Government Technology* 10 (August 1997): 1.
36. Kaveh Ghaemian, "Small-Town Cops Wield Big-City Data," *Government Technology* 9 (September 1996): 38.
37. Tod Newcombe, "Combined Forces," *Crime and the Tech Effect* (Folsom, CA: Government Technology, April 2001), p. 15.
38. *Ibid.*
39. J. Campbell, *Map Use and Analysis*, 2d ed. (Dubuque, IA: William C. Brown, 1993).
40. U.S. Department of Justice, National Institute of Justice, Crime Mapping Research Center, *Mapping Crime: Principle and Practice* (Washington, D.C.: Author, 1999), p. 1.
41. *Ibid.*, p. 2.
42. Lois Pilant, "Computerized Crime Mapping," *The Police Chief* (December 1997): p. 58.
43. Dan Sadler, "Exploring Crime Mapping" (Washington, D.C.: U.S. Department of Justice, National Institute of Justice, Crime Mapping Research Center, 1999), p. 1.
44. Pilant, "Computerized Crime Mapping," p. 58.
45. Chicago Police Department, *CAPS News*, October 1995, pp. 1, 6.
46. Pilant, "Computerized Crime Mapping," pp. 64–65.
47. *Ibid.*, pp. 66–67.
48. Bill McGarigle, "Crime Profilers Gain New Weapons," *Government Technology* 10 (December 1997): 28–29.
49. *Ibid.*
50. Justine Kavanaugh, "Locator System Targets Shooters," *Government Technology* 9 (June 1996): 14–15.
51. Lorraine Green Mazerolle, Cory Watkins, Dennis Rogan, and James Frank, "Using Gunshot Detection Systems in Police Departments: The Impact on Police Response Times and Officer Workloads," *Police Quarterly* 1 (1998): 21–50.
52. "GPS-Loaded Dogs to the Rescue," *Government Technology* 10 (April 1997): 24.
53. Bill McGarigle, "Electronic Mapping Speeds Crime and Traffic Investigations," *Government Technology* 9 (February 1996): 20–21.
54. Justine Kavanaugh, "Drunk Drivers Get a Shot of Technology," *Government Technology* 9 (March 1996): 26.
55. *Ibid.*
56. Http://www.jaycor.com/arrestor.htm, April 11, 1997.

57. Rutt Bridges, "Catching More Criminals in the DNA Web," *Law Enforcement News*, September 15, 2000, p. 9.
58. Drew Robb, "The Long New Arm of the Law," in *Crime and the Tech Effect* (Folsom, CA: Government Technology, April 2001), pp. 44–46.
59. *Ibid.*
60. *Ibid.*, p. 46.
61. Tod Newcombe, "Imaged Prints Go Online, Cops Return to Streets," *Government Technology* 9 (April 1996): 1, 31.
62. Corey Grice, "Technologies, Agencies Converge," *Government Technology* 11 (April 1998): 24, 61.
63. Tod Newcombe, "Adding a New Dimension to Crime Reconstruction," *Government Technology* 9 (August 1996): 32.
64. John McCormick, "Scene of the Crime," *Newsweek*, February 28, 2000, p. 60.
65. Joan Raymond, "Forget the Pipe, Sherlock: Gear for Tomorrow's Detectives," *Newsweek*, June 22, 1998, p. 12.
66. Samuel G. Chapman, *Murdered on Duty: The Killing of Police Officers in America*, 2d ed. (Springfield, IL: Thomas, 1998), p. 33.
67. Patrick Joyce, "Firearms Training: As Close to Real as It Gets," *Government Technology* 8 (July 1995): 14–15.
68. *Ibid.*
69. John McCormick, "On a High-Tech Firing Line," *Newsweek*, December 6, 1999, p. 64.
70. Vanessa O'Connell, "The Next Big Idea: Using 'Fingerprints' of Guns to Solve Cases," *Wall Street Journal*, February 10, 2000, p. A1.
71. *Ibid.*
72. Ray Dussault, "GangNet: A New Tool in the War on Gangs," *Government Technology* (January 1998): 34–35.
73. Pettinari, "Are We There Yet?"
74. George Nicholson and Jeffrey Hogge, "Retooling Criminal Justice: Interbranch Cooperation Needed," *Government Technology* 9 (February 1996): 32.

Challenges of the Future

Key Terms and Concepts

Accelerators
COPS Office

Demographics
National Crime Victim Survey

Where there is no vision, the people perish.—*Proverbs 29:18*

The future must be shaped or it will impose itself as catastrophe.
—*Henry Kissinger*

INTRODUCTION

It has been said that the only thing that is permanent is change. Perhaps more than anything else, the foregoing chapters have demonstrated that even in the relatively tradition-bound domain of policing we certainly live in a dynamic world where society and its problems and challenges are constantly changing. As a result, our lives will be drastically altered as the world continues to change.

Given that, the question at the forefront of our minds is, What will the future bring? That troublesome question becomes even more important and ominous when we consider the world's present state of affairs. As we will see here, there are indicators of future problems and turmoil. Our choice is either to ignore the future until it is upon us or to try to anticipate what the future holds and gear our resources to cope with it.

Predicting the future is not easy. Many variables—such as war, technology, and economic upheaval—can greatly affect otherwise sound predictions and trends. Indeed, much of our planning and predictive efforts revolve around the unstable future of finance. As someone once said, "There's a lot of crime prevention in a T-bone steak." While all predictions are grounded in past trends and future likelihood, unforeseen variables and major events can and do change the course of the future in ways that no one could have anticipated. In other words, the best we can possibly do in this chapter is render an educated guess.[1]

This chapter begins by examining how America is changing, primarily with respect to its people and its crime. Then the focus shifts to what all of the current and predicted demographic and crime-related changes portend for the police, especially in the areas of community oriented policing and problem solving (COPPS), advanced technology, the allocation of resources, the role of the beat officer, and other personnel issues.

A CHANGING NATION

FACES OF AMERICA

The changing **demographics** of our nation, and what that bodes for the future, absolutely commands the attention of our peacekeepers. Nearly 273 million people now live in the United States; a little more than half (51 percent) are female. There are about 2.63 persons per household, and about one-fourth (27.6 percent) of all households are headed by one person.[2] While 56 percent of the population are married and live with their spouse, about 19.4 million adults (9.4 percent) are divorced.[3] These figures can represent threats to societal stability, because the smaller the household, the less the commitment of a population to each other and the less likely deviance will be managed at the household level, thus necessitating governmental intervention. Marriage is our most stable family structure.[4]

Another destabilizing force is the number of people who in a given year change where they reside. The average person moves once every six years, but approximately 17 percent of the population moves each year. Thus, long-term neighborhood stability is the exception rather than the rule. Home ownership—another stability marker—is also a concern: 72 percent of blacks and Hispanics and 38 percent of whites are unable to afford a modestly priced home. Single females fare even worse—57 percent of single white women, 78 percent of single black women, and 73 percent of single Hispanic women are unable to afford a modestly priced home. Being a female black or Hispanic and/or being single often lead to poverty and criminal victimization.[5]

Furthermore, the number of children without a father at home is increasing—these children are more prone to delinquency and other social pathologies. Many problems in crime control are strongly related to father absence: 90 percent of all homeless and runaway youths are from fatherless homes, as are 71 percent of high school dropouts, 70 percent of youths in state institutions, and 75 percent of adolescent patients in substance abuse centers.[6]

We live in a "graying" country as well, where the fastest-growing age group is between 55 and 65.[7] The golden years for baby boomers mean a graying of the population. Nearly half (45.2 percent) of the women 65 years old and over are widowed; of those, 70.1 percent live alone.[8] The oldest old (persons 85 years old and over) are a small but rapidly growing group and are projected to be the fastest-growing part of the elderly population well into the twenty-first century.[9] The rapid growth of the elderly, particularly the oldest old, represents a triumph of efforts to extend human life, but these age groups also require a large share of special services and public support.[10]

The good news is the elderly are less likely than younger people to become victims of violence, personal theft, and household crimes or to be injured during a violent crime—but their injuries, because of their brittle bones, are more severe. Victimization for the elderly can be permanently disabling. Because many are on fixed incomes, they often cannot afford the best medical care. They have a high fear of crime but are less likely to take protective measures than younger people and are more likely to report a crime. They are also targeted more for financial fraud schemes than younger people, which can lead to severe depression and other serious health problems. Their isolation means that a high percentage of incidents in which they are victimized occur in their homes.[11]

Almost one-third of the current population growth is caused by immigration.[12] An estimated 5 million undocumented immigrants reside in the United States as well, and an estimated 275,000 illegal aliens enter and stay in this country each year. Mexico is the leading country of origin of illegal aliens; about 2.7 million (or 54 percent) of the illegal immigrant population comes from that nation.[13]

The impact of immigration to America and the growth of minority group populations in general cannot be overstated. America now accepts nearly 1 million newcomers each year (with a net immigration of about 880,000 persons, which could increase or decrease in future years).[14] Immigrants to this nation, however, are not more prone to crime than native-born Americans. Among men 18 to 40 years old in the United States, immigrants are less likely to go to prison than native-born Americans.[15] As will be seen later, immigrants pose unique challenges for the police.

CRIME, VIOLENCE, AND THE INFLUENCE OF DRUGS AND GUNS

About 29 million violent and property crime victimizations occur each year in this country.[16] A number of factors contribute to these figures: immediate access to firearms, alcohol and substance abuse, drug trafficking, poverty, racial discrimination, and cultural acceptance of violent behavior.[17]

A recent decline in crime has actually made America safer. Between 1994 and 1998, the murder rate dropped 30 percent. The **National Crime Victim Survey (NCVS)** data showed a 7 percent decline in other violent crimes from 1997 to 1998 and a decrease of 26.7 percent from 1993 to 1998, as well as a 12 percent decline in property crimes. Violent crimes are expected to continue to decline.[18]

Why this decline in crime? Possible reasons include that state legislators have imposed tougher sentences on violent criminals and local officials are implementing aggressive and intelligent methods of community policing. The effectiveness of good police work and extended incarceration of hardened criminals is beyond dispute.[19] Notwithstanding this recent decline in offenses, serious challenges remain. New criminal types will dot the national landscape in the future: better educated, upscale, older, and increasingly female.

Still, there are serious problems with respect to violence in the United States. In both the immediate and distant future, police will continue to deal with the "new violence" that has emerged over the past five to ten years. In the past, violence was a means to an end. Revenge, robbery, and jealousy were among the many reasons that people resorted to violence. Today, an entire culture has emerged that sees the use of violence as an end unto itself. The people who make up this culture—gangs, pseudo-gangs, well-armed young people, and others—are not going to change their way of thinking or relax their hostility and aggression simply because there is a fluctuation in drug usage or a switch from cocaine to heroin. For these people, the wanton use of violence—aggression for the sake of aggression—is not considered abhorrent behavior. Taken in combination, these factors present challenges for police leaders. Police will need training and education far beyond what they are given today to understand such violent behavior.[20]

Furthermore, the reduction in crime has had a minimal, if any, effect on reducing citizen fear of crime. Reducing fear, not simply crime statistics, will be a major challenge facing police leaders in the future. Few police leaders and officers know anything about fear. What is it? How does it work? What is the cycle of fear? How can the police intervene to break the cycle? Why is fear contagious? Do problem solving, partnering, and the implementation of crime control strategies and tactics reduce fear? Officers should be trained to assess causes and levels of individual, neighborhood, and community fear.[21]

Drugs are also a major factor in crime and violence; more than 277,000 offenders are in prison for drug-law violations—21 percent of state prisoners and more than 60 percent of federal prisoners. More than 80 percent of state prisoners and 70 percent of federal prisoners have engaged in some form of illicit drug use. One-third of state prison inmates and 22 percent of federal prisoners report they were under the influence of drugs when they committed the crime for which they are in prison.[22]

Most police agencies remain reactive to the drug trade. They sweep street corners, arrest users and low-level dealers, rely on specialists (narcotics units) to assume primary responsibility for enforcement, and participate in regional task forces (generally when funded by the federal government). Police leaders will have to "think out of the box" in the future in response to a changing drug market. A well-planned, strategic approach to the drug trade and problem solving related to drugs will be needed in all agencies. More agencies must become involved in teaching employees about market analysis and forecasting in the drug trade.

In addition to drugs, two more likely crime **accelerators**—guns and alcohol—might still increase the risk of victimization and a general fear of crime. Nearly 200 million guns are in private hands in the United States.[23] Gun violence in the United States is both a criminal justice and a public health problem; firearms are still the weapons most frequently used for murder; they are the weapons of choice in nearly two-thirds of all murders. Strategies and programs to reduce gun violence include interrupting sources of illegal guns, deterring illegal possession and carrying of guns, and responding to illegal gun use.[24]

Alcohol also remains a major factor in crime and violence; almost four in 10 violent crimes involve alcohol.[25] Nearly 2 million offenders (about 36 percent of the total) reported that they were using alcohol at the time of their offense.[26]

WHAT THE FUTURE PORTENDS FOR THE POLICE

What can the police do about these demographic and crime issues? Can they hold the line and reduce the spread of criminality?

Defining solutions to the problems facing future police service is difficult at best. Factors such as locale, political environment, and economics determine how an agency and its employees will view and react to future events. Additionally, as Sheldon Greenberg has noted, for some police leaders, the future is the next fiscal year, while for others it is a three- to five-year span of time toward which they have set into motion a strategic planning process. Still, for others, the future is next Friday and surviving without a crisis until their next day off. Any discussion of the future must include the short term as well as the long term and must give attention to operational issues, administrative issues, the community, and the basic philosophy of policing.[27]

The first priority is for the police to realize that the conventional style of reactive, incident-driven policing employed during the professional era had several drawbacks. That type of police department was hierarchical, impersonal, and rule based, and most important policy decisions were made at the top; line officers made few decisions on their own. Next we analyze where the current policing era—that of community oriented policing and problem solving (COPPS)—fits into the challenges of the future.

COMMUNITY ORIENTED POLICING AND PROBLEM SOLVING

Chapters 1 and 6 examined the current era of policing, COPPS; following are some futuristic issues that remain to be addressed in this connection.

First, many fundamental questions remain concerning COPPS. For example, how many police agencies have made a commitment to community policing? How many agencies have demonstrated the link between community policing and the quality of the communities they serve? How many have embraced community policing to gain a share of available federal dollars? Will community policing endure without federal funding?[28]

From 1994 to 2001, the federal **Office of Community Oriented Policing Services (COPS)**, created by the Violent Crime Control and Law Enforcement Act of 1994, spent about $9 billion to help police agencies implement COPPS, adding officers to the beat and providing technical assistance, technology, equipment, and training. By mid-2000, the COPS office had provided funding to 82 percent of America's police departments by awarding more than 30,000 grants to more than 12,000 departments. This included funding more than 105,000 community policing officers and training more than 90,000 officers and citizens.[29] Funding for the COPS office expired in 2000, however; the future remains uncertain for the office.[30] Any police agency that is currently hanging its COPPS efforts on federal funds may find itself in desperate straits in the future.

Another central issue with respect to COPPS concerns community partnerships. Much of society—including the police—has come to realize that the police cannot function independently to address crime and disorder. The challenge to leaders, now and in the future, is to develop meaningful and lasting partnerships rather than the superficial relationships that exist in many communities. For many police officers, the concept of partnership means simply attending occasional neighborhood association meetings or occasionally visiting neighborhood leaders who live on their beat. Partnerships for the sake of partnership do not endure; they will work only when the mutual benefits to the parties involved are well defined, well understood, and attainable.

Several other issues for the future must be addressed if COPPS is to survive and thrive, including whether police chief executives will change the culture of their agencies, implement the concept, decentralize their department (pushing decision making downward), invest in the necessary technology to locate "hot spots," and develop the necessary mechanisms to support COPPS. These challenges also include recruitment, selection, training, performance appraisals, and reward and promotional systems. It is also necessary that police unions work with administrators to effect the kinds of changes that are needed for COPPS. The answers to these issues could be critical, not only to the future of COPPS but also to policing and society.

Some futurists also see policing generally undergoing a metamorphosis in the near future, including the following changes:

- Ethics will be woven into everything the police do: the hiring process, the FTO program, the decision-making processes. There will be increased emphasis on accountability and integrity within police agencies as policing is elevated to a higher standing and reaches more toward being a true profession. Concomitantly, the majority of officers will be required to possess a college degree.

- Formal awards ceremonies will recognize improving citizens' quality of life as well as felony arrests or other high-risk activities.
- Communications will be greatly improved through internal intranets that contain local and agency operational data, phone books, maps, calendars, calls for service, crime data sheets, speeches, newsletters, news releases, and so on.[31]
- Major cities will no longer require policing experience of the chief police executive (who may be recruited from private industry). The head of the future police agency will essentially be recognized as a CEO with good business sense as a trend toward privatization of certain services becomes more prevalent. Knowledge will continue to increase at lightning speed, forcing the CEO to be involved in trend analysis and forecasting in order to keep ahead of the curve.
- The rigid paramilitary style currently in effect will become obsolete, replaced by work teams consisting of line officers, community members, and business and corporation people.
- The current squad structure will give way to more productive, creative teams of officers who, empowered with more autonomy, will become efficient problem solvers, thus strengthening ties between the police and the citizenry.
- Neighborhoods will more actively participate in the identification, location, and capture of criminals.

On a more negative note, however, some authors point to what they believe are several *unfavorable* social forces that militate against the future of COPPS: Local governments are being pushed toward a more legalistic, crime-control model of policing; the public is less willing to pay more taxes to address fundamental social problems; public policy does not allow the police to focus on the root causes of crime but rather their symptoms—such as criminal conduct—through aggressive strategies rather than through COPPS.[32] Indeed, the landscape of policing is littered with the skeletons of strategies and approaches that died after the departure of a dedicated COPPS chief or sheriff. Although one would hope that such a fate would not befall COPPS, history has shown it to be a possibility that cannot be ignored.

In sum, several questions for the future remain concerning COPPS:

- Will police organizations come to believe that they cannot control crime alone and truly enlist the aid of the community in this endeavor?
- Will police chief executives acquire the innovative drive necessary to change the culture of their departments, implement COPPS, flatten the organizational structure of their department, and see that officers' work is properly evaluated?
- Will police executives have the necessary job security to accommodate COPPS? Should at-will employment of chiefs place COPPS at risk?
- Will those departments work with their communities, other city agencies, businesses, elected officials, and the media to sustain COPPS?
- Will police unions work with administrators to effect change needed for COPPS?
- Will police employees who have not yet done so, from top to bottom, sworn and civilian, realize that the traditional reactive mode of policing has obviously not been successful and cannot work in the future?

- Can those police organizations, from top to bottom, become more customer and value oriented?
- Will police executives and supervisors come to develop the necessary policies and support mechanisms to support COPPS, including recruitment, selection, training, performance appraisals, and reward and promotional systems?
- Will police executives and supervisors begin viewing the patrol officer as a problem-solving specialist? Will they give street officers enough free time and latitude to engage in proactive policing?
- Will police organizations come to view COPPS as a departmentwide and citywide strategy? Will they invest in technology to support problem-oriented policing?
- Will those agencies attempt to bring diversity into their ranks to reflect the changing demographics and cultural customs of our society?

FUTURE HIGH TECHNOLOGY

Certainly the advent of computer crime has changed the world of policing, posing new and extraordinary challenges. An estimated 150 million Americans are now plugged into cyberspace, and thousands more enter the online world each day.[33] This high-tech revolution in our homes and offices has opened a whole new world for the criminal element as well. Pornographers and pedophiles are now on the Web, as well as people who trade stolen credit card numbers, rig auctions, create viruses, and devise baby adoption scams, among many other crimes.[34] Computer crimes will increase dramatically, including such crimes as cyberterrorism, identity theft, credit card fraud, consumer fraud, stock market–related fraud, and industrial espionage; these crimes may well become the next national crime-fighting priorities.[35]

Other computer-related problems will exist as well. For example, from 1995 to 1998 (the last year for which such applications were received), the federal Office of Community Oriented Policing Services poured hundreds of millions of dollars into police agencies to provide grants for equipment.[36] Yet, as discussed in Chapter 14, many departments lack the in-house computer expertise to install or run the software and equipment.

Also, as noted in Chapter 7, at present the police are ill-prepared to investigate computer-related crimes. To meet the future challenges posed by computer crimes, the police must become better educated, more adaptable, and better equipped. Training will have to be virtually continuous to address complex threats to society.[37]

ALLOCATION OF RESOURCES

If the quality of policing is to improve, a greater number of police leaders will need to know more about resource allocation than they currently do. A predicted decline in federal funding for additional police officers, combined with local, state, and federal belt-tightening, will necessitate a greater emphasis on using available resources as efficiently as possible.

Analysis of workload and resources should be part of every supervisor's and executive's routine functioning. Justifying resources should be based on more than perception, statistics such as increased calls for service, and the simple belief that more is better. Once learned, resource allocation is neither overwhelmingly time consuming nor difficult. It becomes second nature.[38]

On the whole, it is unlikely that police budgets will increase substantially in the near future.[39] Some police organizations are already attempting to cope with budgetary problems in some new and unique ways, such as employing lobbyists to present their cases to state legislatures. Some have also begun generating their own revenue by charging for some of their traditionally free public services, such as automobile accident investigations and responses to false alarms. The police can also engage in more proactive kinds of activities and insist that private citizens, institutions, and organizations shoulder greater responsibility for crime control. Some examples of existing collaborative efforts are Drug Abuse Resistance Education (DARE), Mothers Against Drunk Drivers (MADD), Neighborhood Watch, and Court Watch.

ROLE OF THE BEAT OFFICER

The changing role of rank-and-file officers also looms large in future police service. Future generations will have been raised to be at ease and fluent with information technologies and will enter a police service that is much more involved with collective bargaining.

In the past—particularly under the professional model of policing—while undergoing the academy phase of their training, recruits adopted a new identity and a system of discipline in which they learned to take orders and not to question authority. Recruits learned that loyalty to fellow officers, a professional demeanor and bearing, and respect for authority are highly valued qualities. That theme—and the police executive's set of expectations for recruits—must change in the future, however. Only those potential officers who can critically think, plan, and evaluate will be hired. At the same time, chiefs, sheriffs, commanders, and even sergeants will wield less power and control and filter less information; instead, they will move into enhanced roles as coaches, supporters, and resource developers.[40]

People entering future police service will not normally possess military experience and its inherent obedience to authority, but they *will* have higher levels of education and tend to be more independent and less responsive to traditional authoritarian leadership styles. These recruits will have been exposed to more participative, supportive, and humanistic approaches and will want more opportunities to provide input into their work and to address the challenges posed by problem solving. The autocratic leader of the past will not work in the future. The watchwords of the new leadership paradigm are coach, inspire, gain commitment, empower, affirm, flexibility, responsibility, self-management, shared power, autonomous teams, and entrepreneurial units. Therefore, a major need for police leadership will be the surrendering of power to lower organizational employees (a flattened hierarchy).

The tools and functions of police in the future, as seen by futurists, were discussed in Chapter 14.

OTHER PERSONNEL ISSUES

Some traditional police personnel problems are not likely to go away. Such matters as the need for more women and minorities in policing, unionization and job actions, contract and consolidated policing, civilianization, accreditation, higher education, and stress recognition and management for police will not be resolved in the near future. Furthermore, changing societal values, court decisions about the rights of employees, and the Peace Officers' Bill of Rights will make police leadership increasingly challenging. Nor will opportunities decline for officers to engage in graft and corruption, so administrators must develop personnel policies that will protect the integrity of the profession.

SUMMARY

The cup is half full. The cup is half empty. Should we be optimistic or pessimistic about the nation's future? One thing that is for certain is that our society is changing.

This is a very exciting and challenging time in the history of police service. No matter what issues lie ahead, the public will continue to expect a high degree of service from its police. Today's leaders and those who follow will determine whether police agencies embrace their communities or return to being distant and aloof (as under the professional era, discussed in Chapter 1). They will deal with unforeseen problems caused by new drugs, small but hostile groups of extremists, young people who were raised in an environment of violence, and more. They have the opportunity to deal with these issues supported by advanced technology, highly evolved information resources, better trained officers and deputies, and a heightened commitment to interjurisdictional cooperation.

The police will benefit greatly by anticipating what the future holds so that appropriate resources and methods may be brought to bear on the problems ahead. And there is no indication that a significant abatement of today's social problems is looming on the horizon. The police will be affected in many ways by the social and economic changes to come; they can no longer be resistant to change or unmindful of the future.

It is hoped that at some point in the future, Americans can reflect back on the challenges of today and say with total certainty and sincerity that

> The police profession today is the intellectual leadership of the criminal justice profession in the United States. The police are in the lead. They're showing the world how things might better be done.[41]

ITEMS FOR REVIEW

1. Describe in general terms the demographic shifts in America today.
2. Explain the specific changes in crime that are predicted for the future.
3. Delineate how community policing and problem solving will fit into the future crime picture.
4. What are some of the major technological changes that futurists anticipate in policing?
5. Describe how the role of beat police officers will possibly change in the future.
6. What are some problematic police personnel issues that loom on the horizon?

NOTES

1. Those wanting to keep abreast of trends and changes should join the World Future Society; its publication, *The Futurist*, is excellent.
2. U.S. Census Bureau, *Statistics in Brief: Population and Vital Statistics* (Washington, D.C.: Author, 2000).
3. *Ibid.*
4. *Ibid.*
5. *Ibid.*, pp. 1–2.
6. *Ibid.*, p. 11.
7. Cicero Wilson, "Economic Shifts That Will Impact Crime Control and Community Mobilization," in U.S. Department of Justice, National Institute of Justice, *What Can the Federal Government Do to Decrease Crime and Revitalize Communities?* (Washington, D.C.: Author, 1998), p. 4.
8. U.S. Census Bureau, *Current Population Reports: Marital Status and Living Arrangements* (Washington, D.C.: Author, March 1998), p. 2.
9. Frank B. Hobbs, *The Elderly Population* (Washington, D.C.: United States Census Bureau, 1999), p. 2.
10. Konrad M. Kressly, "Golden Years for Baby Boomers: America in the Next Century," *The Harbinger*, January 1998, p. 1.
11. Andrew Karmen, *Crime Victims: An Introduction to Victimology*, 4th ed. (Belmont, CA: Wadsworth, 2001), pp. 263–264.
12. *Ibid.*, p. 2.
13. U.S. Census Bureau, *Illegal Alien Resident Population* (Washington, D.C.: Author, 2000), p. 1.
14. Jennifer Cheeseman Day, *National Population Projections* (Washington, D.C.: United States Census Bureau, 1999), p. 1.
15. Kristin F. Butcher and Anne Morrison Piehl, "Recent Immigrants: Unexpected Implications for Crime and Incarceration," National Bureau of Economic Research, *NBER Working Papers*, No. 6067 (Washington, D.C., June 1997), p. 1.
16. U.S. Department of Justice, Bureau of Justice Statistics Press Release, "National Violent Crime Rate Falls More Than 10 Percent—Violent Victimizations Down One-Third Since 1993," August 27, 2000.
17. Lee P. Brown, "Violent Crime and Community Involvement," *FBI Law Enforcement Bulletin* (May 1992): 2–5.
18. *Ibid.*, pp. 2–3.
19. *Ibid.*, p. 3.
20. Sheldon Greenberg, "Future Issues in Policing: Challenges for Leaders," in *Policing Communities: Understanding Crime and Solving Problems*, eds. Ronald W. Glensor, Mark E. Correia, and Kenneth J. Peak (Los Angeles: Roxbury, 2000), pp. 315–321.
21. *Ibid.*
22. U.S. Department of Justice, Bureau of Justice Statistics Press Release, "More Than Three-Quarters of Prisoners Had Abused Drugs in the Past," January 5, 1999.
23. U.S. Department of Justice, National Institute of Justice, *Guns in America: National Survey on Private Ownership and Use of Firearms* (Washington, D.C.: Author, 1997), pp. 1–2.

24. David Sheppard, "Strategies to Reduce Gun Violence" (U.S. Department of Justice, Office of Juvenile Justice and Delinquency Prevention Fact Sheet #93, February 1999), p. 1.
25. U.S. Department of Justice, Bureau of Justice Statistics, *Alcohol and Crime* (Washington, D.C.: Author, 1998), p. 20.
26. *Ibid.*
27. Greenberg, "Future Issues in Policing," p. 315.
28. *Ibid.*, pp. 318–319.
29. Thomas C. Frazier, "Community Policing Efforts Offer Hope for the Future," *The Police Chief*, August 2000, p. 11.
30. *Ibid.*
31. Dave Pettinari, "Are We There Yet? The Future of Policing/Sheriffing in Pueblo—Or in Anywhere, America," http://www.policefuturists.org/files/yet.html
32. Roy Roberg, John Crank, and Jack Kuykendall, *Police & Society*, 2d ed. (Los Angeles: Roxbury, 2000), pp. 522–523.
33. Margaret Mannix, "The Web's Dark Side," *U.S. News and World Report*, August 28, 2000, p. 36.
34. *Ibid.*
35. Pettinari, "Are We There Yet?," p. 3.
36. Jennifer Nislow, "Big Benefits, Huge Headaches," *Law Enforcement News*, May 15/31, 2000, p. 1.
37. "Cybergame for the Millennium: Cops 'n Robbers Playin' Hide 'n Seek on the Net," *Police Futurist* 8 (1): 4.
38. Greenberg, "Future Issues in Policing," p. 320.
39. Edward J. Tully, "The Near Future: Implications for Law Enforcement," *FBI Law Enforcement Bulletin*, 55 (July 1986): 1.
40. Pettinari, "Are We There Yet?," p. 2.
41. James Q. Wilson, "Six Things Police Leaders Can Do About Juvenile Crime," in *Subject to Debate* (newsletter of the Police Executive Research Forum) September/October 1997, p. 1.

Career Information

This book has examined policing from many perspectives. Those persons who have made the decision to enter into a police career may already be involved with the job market; therefore, this appendix provides some general information in that regard. It discusses some methods for preparing for career opportunities at the federal, state, and local levels; some general advice is also provided about obtaining a police position.

PREPARING FOR JOB HUNTING

A question that is often asked by police recruiters is what the applicant has done to "prepare" himself or herself for a policing career; several measures can be taken to prepare for a prospective career in policing. For example, it is helpful if an applicant can point to studying criminal justice, participating in a police ride-along program or auxiliary police service, having a clean criminal record, staying in good physical shape, and so on.

Also, you may wish to read about the federal, state, and local law enforcement agencies that are described in Chapter 2; furthermore, Chapter 3 discussed police recruitment and the hiring process. Both chapters should serve as a good resource in the job search. Then you can more objectively determine your areas of interest and possible suitability for such employment. The primary limitations to the ability to secure employment are often those characteristics that the applicant can address: mobility, appearance, character, physical ability, academic performance, experience, the quality of the résumé and interview(s), and so forth. And one must generally be patient and heed the old adage that "Rome wasn't built in a day."

Also, at some point an applicant should try to determine whether he or she possesses the background, personality, interest, and physical abilities to pass the entry-level examinations and handle the challenges of police work. Can he or she work well under pressure? Write well? Be action oriented when necessary? Accept cultural differences found in others? Can the applicant enroll in an internship experience at a college or university? Also bear in mind that college and university placement offices may be contacted for job announcements and assistance in preparing résumés, interviewing, and so on. There are also private, for-profit organizations that assist with résumé preparation and developing interviewing skills. Do not underestimate the value of a good résumé.

Applicants must also determine where the job vacancies exist. The National Employment Listing Service, at Sam Houston State University's Criminal Justice Center in Huntsville, Texas, provides job listings on the Internet. Knight Line USA, published semimonthly in Tallahassee, Florida, provides general information about career planning and also provides a job search service. Another excellent resource is *Seeking Employment in Criminal Justice and Related Fields*, by J. Scott Harr and Karen M. Hess.

CAREERS IN FEDERAL LAW ENFORCEMENT

APPLYING FOR A POSITION

Several methods may be employed to learn of openings in federal law enforcement agencies. First, many federal law enforcement agencies have agents who are assigned specifically as recruitment coordinators; they may be contacted in their respective agencies for general information about the application process and the hiring outlook. Some people have also been successful in finding federal law enforcement positions by consulting the Internet. One can also look for announcements at a state employment office.

Before considering a career in any federal law enforcement agency, an applicant must first complete and submit an application form—the SF-171 form—for federal employment. The federal Office of Personnel Management (OPM) coordinates the testing of most federal law enforcement agencies, although requirements for federal positions are determined by the individual agencies (the FBI and some other agencies do not utilize the OPM). Once the applicant has tested and passed, he or she receives an employment rating for the particular agency. The applicant must pass the test with a minimum score of 70. The recruiting agency can then proceed to interview the best candidates, after which a background investigation is initiated.

The minimum salary or Government Service (G.S.) grade of hiring depends on the applicant's education, experience, and training, although some agencies do not hire above a specific level (generally GS-5, 7, or 9).

CAREERS WITH THE STATE POLICE

As discussed in Chapter 2, most states have a state police or highway patrol agency; many have an investigative agency. A number of other specialized investigative units also exist at the state level, such as a separate office of alcoholic beverage control, a fire marshal, a state revenue office, and an office of wildlife or natural resources.

Nearly all of the state police, highway patrol, and investigative agencies require a high school degree or its equivalent as a minimal educational requirement for trooper, investigator, or agent positions.

Persons who are interested in state employment should watch for hiring notices on television, in the newspapers, or on the radio or for hiring announcements at their state employment office; or they may contact a state personnel office and ask for information concerning a particular state agency.

CAREERS IN LOCAL POLICING

Most of this book's previous chapters have examined the roles and functions of local police departments and sheriff's offices. The 17,000 municipal police agencies constitute the largest segment of police personnel in the nation. They also offer a broad spectrum of specialized assignments. At the county level, policing is primarily in the hands of an elected sheriff; county prosecutors may have investigators assigned to that office by the county sheriff or have a separate investigative unit hired by and reporting to that office. Counties may have other specialized police personnel, such as county park police or part-time deputies.

Persons who are interested in local police employment should contact the relevant human resources department for the city or county agency in which they are interested or contact the agency directly (many smaller agencies do not have a human resources department or advertise position vacancies). They should also watch for hiring notices on television, in the newspapers, or on the radio.

Work Cited

J. Scott Harr and Karen M. Hess, *Seeking Employment in Criminal Justice and Related Fields,* 3d ed. (Belmont, CA: Wadsworth, 2000).

Police-Related Information Sources

In addition to the groups listed in Appendix A, there are a host of other organizations and agencies that provide information on justice-related topics. A partial list is as follows:

Administrative Office of the Courts
202-273-2777
http://www.uscourts.gov

Alcohol and Drug Information Clearinghouse
301-468-2600
800-726-6686
infor@health.org
http://www.health.org

American Academy of Forensic Sciences
719-636-1100
http://www.aafs.org

American Bar Association
312-988-5000
800-285-2221
http://www.abanet.org

American Civil Liberties Union
212-549-2500
elistclu@aol.com
http://www.aclu.org

Association of Certified Fraud Examiners
http://www.acfe.org

Association of Trial Lawyers
202-965-3500
800-424-2727
help@atl.ahq.org
http://www.atlanet.org

Boot camp information
http://dev.abanet.org/crimjust/juvekust/
    cjbootcamp.html

Boys and Girls Clubs of American Gang
    Prevention Program
404-815-5700
fsanchez@bgca.org
http://www.bgca.org

Bureau of Alcohol, Tobacco and Firearms
202-927-7777
http://www.atf.treas.gov

Bureau of Justice Assistance
800-668-4252
askncjrs@ncjrs.org
http://www.ncjrs.org

Bureau of Justice Statistics
202-633-3047
800-732-3277
ncjrs@ncjrs.org
http://www.ncjrs.org
http://www.ojp.usdoj.bjs

Child Abuse and Neglect Clearinghouse
800-394-3366

Childhelp USA/Forrester National Child
  Abuse Hotline
800-422-4453

Children and Family Justice Center
312-503-1235

Community court forum
http://www.communitycourts.org

Community policing information
http://www.communitypolicing.org/
  index.html

COPNET (links to law enforcement agencies
  nationwide)
http://police.sas.ab.ca

Crime Statistics Site
http://www.crime.org/links.html

Death Penalty Information Center
202-347-2531
dpic@essential.org
http://www.essential.org/dpic

Domestic Violence Resource Center
800-537-2238

Drug court information
http://www.calyx.net/~schaffer/GOVPUBS/
  dcourt1.html

Drug Enforcement Administration
202-307-1000
http://www.usdoj.gov/dea

Drug information
800-578-3472
http://www.whitehousedrugpolicy.gov
http://www.indesmith.org

Drug Policy Foundation
202-537-5005

Drugs and Crime Data Center
800-666-3332

Federal Bureau of Investigation
202-324-3000
http://www.fbi.gov
postmaster@fbi.gov

Federal jobs in general
http://www.usajobs.opm.gov

Federal Judicial Center
http://www.fjc.gov

Forensic science information
http://ash.lab.rl.fws.gov

Government agency links
http://www.fjc.gov/govlinks.html

Handgun Control, Inc.
202-898-0792
http://www.handguncontrol.org

Immigration and Naturalization Service
202-514-4301
800-546-3224
http://www.usdoj.gov.ins

International Association of Chiefs of Police
703-836-6767
800-843-4227
http://www.theiacp.org

International perspectives on justice
intjj@cunyvm.cuny.edu
http://pap01.adt.jjay.cuny.edu

Jefferson Institute for Justice Studies
202-659-2882
jjacoby@DC.net

Justice Information Center
http://www.ncjrs.org

Justice Research and Statistics Association
202-624-8560

Justice system jobs in general
http://www.shsu.edu/~icc_nels

Justice Technology Information Network
  (JUSTNET)
nlectc@aspensys.com
http://www.nlectc.org

Law enforcement
http://www.officer.com/jobs.html

Law Enforcement Agency Information
http://www.officer.com/agencies.html

Mothers Against Drunk Driving
  (MADD)
214-744-6233
http://www.madd.org

National Alcohol and Drug Information
  Clearinghouse
301-468-2600
http://www.health.org

National Archive of Criminal Justice Data
800-999-0960
http://www.icpsr.unmich.edu/NACJD/
  index.html

National Association of Attorney Generals
202-434-8000

National Association of Criminal Defense
  Lawyers
202-872-8600
http://www.criminaljustice.org

National Center for Missing and Exploited
  Children
800-843-5678
http://www.missingkids.com/cybertip

National Center for State Courts
757-259-1818
http://www.ncsc.dni.us

National Clearinghouse for Alcohol and Drug
  Information
800-729-6686
http://www.health.org

National Clearinghouse on Child Abuse and
  Neglect
800-394-3366

National Consortium on Violence Research
412-268-8269
ab0q@andrew.cmu.edu
http://www.heinz.cmu.edu/ncovr

National Council on Crime and Delinquency
  (NCCD)
415-896-6223
202-638-0556

National Crime Prevention Council
202-466-6272
800-627-2901
http://www.ncpc.org

National Criminal Justice Commission
http://www.ncianet.org/ncia

National Criminal Justice Reference Service
  (NCJRS)
800-851-3420
askncjrs@ncjrs.org
look@ncjrs.org
http://www.ncjrs.org

National District Attorney Assocation
703-549-9222
http://www.ndaa-apri.org

National Domestic Violence Hotline
512-453-8117
800-799-7233
http://www.inetport.com/ndvh

National Drugs and Crime Clearinghouse
800-666-3332
askncjrs@ncjrs.org
http://www.ncjrs.org

National Fraud Information Hotline
800-876-7060
http://www.fraud.org

National Institute of Drug Abuse (NIDA)
301-443-1124
http://www.nida.hin.gov

National Institute of Justice (NIJ)
800-851-3420
hillsman@justice.usdoj.gov
http://www.ojp.usdoj.gov/nij

National Institute of Justice Crime Mapping
  Research Center
http://www.nlectc.org/cmrc

National Institute of Justice Office for Victims
  of Crime
http://www.jop.usdoj.gov/ovc

National Law Enforcement and Corrections
  Technology Center
http://www.nlectc.org

National Organization for Victim Assistance
  (NOVA)
202-232-6682
800-897-8794
nova@access.diget.net
http://www.access.digex.net/~nova

National Research Council
202-334-3577
http://www.nas.edu

National Runaway Hotline
800-231-6946
http://www.nrscrisisline.org

National School Safety Center
805-373-9977

National Security Agency
301-688-6311
http://www.nsa.gov

National Victim Center (NVC)
817-877-3355
http://www.nvc.org

National Victims of Crime Resource Center
800-627-6872
http://www.ojp.usdoj.gov.ovc

National White Collar Crime Center
804-323-3563
800-221-4424
http://www.iir.com/nwccc/nwccc.htm

National Youth Gang Center
904-385-0600
nygc@iir.com
http://www.iir.com/nygc.htm

Non-government criminal justice organizations
http://www.ncianet.org/ncia/ocj.html

Office of Community Oriented Policing
  Services (COPS)
202-616-1728
http://www.usdoj.gov/cops

Office of International Criminal Justice (OICJ)
312-996-0159
http://www.acsp.uic.edu

Office of National Drug and Control Policy
  (Drugs and Crime Clearinghouse)
800-666-3332
askncjrs@aspensys.com
http://www.ncjrs.org

Office of Victims of Crime Resource Center
  (OVCRC)
800-627-6872
http://www.ojp.usdoj.gov/ovc

Partnership Against Violence Network
  (PAVNET)
301-504-5462
http://www.pavnet.org

Partnership for Responsible Drug Information
212-362-1964
adw7@columbia.edu

Police Executive Research Forum (PERF)
202-466-7820
http://www.policeforum.org

Police Foundation
202-833-1460
pfinfo@policefoundation.org

Safe Street Alliance
202-822-8100
safestsl@aol.com

State and local governments on the net
http://www.piperinfo.com/state/states.html

Uniform Crime Reports
http://www.getsafe.com/fbi/pre97tbs.html

United Nations Crime and Justice Information
  Network
http://www.ifs.univie.ac.at/~uncjin/
  uncjin.html

United Nations Online Crime and Justice
  Clearinghouse
http://www.unojust.org

U.S. Customs Service
202-927-3736
http://www.ustreas.gov/treasury/bureaus/
  customs.customs.html

U.S. Department of Justice
202-514-2000
askbjs@ojp.usdoj.gov
http://www.usdoj.gov

U.S. Department of State
202-647-4000
http://www.state.gov

U.S. Department of the Treasury
202-622-2000
http://www.ustreas.gov

U.S. Marshals Service
202-307-9100
800-336-0102
http://www.usdoj.gov/marshals

U.S. Postal Inspectors Service
202-268-2000
http://www.blue.ups.gov

U.S. Secret Service
202-435-7575
http://www.ustreas.gov/treasury/bureaus/
    usss/usss.html

U.S. Supreme Court
202-479-3211
http://www.law.cornell.edu/supct

Vera Institute of Justice
212-334-1300
http://broadway.vera.org

Victim's Resource Center
800-627-6872
http://www.ojp.usdoj.gov/ovc

The National Criminal Justice Reference produces a newsletter twice each month that covers a variety of justice-related topics. To subscribe to this free bimonthly newsletter on-line, send this message:

"subscribe justinfo" and give your name. Send this request to: lisproc@ncjrs.org

The U.S. Department of Justice provides a variety of justice-related printed materials free of charge. To subscribe to this service, contact

NCJRS User Services
Box 6000
Rockville, MD 20849-6000
800-851-3420
askncjrs@ncjrs.aspensys.com
http://www.ncjrs.org

The Police Corps

The Police Corps is a federal program administered by the U.S. Department of Justice, Office of the Police Corps and Law Enforcement Education (OPCLEE). The program is designed to address violent crime by increasing the number of police officers and sheriff's deputies with advanced education and training. The program has three components:

- It provides scholarships on a competitive basis to students who agree to earn their bachelor's degrees, complete approved Police Corps training, and then serve for four years on patrol, as assigned, with law enforcement agencies in great need.
- It provides funds to states to develop and provide 16 to 24 weeks of rigorous Police Corps training. Undergraduates must attend college full time and may receive up to $7,500 per academic year to cover the expenses of study toward a bachelor's or master's degree (graduate students complete their service in advance).
- It provides state and local agencies that hire Police Corps officers $10,000 per year for each of an officer's first four years of service.

Individuals apply to the state where they are willing to serve. Allowable educational expenses for full-time students include reasonable room and board. A student may receive up to $30,000 under the program. To be eligible, a student must attend a public or nonprofit four-year college or university. Participants may choose to study criminal justice or may pursue degrees in other fields, regardless of family income or resources. They must also possess the necessary mental and physical capabilities and moral characteristics to be an effective police officer, be of good character, meet the standards of the police agency with which they will serve, and demonstrate sincere motivation and dedication to law enforcement and public service.

Police Corps participants have all the rights and responsibilities of—and are subject to all rules and regulations that apply to—other members of the police departments with which they serve. If a Police Corps participant does not satisfactorily complete his or her education, training, and service obligations, he or she must repay all scholarships and reimbursements received through the program, plus interest.

Because not all states participate in the Police Corps program and some of the conditions may vary from state to state, individuals and police agencies interested in learning about the Police Corps may contact the Office of the Police Corps and Law Enforcement Education at 810 Seventh St., NW, Washington, D.C. Information on participating states and agency contacts may also be acquired by calling the U.S. Department of Justice Response Center at (800) 421-6770, or at the OPCLEE Web site: www.ojp.usdoj.gov/opclee.

Name Index

Subject Index